T3-AJK-920

CRC HANDBOOK of PESTICIDES

Editor

G.W.A. Milne

CRC Press
Boca Raton Ann Arbor London Tokyo

Library of Congress Cataloging-in-Publication Data

CRC Handbook of Pesticides / editor, G.W.A. Milne.
p. cm.
Includes bibliographical references (p.) and index
ISBN 0-8493-2447-5
1. Pesticides--Handbooks, manuals, etc. 2. Pesticides--Toxicology--Handbooks, manuals, etc. I. Milne, George W. A., 1937- . II. Title: Handbook of Pesticides.
SB961.H36 1994
615.9' 02--dc20

94-39758
CIP

International Standard Book Number 0-8493-2447-5
Library of Congress Card Number 94-39758
Printed in the United States of America 1 2 3 4 5 6 7 8 9 0
Printed on acid-free paper

THE EDITOR

George W. A. Milne is a Research Chemist in the National Cancer Institute, one of the National Institutes of Health in Bethesda, Maryland. He obtained his B.Sc. (1957), M.Sc. (1958), and Ph.D. (1960) degrees from the University of Manchester in England and spent two postdoctoral years at the University of Wisconsin before joining NIH in 1962. His research interests have included organic and natural products chemistry and analytical chemistry, and he spent over ten years working on the application of mass spectrometry to identification of organic compounds. This work led to the development in 1972 of the NIH/EPA Mass Spectral Database, which was the first computer-searchable mass spectral library. Since the mid-1970s, Dr. Milne has worked increasingly in the application of computers in chemistry and currently heads a group which is using molecular modeling in the design of drugs for the treatment of cancer and AIDS. He has published over 150 papers and 2 books and since 1989 has been the Editor-in-Chief of the American Chemical Society's *Journal of Chemical Information and Computer Sciences.*

INTRODUCTION

A pesticide may be defined as a chemical used to eradicate pests, a term which includes unwanted plants, such as weeds; insects, such as ants or boll weevils; or animals, such as rats or mice. Chemicals designed for use against plants or insects are widely used in agriculture and their use is generally termed crop protection.

Chemicals that are used as pesticides are generally toxic to various species, including man. They often exhibit both short-term, or acute toxicity, or long-term (chronic) toxicity and because of this their use is regulated by the U.S. Government. All pesticide chemicals are assigned Food Crop Tolerances which are promulgated under Section 408 of the Food, Drug and Cosmetic Act administered by the Environmental Protection Agency.

As of July 1994, some 892 pesticide chemicals, or "active ingredients", have been registered by the EPA. The 386 chemicals listed in this compilation are those most commonly used as pesticide chemicals. For each chemical listed, there is given its structural and molecular formula, its Chemical Abstracts Registry Number (CAS RN)[1] and its number, if available, in the Merck Index.[2] All pesticide chemicals are associated with many chemical names, trivial names, synonyms and trade names, in English and foreign languages, especially the European languages. All known names for each chemical have been collected in this compilation which has a total of approximately 7340 separate names, for an average of 19 names per chemical. These are presented in each record and are also arranged into a chemical name index.

The known names for the chemical are given and are followed by a summary of its physical properties, specifically, its melting point, boiling point, density, solubility and, when available, its octanol/water partition coefficient, which is generally regarded as an indicator of its potential for absorption in the body.

In each case, data pertaining to the compound's acute toxicity in various species are given. These have been drawn from the Registry of Toxic Effects of Chemical Substances,[3] a publication of the National Institute of Occupational Safety and Health. Finally, for each of the pesticide chemicals there is a reference to the appropriate Section of the Code of Federal Regulations, 40 CFR 180, where the various crop tolerances may be found. Changes and updates to these Crop Tolerance data will be found in the *Pesticide Chemical News Guide*,[4] a publication of CRC Press, Inc.

EXPLANATION OF THE DATA FIELDS

CAS RN: The Chemical Abstracts Registry Number for the compound, established by the Chemical Abstracts Service, Columbus, OH. This number may have up to 8 digits and is given in the form 12345-67-8, where the final digit (8) is an arithmetic check digit.

Molecular Formula: The molecular formula written in the "Hill" order in which C appears first, and H second (if present) and then the other elements are listed in alphabetical order. Salts and other molecular complexes are designated by the appropriate "dot-disconnected" molecular formula (*e.g.*, $H_2O_4S.Zn$ for zinc sulfate).

Chemical Structure: The chemical structure for the compound, given in the standard form. Stereochemistry at chiral centers is given where known. In some cases (*e.g.*, toxaphene), no structure is given because the material is a variable mixture of incompletely characterized components.

Chemical Name: All known names for the compound. These include systematic chemical names. The notations "8CI" and "9CI" indicate that the name was developed in accordance with nomenclature rules associated with the Eight or Ninth Collective Index of the Chemical Abstracts Service. The suffix "VAN" denotes a "valid, ambiguous name", and "ACN" denotes an "accepted common name". Names in foreign languages are included and the language is identified.

Merck Index No: For compounds included in the 11th Edition of the Merck Index,[2] the Monograph Number used in the Index is given. This number is not a unique identifier for a single compound because several derivatives or isomers of a compound may be included in the same Monograph.

MP: The melting point, in °C. Most values are given to the nearest degree, or over a range of temperatures. Where several different values have been reported, they are reproduced here.

BP: The boiling point in °C. Boiling points are often measured at reduced pressure and in such cases, the operative pressure (in mmHg) is given as a superscript. Thus 125^{o25} indicates a boiling point of 125°C at a pressure of 25 mmHg, and 125° means 125°C at ambient pressure, *i.e.*, 760 mmHg.

Density: The density of the compound is given with implicit units of g/ml. The measurement temperature, when available, is given as a superscript. Thus 1.234^{25} means a density of 1.234 g/ml at 25°C, while 1.234 means a density of 1.234 at an unspecified temperature. Where specific gravity, as opposed to density, was measured, this is noted.

Solubility: The solubility of the compound in specified solvents. The units reported in the original publication are used and the temperature is given when available. Thus, "48 mg/l water 25°" means 48 mg dissolve in 1 l of water at 25°C. Where available, the pH of the measurement is recorded.

Octanol/water PC: The partition coefficient of the compound between water and octanol. This is a unitless ratio. The temperature and pH of measurement are given when available, but otherwise, these measurements should be assumed to have been recorded at 25°C and pH 7.

LD_{50}: The acute toxicity data for the compound are listed under this heading. The majority of these data are LD_{50} data, *i.e.*, the dose that is lethal to 50% of the individuals tested, but other values, such as LCLo, TDLo, and LC_{50}, are cited. These are defined below. The species and the route of administration are given in abbreviated form. The abbreviations are also defined below.

CFR: The reference to the Section of the Code of Federal Regulations in which the Food Crop Tolerances will be found.

LIST OF ABBREVIATIONS

Species

bwd - wild bird species
cat - cat
chd - child
ckn - chicken
dck - duck
dog - dog
dom - domestic animal (*e.g.*, cat, dog)
frg - frog
ger - gerbil
gpg - guinea pig
ham - hamster
hmn - human
inf - infant
mam - mammal
man - man
mky - monkey

mus - mouse
pgn - pigeon
pig - pig
qal - quail
rat - rat
rbt - rabbit
trk - turkey
unk - unknown
wmn - woman

Routes of Administration

ice - intracerebral
ihl - inhalation
ims - intramuscular
ipr - intraperitoneal
ivn - intravenous
ocu - intraocular
orl - oral
par - parenteral
scu - subcutaneous
skn - topically applied to the skin
unr - unrecorded

Toxicity Measurements

LC_{50} - concentration lethal to 50% of the animals dosed
LCLo - lowest lethal concentration
LD_{50} - dose lethal to 50% of the animals dosed
LDLo - lowest lethal dose
TC_{50} - concentration toxic to 50% of the animals dosed
TCLo - lowest toxic concentration
TD_{50} - dose toxic to 50% of the animals dosed
TDLo - lowest toxic dose

REFERENCES

1. The Chemical Abstracts Service, a Division of the American Chemical Society, has since 1965 registered all new chemical compounds with a Chemical Abstracts Registry Number (CAS RN). Every distinct chemical structure is assigned a unique CAS RN and this number is useful as a means of locating information pertaining to the compound.

2. The Merck Index, 11th Edition, 1989, Merck & Co., Rahway, NJ.

3. Registry of Toxic Effects of Chemical Substances (RTECS) is a collection of toxicological data taken from the open literature and compiled and published by the National Institute for Occupational Safety and Health. RTECS is available online via vendors such as Dialog (Palo Alto, CA) and Toxline (NLM, Bethesda, MD).

4. *The Pesticide Chemical News Guide* is published by Food Chemical News, Inc., 1101 Pennsylvania Avenue SE, Washington, DC 20003.

CAS RN 3383-96-8
$C_{16}H_{20}O_6P_2S_3$

Abate
Phosphorothioic acid, O,O'-(thiodi-4,1-phenylene) O,O,O',O'-tetramethyl ester (9CI)
Phosphorothioic acid, O,O'-(thiodi-p-phenylene) O,O,O',O'-tetramethyl ester (8CI)
Difenphos
Temephos

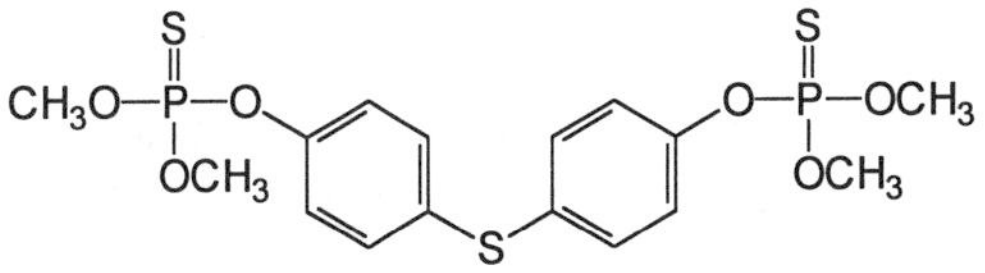

Merck Index No. 9075
MP: 30.0 - 30.5°
BP:
Density: 1.32
Solubility: Insol. water, hexane, sol. acetonitrile, carbon tetrachloride, diethyl ether, dichloromethane, toluene
Octanol/water PC: 80,900
LD_{50}: rat orl 1000 mg/kg; rat orl 8600, 13000 mg/kg; rat skn 1370 mg/kg; rat ipr 912 mg/kg; rat scu 2302 mg/kg; rat unr 8600 mg/kg; mus orl 223 mg/kg; mus ipr 683 mg/kg; rbt orl 313 mg/kg; rbt skn 970 mg/kg; pgn orl 50 mg/kg; qal orl 75 mg/kg; dck orl 79 mg/kg; mam orl 4700 mg/kg; mam unr 4000 mg/kg; bwd orl 32 mg/kg.
CFR: 40CFR180.170

CAS RN 30560-19-1
$C_4H_{10}NO_3PS$

Acephate
Phosphoramidothioic acid, acetyl-, O,S-dimethyl ester (8CI9CI)
Acetylphosphoramidothioic acid ester
Orthene

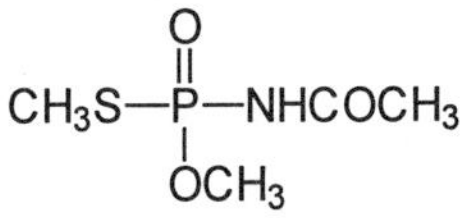

Merck Index No. 26
MP: 82 - 93° (64 - 68° impure)
BP:
Density: 1.35
Solubility: 790 g/l water 20°, 151 g/l acetone, >100 g/l ethanol, 35 g/l ethyl acetate, 16 g/l benzene, 0.1 g/l hexane
Octanol/water PC:
LD_{50}: rat orl 750 mg/kg; rat orl 700 mg/kg; rat skn >2500 mg/kg; mus orl 233 mg/kg; mus ihl LCLo 2200 mg/m³/5H; dog orl LDLo 681 mg/kg; rbt skn 2000 mg/kg; ckn orl 852 mg/kg; ckn unr 568 mg/kg; dck orl 350 mg/kg; mam orl 321 mg/kg; brd orl 106 mg/kg; bwd orl 140 mg/kg.
CFR: 40CFR180.108

CAS RN 34256-82-1
$C_{14}H_{20}ClNO_2$

Acetochlor (ACN)
Acetamide, 2-chloro-N-(ethoxymethyl)-N-(2-ethyl-6-methylphenyl)- (9CI)
o-Acetotoluidide, 2-chloro-N-(ethoxymethyl)-6'-ethyl- (8CI)
2-Chloro-N-(ethoxymethyl)-6'-ethyl-*o*-acetotoluidide (ACN)

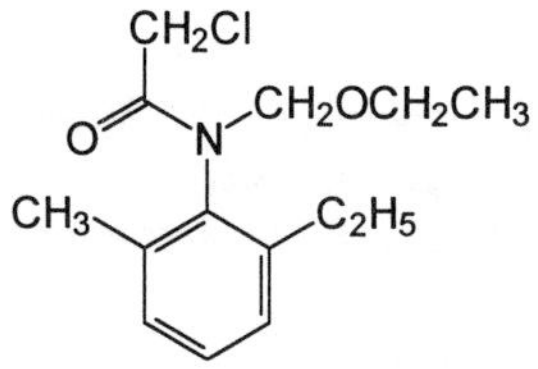

Merck Index No.
MP: 0°
BP:
Density:
Solubility: 223 mg/l water 25°, sol. diethyl ether, acetone, benzene, chloroform, ethanol, ethyl acetate, toluene
Octanol/water PC:
LD_{50}: rat orl 3400 mg/kg; rbt orl 600 mg/kg
CFR: 40CFR180. temp. 1982

CAS RN 50594-66-6
$C_{14}H_7ClF_3NO_5$

Acifluorfen (ACN)
Benzoic acid, 5-[2-chloro-4-(trifluoromethyl)phenoxy]-2-nitro- (9CI)
2-Nitro-5-[2-chloro-4-(trifluoromethyl)phenoxy]benzoic acid
5-[2-Chloro-4-(trifluoromethyl)phenoxy]-2-nitrobenzoic acid

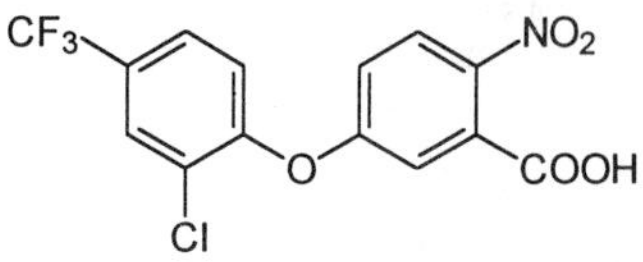

Merck Index No. 105
MP: 142-160°, 151.5 - 157°
BP:
Density:
Solubility: 120mg/l water 23-25°, 600 g/kg acetone, 500 g/kg ethanol, 50 g/kg dichloromethane, <10g/kg xylene, kerosene>250g/l water (sodium salt)
Octanol/water PC:
LD_{50}: rat orl 1300 mg/kg; rat orl 1370 mg/kg; rat ihl LC_{50} >6900 mg/m^3/4H; mus orl 1370 mg/kg; rbt skn 3680 mg/kg; qal orl 325 mg/kg; dcl orl 2821 mg/kg.
CFR: 40CFR180.383

CAS RN 15972-60-8
$C_{14}H_{20}ClNO_2$

Alachlor (ACN)
Acetamide, 2-chloro-N-(2,6-diethylphenyl)-N-(methoxymethyl)- (9CI)
Acetanilide, 2-chloro-2',6'-diethyl-N-(methoxymethyl)- (8CI)
Alanex

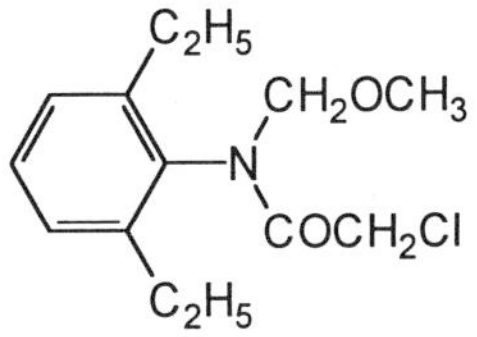

C_2H_5 CH_2OCH_3 N $COCH_2Cl$ C_2H_5

Merck Index No. 193
MP: 40°, 40 - 41°
$BP_{0.02}$: 100°, $BP_{0.3}$ 135°
d_{25}: 1.133
Solubility: 140 mg/l water 23°, sol. diethyl ether, acetone, benzene, chloroform, ethanol, ethyl acetate, sp. sol. heptane
Octanol/water PC:
LD_{50}: rat orl 1200 mg/kg; rat orl 930 mg/kg; rat skn >2000 mg/kg; mus orl 462 mg/kg; rbt inh LC_{50} >5100 mg/m^3; rbt skn 3500 mg/kg; mam orl 3000 mg/kg; mam unr 1200 mg/kg.
CFR: 40CFR180.249

CAS RN 1596-84-5
$C_6H_{12}N_2O_3$

Alar
Butanedioic acid, mono(2,2-dimethylhydrazide) (9CI)
Succinic acid, mono(2,2-dimethylhydrazide) (8CI)
Aminozide
Daminozide

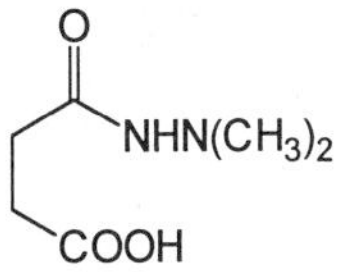

O $NHN(CH_3)_2$ COOH

Merck Index No. 2810
MP: 154 - 155°, 157 - 164°
BP:
Density:
Solubility: 10% water, 5% methanol, 2.5% acetone, 100g/kg water 25, 50 g/kg methanol, 25 g/kg acetone, insol. hydrocarbons
Octanol/water PC: 0.031 (21°)
LD_{50}: rat orl 8400 mg/kg; mus orl 6300 mg/kg; rat ihl >147 gm/m^3; mus ipr 1325 mg/kg; rbt skn >5g/kg.
CFR: 40CFR180.246

CAS RN 20859-73-8
AlP

Aluminium phosphide
Aluminum phosphide (AlP) (8CI9CI)
Al-Phos
Aluminium phosphide (AlP)

Al≡P

Merck Index No. 363
MP: >1000°
BP:
Density: 2.85[15]
Solubility:
Octanol/water PC:
LD_{50}: man ihl LCLo 2800 mg/m^3
CFR: 40CFR180.225

CAS RN 504-24-5
$C_5H_6N_2$

***p*-Aminopyridine**
4-Pyridinamine (9CI)
Pyridine, 4-amino- (8CI)
γ-Aminopyridine

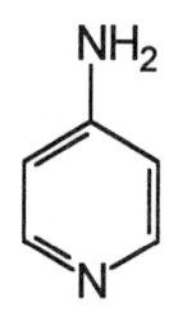

Merck Index No.
MP: 158-159°
BP: 180°[13]
Density:
Solubility: 76600 ppm water 25°, 83000 ppm water 25°
Octanol/water PC:
LD_{50}: rat orl 21 mg/kg; rat ipr 6.5 mg/kg; rat scu 19 mg/kg; mus orl LDLo 42 mg/kg; mus ipr 10 mg/kg; mus scu LDLo 5mg/kg; mus ivn 7 ng/kg; mus ice 4 mg/kg; dog orl 3700 mg/kg; rbt ivn 5.5 mg/kg; pgn orl 6 mg/kg; orl qal 7.65 mg/kg; orl dck 4200 mg/kg; orl bwd 2.37 mg/kg.
CFR: 40CFR180.312

CAS RN 2227-13-6
$C_{12}H_6Cl_4S$

Tetrasul
Sulfide, *p*-chlorophenyl 2,4,5-trichlorophenyl (8CI)
p-Chlorophenyl 2,4,5-trichlorophenyl sulfide (ACN)
Animert
Animert V-10

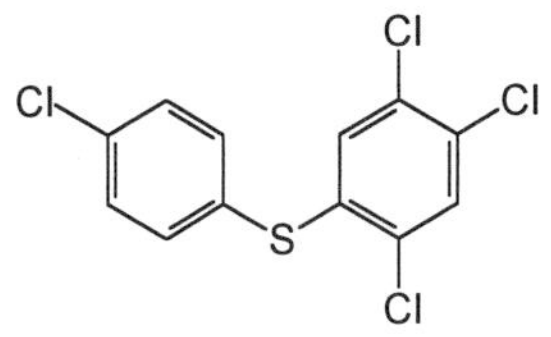

Merck Index No.
MP:
BP:
Density:
Solubility:
Octanol/water PC:
LD_{50}: rat orl 3960 mg/kg, rat ipr 6810 mg/kg, mus orl 5010 mg/kg, mus ipr >6810 mg/kg; rbt orl LDLo 1350 mg/kg; rbt skn 2000 mg/kg; gpg orl 500 mg/kg; gpg skn 8200 mg/kg; gpg ipr 550 mg/kg.
CFR: 40CFR180.256

CAS RN 9006-42-2
$C_{16}H_{33}N_{11}S_{16}Zn_3$

Metiram (9CI)
Polyram (8CI)
Amarex
Carbatene
FMC-9102

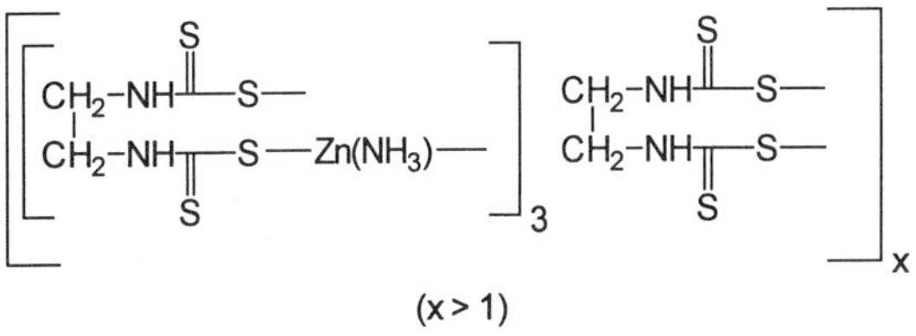

(x > 1)

Merck Index No.
MW:
MP: 140° (dec)
BP:
Density:
Solubility: insol. water, sol. pyridine, pract. insol. ethanol, acetone, benzene
Octanol/water PC: 2 (pH 7)
LD_{50}: rat orl 2850 mg/kg; rat ihl LC_{50} > 5700 mg/m^3/4H; rat skn >2000 mg/kg; rat unr 3250 mg/kg; mus orl 2630 mg/kg; rbt orl 620 mg/kg; gpg orl 2400 mg/kg; gpg unr 3750 mg/kg; ckn orl 7660 mg/kg; dom orl 1000 mg/kg; mam unr 3290 mg/kg.
CFR: 40CFR180.217

CAS RN 140-57-8
$C_{15}H_{23}ClO_4S$

Aramite
Sulfurous acid, 2-chloroethyl 2-[4-(1,1-dimethylethyl)phenoxy]-1-methylethyl ester (9CI)
Sulfurous acid, 2-(*p-tert*-butylphenoxy)-1-methylethyl 2-chloroethylester (8CI)
β-Chloroethyl-β-(*p-t*-butylphenoxy)-α-methylethyl sulfite

Merck Index No. 794
MP: -31.7°
BP: $175^{\circ 0.1}$, 200 - $210^{\circ 7}$
Density: 1.1450 - 1.1620
Solubility: insol. water
Octanol/water PC:
LD_{50}: rat orl 3900 mg/kg; hmn orl LDLo 429 mg/kg; mus ipr LDLo 200 mg/kg; gpg orl 3900 mg/kg.
CFR: 40CFR180.107 (revoked 5/4/88)

CAS RN 1861-40-1
$C_{13}H_{16}F_3N_3O_4$

Balan
Benzenamine, N-butyl-N-ethyl-2,6-dinitro-4-(trifluoromethyl)- (9CI)
p-Toluidine, N-butyl-N-ethyl-α,α,α-trifluoro-2,6-dinitro- (8CI)
α,α,α-trifluoro-2,6-dinitro-N,N-ethylbutyl-*p*-toluidine
p-Toluidine, N-butyl-2,6-dinitro-N-ethyl-α,α,α-trifluoro-
Balfin
Banafine
Benalan
Benefex
Benefin
Benephin
Benfluralin
Bethrodine
Binnell
Blulan
Bonalan
Carpidor
Emblem

Merck Index No. 1048
MP: 65-66.5°
BP: 121-$122^{\circ 0.5}$, BP: 148-$149^{\circ 7}$
Density:
Solubility: <1mg/ml water, 650g/l acetone, 600g/l dioxane, 580g/l methyl ethyl ketone, >500g/l chloroform, 450g/l dimethylformamide, 450g/l xylene, 40g/l methanol, 24g/l ethanol
LD_{50}: rat orl >10000mg/kg
Octanol/water PC: 195000 (25°, pH 7)
CFR: 40CFR180.208

CAS RN 1918-00-9
$C_8H_6Cl_2O_3$

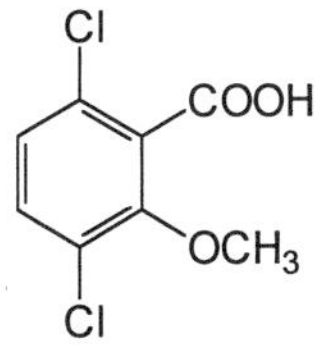

Banvel D
Benzoic acid, 3,6-dichloro-2-methoxy- (9CI)
o-Anisic acid, 3,6-dichloro- (8CI)
o-Anisic, 3,6-dichloro-
Acido (3,6-dicloro-2-metossi)-benzoico (Italian)
Banes
Banex
Banlen
Banuel D
Banvel
Brush buster
Compound B dicamba
Dianat
Dianate
Dicamba (ACN)
Dicambe
Mediben
Mondak

Merck Index No. 3026
MP: 114 - 116°
BP:
Density: 1.57^{25}
Solubility: 6.5g/l water 25°, 922 g/l ethanol, 916 g/l cyclohexanone, 810 g/l acetone, 260 g/l dichloromethane, 130 g/l toluene, 78 g/l xylene
Octanol/water PC:
LD_{50}: rat orl 1040 mg/kg
CFR: 40CFR180.227

CAS RN 12069-69-1
$CH_2Cu_2O_5$

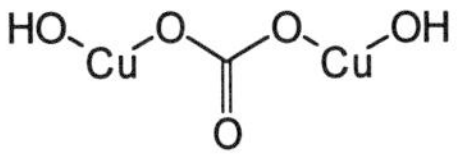

Basic copper carbonate
Copper, [μ-[carbonato(2-)-O:O']]dihydroxydi- (9CI)
Copper, (carbonato)dihydroxydi- (8CI)
(Carbonato)dihydroxydicopper
Artificial malachite
Basic copper(II) carbonate
Basic cupric carbonate
Carbonic acid, copper(2+) salt (1:1), basic
Cheshunt compound
Copper basic carbonate
Copper carbonate (ACN)
Copper carbonate (basic) ($CuCO_3.Cu(OH)_2$)
Copper carbonate hydroxide ($Cu_2CO_3(OH)_2$)
Copper carbonate hydroxide [$CuCO_3.Cu(OH)_2$]
Copper carbonate [$CuCO_3.Cu(OH)_2$]
Copper carbonate, basic

Merck Index No. 2634
MP: 200°(dec)
BP:
Density: 3.7 - 4.0
Solubility: Insol. water
Octanol/water PC:
LD_{50}: rat orl 1350 mg/kg; rbt orl 159 mg/kg; pgn orl LDLo 1000 mg/kg; dck orl LDLo 900 mg/kg; brd orl 900 mg/kg.
CFR: 40CFR180.136

CAS RN 43121-43-3
$C_{14}H_{16}ClN_3O_2$

Bayleton
2-Butanone, 1-(4-chlorophenoxy)-3,3-dimethyl-1-(1H-1,2,4-triazol-1-yl)- (9CI)
Amiral
Bay-meb-6447
MEB 6447
Triadimefon
1-(4-Chlorophenoxy)-3,3-dimethyl-1-(1,2,4-triazol-1-yl)-butan-2-one
1-(4-Chlorophenoxy)-3,3-dimethyl-1-(1H-1,2,4-triazol-1-yl)-2-butanone
1H-1,2,4-Triazole, 1-[(tert-butylcarbonyl-4-chlorophenoxy)methyl]-
2-Butanone, 1-(4-chlorophenoxy)-3,3-dimethyl-1-(1,2,4-triazol-1-yl)-

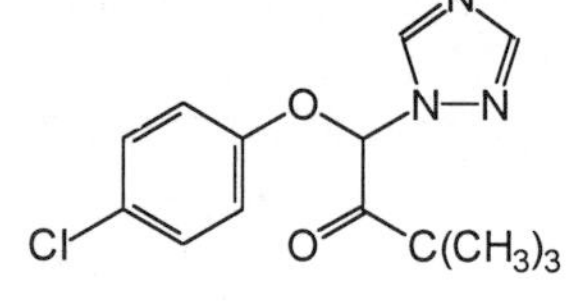

Merck Index No. 9507
MP: 82°, 82.3°
BP:
Density: 1.22
Solubility: 260mg/l water 20°, >200g/l dichloromethane, >200g/l toluene, 100-200g/l isopropanol, 10-20g/l hexane
Octanol/water PC: 1510
LD_{50}: rat orl 568, 363mg/kg; rat ihl LC_{50} >450 mg/m^3/4H; rat skn 5000 mg/kg; mus orl 1000 mg/kg; dog orl >500 mg/kg; rbt orl 500 mg/kg; qal orl 1750 mg/kg.
CFR: 40CFR180.410

CAS RN 22781-23-3
$C_{11}H_{13}NO_4$

Bendiocarb (ACN)
1,3-Benzodioxol-4-ol, 2,2-dimethyl-, methylcarbamate (9CI)
Carbamic acid, methyl-, 2,3-(isopropylidenedioxy)phenyl ester (8CI)
Carbamic acid, methyl-, 2,3-(dimethylmethylenedioxy)phenyl ester
Ficam W
Ficam 80 W
Fisons NC 6897
Garvox
Multamat
NC-6897
2,2-Dimethyl-1,3-benzdioxol-4-yl N-methylcarbamate
2,2-Dimethyl-1,3-benzodioxol-4-ol methylcarbamate (ACN)

Merck Index No. 1044
MP: 129 - 130°
BP:
Density: 1.25
Solubility: 40mg/l water 20°, 200g/kg acetone, 200g/kg dichloromethane, 200 g/kg dioxane, 40 g/kg benzene, 40 g/kg ethanol, 10 g/kg o-xylene, 0.35 g/kg hexane, <1 g/kg kerosene
Octanol/water PC: 50
LD_{50}: rat orl 40 mg/kg; rat skn 566 mg/kg; mus orl 45 mg/kg; rbt orl 35 mg/kg; gpg orl 35 mg/kg; qal orl 21 mg/kg; mam orl 35 - 100mg/kg.
CFR: 40CFR180. temp 1979

CAS RN 17804-35-2
$C_{14}H_{18}N_4O_3$

Benomyl (ACN)
Carbamic acid, [1-[(butylamino)carbonyl]-1H-benzimidazol-2-yl]-, methyl ester (9CI)
2-Benzimidazolecarbamic acid, 1-(butylcarbamoyl)-, methyl ester (8CI)
Arilate
Benlate
Benlate 50W
Butylcarbamoylbenzimidazole-2 methyl carbamate
BBC
F 1991
Fundazol
Fungicide D-1991
Fungicide 1991
F1991
Methyl N-(1-butylcarbamoyl-2-benzimidazole)carbamate

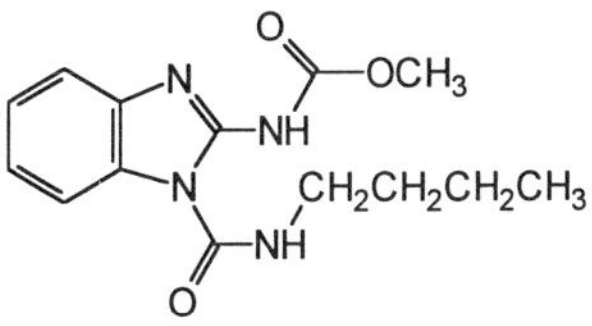

Merck Index No. 1053
MP: dec
BP:
Density:
Solubility: 2 mg/l water 25°, 94 g/kg chloroform, 53 g/kg dimethylformamide, 18 g/kg acetone, 10 g/kg xylene, 4 g/kg ethanol
Octanol/water PC:
LD_{50}: rat orl >9590mg/kg; rat orl 10000 mg/kg; rat ihl LC_{50} >2000 mg/m^3/4H; rat skn >1000 mg/kg; rat unr 9920 mg/kg; mus orl 5600 mg/kg; rbt skn >10000 mg/kg; mam unr 10000 mg/kg; bwd orl 100 mg/kg.
CFR: 40CFR180.294

CAS RN 741-58-2
$C_{14}H_{24}NO_4PS_3$

Bensulide
Phosphorodithioic acid, O,O-bis(1-methylethyl) S-[2-[(phenylsulfonyl)amino]ethyl] ester
Bensulfide
Betasan (obsolete)
Disan (pesticide)
Exporsan
Exposan

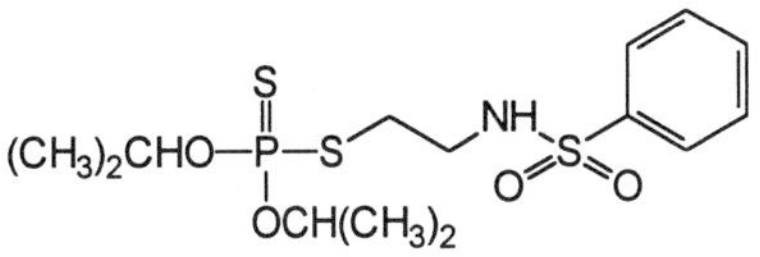

Merck Index No.
MP: 34.4°
BP:
Density: 1.23; 1.224
Solubility: 25mg/l water 25°, 300 mg/l kerosene 20, misc. acetone, ethanol, Methyl isobutyl ketone, xylene
Octanol/water PC: 16,500
LD_{50}: rat orl 770 mg/kg; rat orl 271 mg/kg; rat skn 3950 mg/kg; rat unr 770 mg/kg; mus orl 1540 mg/kg; mus ipr 630 mg/kg; mus scu 630 mg/kg; rbt skn 2000 mg/kg; qal orl 1386 mg/kg.
CFR: 40CFR180.241

CAS RN 25057-89-0
$C_{10}H_{12}N_2O_3S$

Bentazon (ACN)
1H-2,1,3-Benzothiadiazin-4(3H)-one, 3-(1-methylethyl)-, 2,2-dioxide (9CI)
1H-2,1,3-Benzothiadiazin-4(3H)-one, 3-isopropyl-, 2,2-dioxide (8CI)
Basagran
Bendioxide
Bentazone
BAS 351 H
Poast
1H-2,1,3-Benzothiadiazin-4(3H)-one, 3-isopropyl- 2,2-dioxide
3-(1-Methylethyl)-(1H)-2,1,3-benzothiadiazin-4(3H)-one 2,2-dioxide
3-(1-Methylethyl)-1H-2,1,3-benzothiazain-4(3H)-one 2,2-dioxide
3-Isopropyl-1H-benzo-2,1,3-thiadiazin-4-one-2,2-dioxide

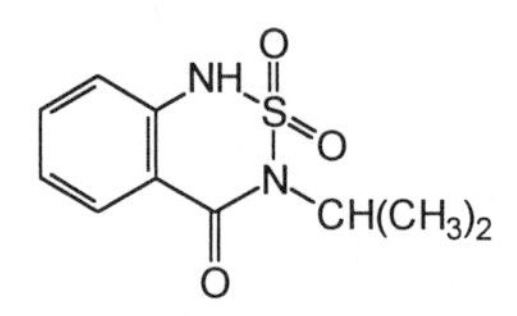

Merck Index No. 1060
MP: 137 - 139°
BP:
Density:
Solubility: 500 mg/l water 20°, 1507 g/kg acetone, 861g/kg ethanol, 650 g/kg ethyl acetate, 616 g/kg diethyl ether, 180 g/kg chloroform, 33 g/kg benzene, 0.2 g/kg cyclohexane
Octanol/water PC: 0.35
LD_{50}: rat orl 1100mg/kg; rat ihl LC_{50} 5100 mg/m^3/4H; rat skn 2500 mg/kg; rat par 1100 mg/kg; dog orl 450 mg/kg; cat orl 500 mg/kg; rbt orl 750 mg/kg; qal orl 720 mg/kg; dcl orl 15000 mg/kg.
CFR: 40CFR180.355

CAS RN 309-00-2
$C_{12}H_8Cl_6$

Aldrin
1,2,3,4,10,10-hexachloro-1α,4α4αβ,5α,8α,8αβ-hexahydro-1:4,5:8-dimethanonaphthalene
HHDN
Octalene

Merck Index No. 219
MP: 104.0 - 104.5°
BP: $145^{\circ 2}$
Density:
Solubility:<0.05 mg/l water; mod. sol paraffins, esters, ketones; spar. sol . alcohols
Octanol/water PC:
LD_{50}: hmn orl TDLo 14 mg/kg; chd orl LDLo 1.25 mg/kg; rat orl 39, 60 mg/kg; rat ihl LCLo 5.8 mg/m^3/4H; rat skn 98 mg/kg; rat ipr 150 mg/kg; rat scu 62 mg/kg; mus orl 44 mg/kg; mus ipr 50 mg/kg; mus ivn 21 mg/kg; dog orl 65 mg/kg; dog ipr 1999 mg/kg; dog unr LDLo 95 mg/kg; cat orl 15 mg/kg; cat skn 75 mg/kg; rbt orl 50 mg/kg; rbt skn 15 mg/kg; rbt scu LDLo 100 mg/kg; gpg orl 33 mg/kg; ham orl 100 mg/kg; pgn orl 56.2 mg/kg; ckn orl 10 mg/kg; qal orl 42.1 mg/kg; dck orl 520 mg/kg; mam orl 39 mg/kg; mam unr 40 mg/kg; bwd orl 7.2 mg/kg.
CFR: 40CFR180.135 (removed 12/24/86)

CAS RN 584-79-2
$C_{19}H_{26}O_3$

Allethrin
2-Methyl-4-oxo-3-(2-propenyl)-2-cyclopenten-1-yl-2,2-dimethyl-3-(2-methyl-1-propenyl) cyclopropanecarboxylate
Pyrocide
(RS)-3-allyl-2-methyl-4-oxocyclopent-2-enyl-(1RS)-*cis,trans*-chrysanthemate

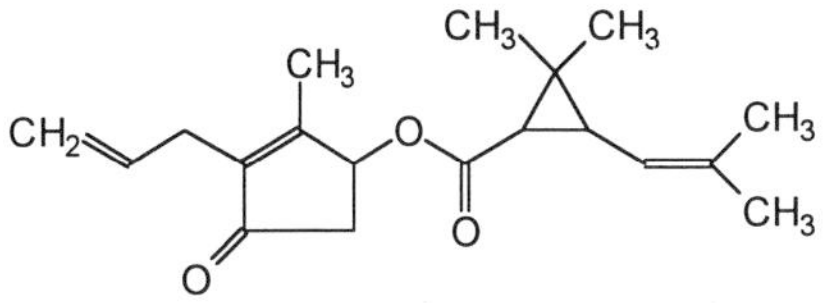

Merck Index No.
MP:
BP: 160°
Density: 1.010 SG
Solubility: sol. alcohols, pet. ether, kerosene, carbon tetrachloride, dichloromethane, nitromethane
Octanol/water PC:
LD_{50}: rat orl 685 mg/kg; rat orl 1100 mg/kg; rat ihl LCLo 13800 mg/m^3/4H; rat skn >11200 mg/kg; rat ivn LDLo 4 mg/kg; rat unr 680 mg/kg; mus orl 370 mg/kg; mus skn 1200 mg/kg; mus ipr 38 mg/kg; mus ice 4 mg/kg; rbt orl 4290 mg/kg; rbt skn 11332 mg/kg; rbt ipr 11200 mg/kg; qal orl 2030 mg/kg.
CFR: 40CFR180.113

CAS RN 67485-29-4
$C_{25}H_{24}F_6N_4$

Amdro
Hydramethylnon
Tetrahydro-5,5-dimethyl-2-(1H)-pyrimidinone[3[4-(trifluoromethyl-styryl)-cinnamylidene- hydrazone
Amidinohydrazone

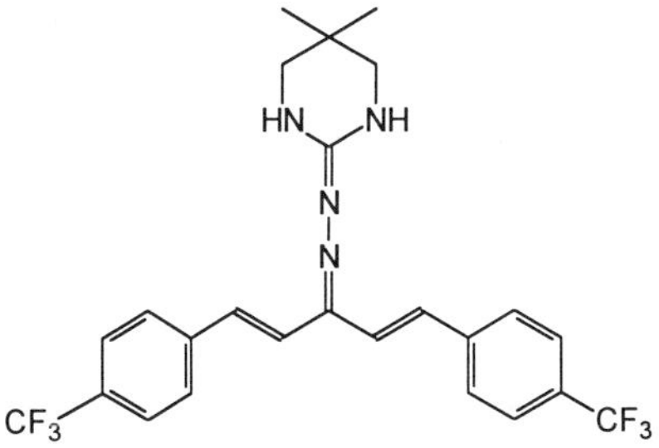

Merck Index No. 4684
MP: 185-190°, 189 - 191°
BP:
Density:
Solubility: 0.005 - 0.007 mg/l water; 360 g/l acetone; 72 g/l ethanol; 230 g/l methanol
Octanol/water PC: 206
LD_{50}: rat orl 1131 mg/kg; rat orl 1300 mg/kg; rat ihl LC_{50} >5000 mg/m^3/4H; rbt skn >5000 mg/kg; qal orl 1828 mg/kg; dck orl >2510 mg/kg.
40CFR180.395

CAS RN 834-12-8
$C_9H_{17}N_5S$

Ametryn
N-Ethyl-N'-(1-methylethyl)-6-(methylthio)1,3,5-triazine-2,4-diamine
Amephyt
Ametrex

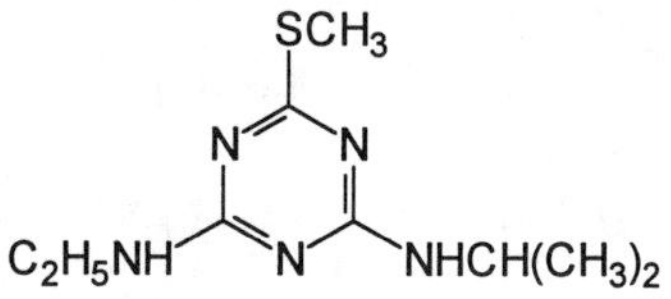

Merck Index No. 402
MP: 88 - 89°
BP:
Density:
Solubility: 185 mg/l water 20°, 600 g/l dichloromethane, 500 g/l acetone, 450 g/l methanol, 400 g/l toluene, 14 g/l hexane.
Octanol/water PC: 676
LD_{50}: rat orl 508 mg/kg; rat orl 590 mg/kg; rat ihl LC_{50} >2200 mg/m^3/4H; rat skn >3000 mg/kg; mus orl 965 mg/kg; rbt skn 8160 mg/kg.
40CFR180.258

CAS RN 69377-81-7
$C_7H_5Cl_2FN_2O_3$

Fluroxypyr
[(4-amino-3,5-dichloro-6-fluoro-2-pyridyl)oxy]acetic acid
Starane
Dowco 433

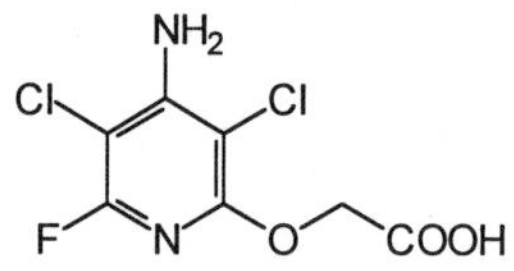

Merck Index No. 4128
MP: 232-233°
BP:
Density:
Solubility: 91 mg/l water; 64.7 g/l acetone; 43.7 g/l methanol; 11.8 g/l ethyl acetate; 11.8 g/l isopropanol; 0.1 g/l dichloromethane; 0.9 g/l toluene; 0.3 g/l xylene
LD_{50}: rat orl 2405 mg/kg; rat ipr 458 mg/kg; rat ipr 519 mg/kg; rat ihl LC_{50} >296 mg/m^3/4H; rat skn >2000 mg/kg; rat ipr 458 mg/kg; rbt skn >5000 mg/kg; dck unr >2000 mg/kg.
Octanol/water PC:
CFR: 40CFR180. (Pending, 1987)

CAS RN 33089-61-1
$C_{19}H_{23}N_3$

Amitraz
acarac
N-methyl bis(2,4-xylyliminomethyl)amine
N-methyl-N'-2,4-xylyl-N-(N-2,4-xylylformimidoyl)formamidine

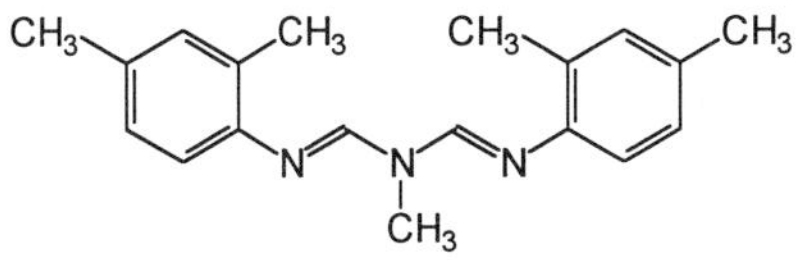

Merck Index No. 503
MP: 86-87°
BP:
Density: 1.128
Solubility: 1 mg/l water; >300 g/l acetone; toluene, xylene
Octanol/water PC: 316,000
LD_{50}: rat orl 400 mg/kg; rat orl 1600mg/kg; rat ihl LC_{50} 65000 mg/m^3/6H;rat skn >1600 mg/kg; rat ipr 800 mg/kg; rat unr 600 mg/kg; mus orl 1600 mg/kg; mus orl 400mg/kg; dog orl 100 mg/kg; rbt orl >100 mg/kg; rbt skn >200 mg/kg; gpg orl >400 mg/kg; qal orl 788 mg/kg; dck orl 7000 mg/kg.
40CFR180.287

CAS RN 101-05-3
$C_9H_5Cl_3N_4$

Anilazine
Triazine
2,4-dichloro-6-(*o*-chloroanilino)-*s*-triazine
4,6-dichloro-N-(2-chlorophenyl-1,3,5-triazin-2-amine

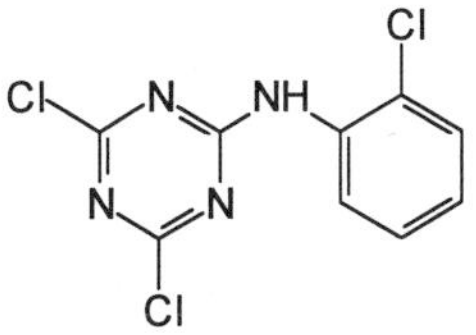

Merck Index No. 685
MP: 159-160°
BP:
Density: 1.8
Solubility: 8 mg/l water; 100 g/l acetone; 60 g/l C6H5Cl; 50 g/l toluene; 40 g/l xylene
Octanol/water PC: 80.6 (pH 7)
LD_{50}: rat orl >5000mg/kg; rat orl 2700 mg/kg; rat ihl LC_{50} >228 mg/m^3/1H; rat skn >5000 mg/kg; rat ipr 16 mg/kg; mus orl 6020 mg/kg; mus ipr 30 mg/kg; dog orl >5000 mg/kg; rbt orl 400 mg/kg; ckn unr LDLo 3750 mg/kg; qal unr 2500 mg/kg.
40CFR180.158

CAS RN 75-60-5
$C_2H_7AsO_2$

Ansar
Cacodylic acid
Dimethylarsinic acid
bolls-eye

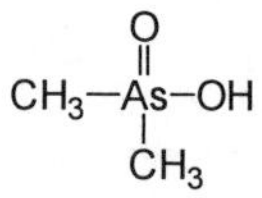

Merck Index No. 1603
MP: 192-198°; 195-196°
BP:
Density:
Solubility: 2 kg/kg water; sol. in lower alcohols; insol. diethyl ether
Octanol/water PC:
LD_{50}: rat orl 1350mg/kg; rat orl 644 mg/kg; rat ihl LCLo >2600 mg/m^3/2H; mus ipr 500 mg/kg; mus unr 185 mg/kg.
40CFR180.311

CAS RN 100728-84-5 (81405-85-8)
$C_{15}H_{18}N_2O_3$

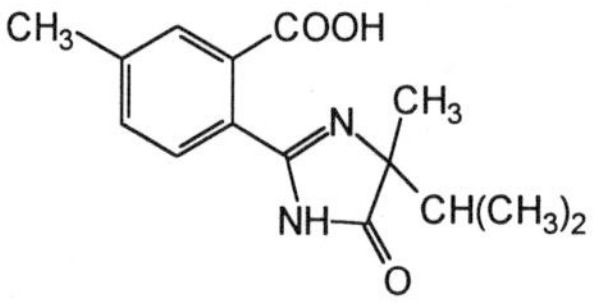

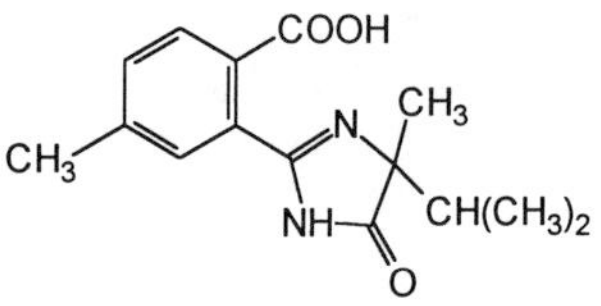

Imazamethabenz
Assert
Dagger
6-(4-isopropyl-4-methyl-5-oxo-2-imidazolin-2-yl)-*m*-toluic acid

Merck Index No. 4825
MP: 108 - 117°; 113 - 122°; 144 - 153° (3 stages)
BP:
Density: 0.22
Solubility: 1370 mg/kg m-isomer, water; 857 mg/kg p-isomer, water; 230 g/kg acetone; 216 g/kg DMSO; 0.6 g/kg hexane; 183 g/kg isopropanol; 309 g/kg methanol; 172 g/kg dichloromethane; 45 g/kg toluene
Octanol/water PC: 35 (p-isomer); 66 (m-isomer)
LD_{50}: rat orl >5000 mg/kg; rat ihl LC_{50} >5800 mg/m^3; rbt skn 2000 mg/kg; qal orl >2150 mg/kg; dck orl >2150 mg/kg.
40CFR180.437

CAS RN 3337-71-1
$C_8H_{10}N_2O_4S$

Asulam
Asilan
Asulox
Methyl [(4-aminophenyl)sulfonyl]carbamate

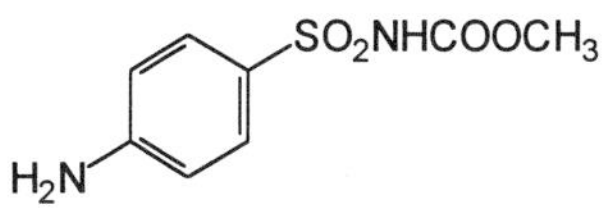

Merck Index No.
MP: 143-144°
BP:
Density:
Solubility: 5 g/l water; >800 g/l dimethylformamide; 340 g/l acetone; 280 g/l methanol; 280 g/l MEK; 120 g/l ethanol; <20 g/l hydrocarbons, chlorinated hydrocarbons
Octanol/water PC:
LD_{50}: rat orl 2000 mg/kg; rat ihl 18900 mg/m^3; rat skn >10000 mg/kg; mus orl >5000 mg/kg; mus unr 5000 mg/kg; dog orl >5000 mg/kg; pgn orl >2000 mg/kg; ckn orl >2000 mg/kg; qal orl >2000 mg/kg.
40CFR180.360

CAS RN 1912-24-9
$C_8H_{14}ClN_5$

Atrazine
6-chloro-N-ethyl-N'-(1-methylethyl)-1,3,5-triazine-2,4-diamine
Aatrex
Aktikon

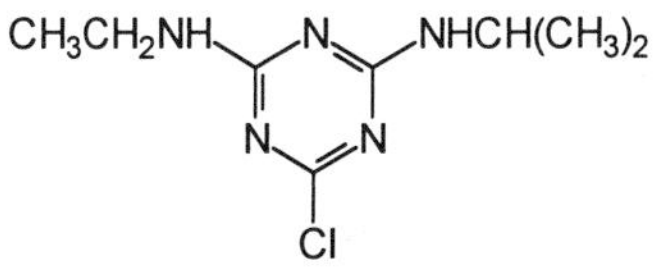

Merck Index No. 886
MP: 171 - 174°, 176°
BP:
Density:
Solubility: 28 mg/l water; 183 g/kg DMSO; 52 g/kg chloroform; 28 g/kg ethyl acetate; 18 g/kg methanol; 12 g/kg diethyl ether; 0.36 g/kg pentane
Octanol/water PC: 219
LD_{50}: rat orl 672 mg/kg; rat ihl LC_{50} 5200 mg/m^3/4H; rat skn >12500 mg/kg; rat ipr 235 mg/kg; mus orl 850 mg/kg; mus orl 1750mg/kg; mus ipr 626 mg/kg; rbt orl 750 mg/kg; rbt skn 7500 mg/kg; ham orl 1000 mg/kg; mam unr 1400 mg/kg.
40CFR180.220

CAS RN 71751-41-2
$C_{48}H_{72}O_{14}$

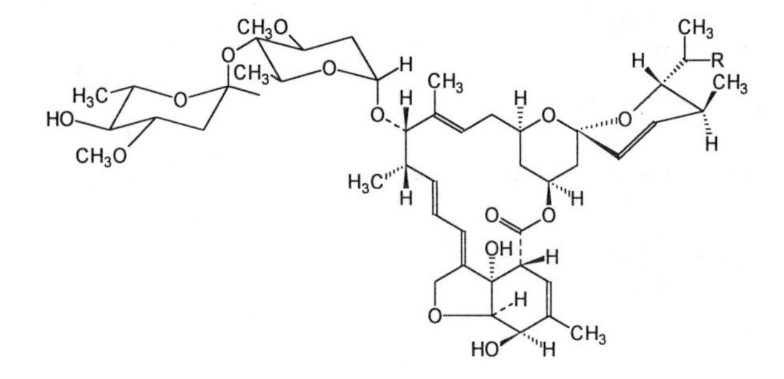

Avermectin B1a
Abamectin
Affirm
Agri-mek

Merck Index No. 1
MP: 150-155°
BP:
Density:
Solubility: 350 g/l toluene; 100 g/l acetone; 70 g/l isopropanol; 25 g/l chloroform; 20 g/l ethanol; 19.5 g/l methanol; 10 g/l nBuOH; 6 g/l cyclohexane
Octanol/water PC:
LD_{50}: rat orl 10mg/kg; rat ihl LC_{50} 1100 mg/m^3/4H; mus orl 13.5 mg/kg; mky orl 17 mg/kg; rbt skn >2000 mg/kg; dck unr 84.6 mg/kg; brd unr >2000 mg/kg.
40CFR180.449

CAS RN 86-50-0
$C_{10}H_{12}N_3O_3PS_2$

Azinphos-methyl
O,O-Dimethyl S-[(4-oxo-1,2,3-benzotriazin-3(4H)-yl)methyl]phosphorodithioate
Azinphosmethyl
Metiltriazotion

Merck Index No. 926
MP: 73-74°
BP:
Density: 1.44 SG
Solubility: 30 mg/l water; >20 g/l toluene, dichloromethane
Octanol/water PC:
LD_{50}: rat orl 7 mg/kg; rat orl 11mg/kg; rat ihl LC_{50} 69 mg/m^3/1H; rat der 220mg/kg; rat ipr 4.9 mg/kg; rat ivn 7.5 mg/kg; rat unr 15 mg/kg; mus orl 8.6 mg/kg; mus skn 65 mg/kg; dor orl 10 mg/kg; gpg orl 80 mg/kg; gpg skn >280 mg/kg; gpg ipr 40 mg/kg; ckn orl 277 mg/kg; qal unr 32.2 mg/kg; dck orl 136 mg/kg; bwd orl 8.5 mg/kg; bwd orl 8 mg/kg.
40CFR180.154

CAS RN 6923-22-4
$C_7H_{14}NO_5P$

Monocrotophos
Azodrin
Bilobran
(E)-Dimethyl 1-methyl-3-(methylamino)-3-oxo-1-propenyl phosphate

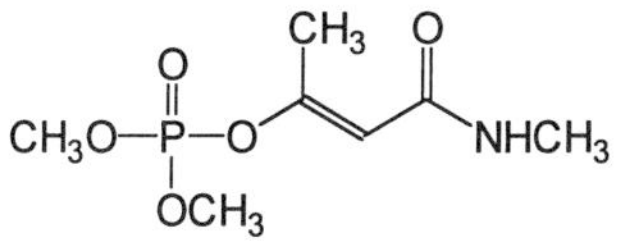

Merck Index No. 6161
MP: 54-55°
BP: $125^{o0.0005}$
Density: 1.33
Solubility: 1 kg/kg water; 1000 g/kg methanol; 800 g/kg dichloromethane; 700 g/kg acetone; 250 g/kg octanol; 60 g/kg toluene
Octanol/water PC:
LD_{50}: rat orl 8 mg/kg; rat ihl LC_{50} 63 mg/m^3/4H; rat skn 112 mg/kg; rat ipr LDLo 120 mg/kg; rat scu 6.964 mg/kg; rat ivn 9.2 mg/kg; mus orl 15 mg/kg; mus ipr 3.8 mg/kg; mus scu 8.71 mg/kg; mus ivn 9.2 mg/kg; rbt skn 354 mg/kg; pgn orl 3 mg/kg; ckn orl 3.54 mg/kg; qal orl 4 mg/kg; dck orl 3.36 mg/kg; bwd orl 0.80 mg/kg; bwd skn 4.2 mg/kg;
40CFR180.296

CAS RN 5251-93-4
$C_9H_9NO_4$

Benzadox (ACN)
Acetic acid, [(benzoylamino)oxy]- (9CI)
Acetic acid, (benzamidooxy)- (8CI)
(Benzamidooxy)acetic acid (ACN)
[(Benzoylamino)oxy]acetic acid
Benzamidooxyacetic acid
Benzamidoxyacetic acid
Hydroxylamine, N-benzoyl-O-(carboxymethyl)-
S 6173
S-7,173
S-7173
Topcide
Topicide

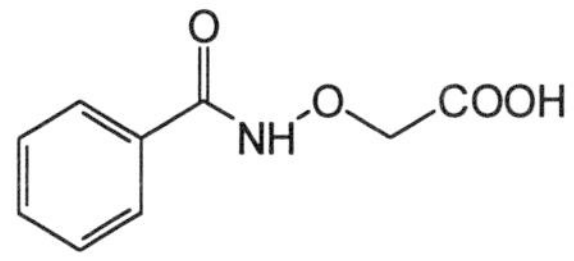

Merck Index No.
MP: 145°, 138 - 139°, 121 - 123°
BP:
Density:
Solubility:
Octanol/water PC:
LD_{50}: chd orl TDLo 180 mg/kg; chd orl TDLo 111 mg/kg; man skn TDLo 20 mg/kg/6W-I; rat orl 76 mg/kg; rat skn 414 mg/kg; rat ipr 35 mg/kg; rat unr 125 mg/kg; mus orl 44 mg/kg; mus ipr 125 mg/kg; dog orl 40 mg/kg; dog ivn LDLo 7.50 mg/kg; cat orl 25 mg/kg; cat unr LDLo 50 mg/kg; rbt orl 60 mg/kg; rbt skn 50 mg/kg; rbt ivn 4.50 mg/kg; gpg orl 127 mg/kg; ham orl 360 mg/kg; ham ipr 640 mg/kg; ham unr 300 mg/kg.
40CFR180.270 (revoked 5/4/88)

CAS RN 66-76-2
$C_{19}H_{12}O_6$

BHC
2H-1-Benzopyran-2-one, 3,3'-methylenebis[4-hydroxy- (9CI)
Coumarin, 3,3'-methylenebis[4-hydroxy- (8CI)
Acadyl
Acavyl
Antitrombosin
Baracoumin
Bis(4-hydroxycoumarin-3-yl)methane
Bis-3,3'-(4-hydroxycoumarinyl)methane
Bishydroxycoumarin
Cuma
Cumid
CUMA
Di(4-hydroxy-3-coumarinyl)methane
Di-4-hydroxy-3,3'-methylenedicoumarin
Dicoumal
Dicoumarin
Dicoumarol
Dicoumarolum
Dicuman

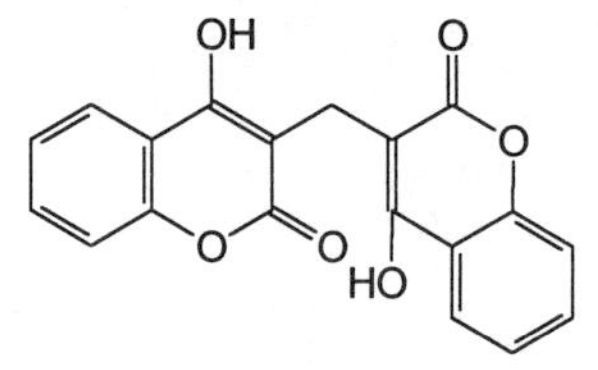

Merck Index No. 3080
MP: 287 - 293°
BP:
Density:
Solubility: sol. in alkaline solutions, pyridine and other organic bases, sl. sol. benzene, chloroform, insol. water, alcohol, ether.
Octanol/water PC:
LD_{50}: rat orl 250 mg/kg; rat orl 541.6mg/kg; rat ivn 52 mg/kg; mus orl 233 mg/kg; mus ipr 91 mg/kg; mus scu 50 mg/kg; mus ivn 42 mg/kg; dog ivn 40 mg/kg; rbt orl 75 mg/kg; rbt ivn 22 mg/kg; gpg ivn 59 mg/kg.
CFR: 40CFR180.140 (removed 7-16-86)

CAS RN 1214-39-7
$C_{12}H_{11}N_5$

N-Benzyladenine
1H-Purin-6-amine, N-(phenylmethyl)- (9CI)
Adenine, N-benzyl- (8CI)
Adenine, 6-benzyl-
BA
BA (growth stimulant)
BA (growth stimulator)
BAP
BAP (cytokinin)
BAP (growth stimulant)
N^6-Benzyladenine
N(6)-(Benzylamino)purine
N-(Phenylmethyl)-1H-purin-6-amine
N-[Phenylmethyl]-1H-purin-6-amine (ACN)
N6-Benzyladenine
Promalin
Purine, 6-amino-N-phenylmethyl-
SD 4901
SQ 4609
Verdan senescence inhibitor

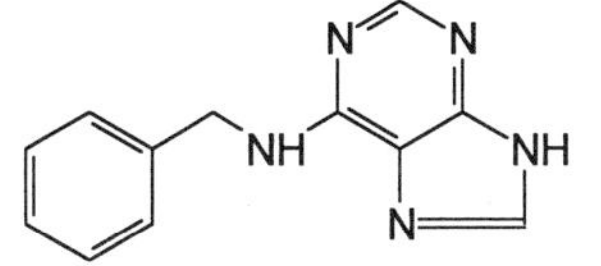

Merck Index No.
MP: 229 - 231°, 226 - 228°
BP:
Density:
Solubility:
Octanol/water PC:
LD_{50}: rat orl 2125 mg/kg; mus orl 1300 mg/kg; mus skn >5000 mg/kg; mus scu >2300 mg/kg.
CFR: 40CFR180.376

CAS RN 42576-02-3
$C_{14}H_9Cl_2NO_5$

Bifenox (ACN)
Benzoic acid, 5-(2,4-dichlorophenoxy)-2-nitro-, methyl ester (9CI)
Methyl 5-(2,4-dichlorophenoxy)-2-nitrobenzoate (ACN)
Methyl 5-(2',4'-dichlorophenoxy)-2-nitrobenzoate
Modown
MC 4379
MC-4379
2,4-Dichlorophenyl 3-methoxycarbonyl-4-nitrophenyl ether

COOCH3
O2N
Cl
O
Cl

Merck Index No. 1228
MP: 84-86°
BP:
Density:
Solubility: 0.35 mg/l water 25°, 400 g/l acetone, 400 g/l C6H5Cl, 300 g/l xylene, <50g/l ethanol
Octanol/water PC: 31,700
LD_{50}: rat orl 6400 mg/kg; rat ihl LC50 >200000 mg/kg; mus orl 4556 mg/kg; rbt skn >20000 mg/kg.
40CFR180.351

CAS RN 485-31-4
$C_{15}H_{18}N_2O_6$

Binapacryl (ACN)
2-Butenoic acid, 3-methyl-, 2-(1-methylpropyl)-4,6-dinitrophenyl ester (9CI)
Crotonic acid, 3-methyl-, 2-sec-butyl-4,6-dinitrophenyl ester (8CI)
(6-(1-Methyl-propyl)-2,4-dinitro-fenyl)-3,3-dimethyl-acrylaat (Dutch)
(6-(1-Methyl-propyl)-2,4-dinitro-phenyl)-3,3-dimethyl-acrylat (German)
(6-(1-Metil-propil)-2,4-dinitro-fenil)-3,3-dimetil-acrilato (Italian)
Acricid
Ambox
Dapacryl
Dinapacryl
Endosan
ENT 25793
FMC 9044
HOE 2784
HOE 2784 oa
Morocide
Morotsid
Morrocid

CH3
CH2CH3
CH3
O
CH3
O
O2N
NO2

Merck Index No. 1237
MP: 70°
BP:
Density: 1.25 - 1.28
Solubility: v. sol. acetone, xylene, sol. ethanol, kerosene, insol. water
Octanol/water PC:
LD_{50}: rat orl 58, 63mg/kg
CFR: 40CFR180.319

CAS RN 2439-99-8
$C_4H_{11}NO_8P_2$

Glyphosine (ACN)
Glycine, N,N-bis(phosphonomethyl)- (8CI9CI)
C 9552
CP 41845
Glycine-N,N-bis(methylenephosphonic acid)
Glycine, N,N-bis(phosphonomethyl)
Glycinedimethanephosphonic acid
MON 845
MON-O45
N,N-Bis(phosphonomethyl)glycine (ACN)
Nitrilomonoacetic acid dimethylenephosphonic acid
Nitrilomonomethylcarbonyldimethylphosphonic acid
Phosphonic acid, [[(carboxymethyl)imino]dimethylene]di-
Polaris

COOH
O O
HO—P—N—P—OH
OH OH

Merck Index No. 4409
MP:
BP:
Density:
Solubility: 248g/l water 20°
Octanol/water PC:
LD_{50}: rat orl 3925 mg/kg; mam orl 3925mg/kg
CFR: 40CFR180.354 (revoked 5-4-88)

CAS RN 7440-42-8
B

Boron (8CI9CI)
Boron atom
Neutron absorber

B

Merck Index No. 1345
MP: ca. 2200°
BP: $2140^{\circ 0.0000156}$
Density: 2.31, 2.350
Solubility: insol. water, HCl, HF, sol. boiling HNO_3, H_2SO_4, molten metals
Octanol/water PC:
LD_{50}: rat orl 650 mg/kg; mus orl 560 mg/kg; mus ipr LDLo 1000 mg/kg; dog orl 310 mg/kg; cat orl 250 mg/kg; rbt orl 310 mg/kg; gpg orl 310 mg/kg; mam orl 300 mg/kg.
CFR: 40CFR180.271

CAS RN 99-30-9
$C_6H_4Cl_2N_2O_2$

Botran
Benzenamine, 2,6-dichloro-4-nitro- (9CI)
Aniline, 2,6-dichloro-4-nitro- (8CI)
Allisan
AL-50
Bortran
CDNA
CNA
Dichloran
Dichloran (amine fungicide)
Dicloran
Ditranil
DCNA
DCNA (fungicide)
Resisan
RD-6584
U-2069
2,6-Dichlor-4-nitroanilin (Czech)
2,6-Dichloro-4-nitroaniline (ACN)

Merck Index No:
MP: 195°
BP:
SG:
Solubility: 6.3mg/l water 20°, 34g/l acetone, 40g/l dioxane, 12g/l chloroform, 19g/l ethyl acetate, 4.6g/l benzene, 3.6g/l xylene, 0.06g/k cyclohexane
Octanol/water PC: 63
LD_{50}: rat orl 2400 mg/kg; rat ihl LC_{50} >21600 mg/m^3/1H; mus orl 1500 mg/kg; mus skn >5000 mg/kg; mus ivn 56 mg/kg; mus unr 4490 mg/kg; rbt skn >2000 mg/kg; gpg orl 1450 mg/kg; dck orl >2000 mg/kg; mam unr 1500 mg/kg.
CFR: 40CFR180.200

CAS RN 1897-45-6
$C_8Cl_4N_2$

Chlorothalonil (ACN)
1,3-Benzenedicarbonitrile, 2,4,5,6-tetrachloro- (9CI)
Isophthalonitrile, tetrachloro- (8CI)
m-Phthalodinitrile, tetrachloro-
m-Tetrachlorophthalodinitrile
m-Tetrachlorophthalonitrile
m-TCPN
Bravo
Bravo 6F
Bravo-W-75
Chloroalonil
Chlorthalonil
Daconil
Daconil 2787
DAC 2787
Exotherm
Forturf
Isophthalonitrile, 2,4,5,6-tetrachloro-
Nopcocide N 96
NCI-C00102

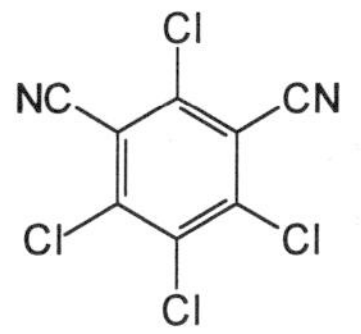

Merck Index No. 2167
MP: 250-251°
BP: 350°
Density:
Solubility: 0.6mg/l water 25°, 80g/kg xylene, 30g/kg cyclohexanone, 30g/kg dimethylformamide, 20g/kg acetone, 20g/kg DMSO, <10g/kg kerosene
Octanol/water PC:
LD_{50}: rat orl >10g/kg
CFR: 40CFR180.275

CAS RN 314-40-9
$C_9H_{13}BrN_2O_2$

Bromacil (ACN)
2,4(1H,3H)-Pyrimidinedione, 5-bromo-6-methyl-3-(1-methylpropyl)- (9CI)
Uracil, 5-bromo-3-*sec*-butyl-6-methyl- (8CI)
Borea
Borocil IV
Bromazil
Cynogan
Herbicide 976
Hyvar x-ws
Hyvar X
Hyvar X bromacil
Hyvar X Weed Killer
Hyvar X-L
Hyvarex
Krovar II
Nalkil
Uragan
Urox B
Urox HX or B
Urox-Hx

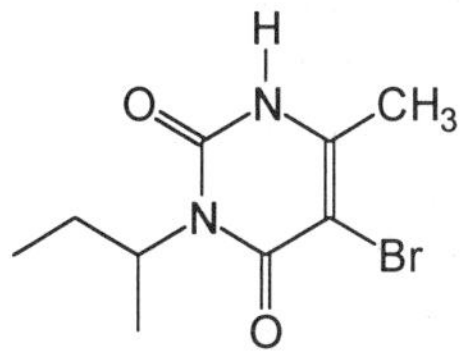

Merck Index No. 1370
MP: 158-159°
BP:
Density: 1.55
Solubility: 815mg/l water 25°, 134g/l ethanol, 167g/l acetone, 71g/l acetonitrile, 32g/l xylene, 88g/l 3% aq. NaOH
Octanol/water PC:
LD_{50}: rat orl 5200mg/kg
CFR: 40CFR180.210

CAS RN 1689-84-5
$C_7H_3Br_2NO$

Bromoxynil (ACN)
Benzonitrile, 3,5-dibromo-4-hydroxy- (8CI9CI)
Brittox
Brominal
Brominal M
Brominal Plus
Brominex
Brominil
Broxynil
Buctril
Buctril industrial
Butilchlorofos
Chipco buctril
Chipco Crab-Kleen
Chipro buctril
ENT 20852
M&B 10,064
MB 10064
MB-10064
Nu-lawn weeder

Br
OH
NC
Br

Merck Index No.
1431
MP: 194-195°
BP:
Density:
Solubility: 130mg/l water 25°, 90g/l methanol, 70g/l ethanol, 170g/l acetone, 170g/l cyclohexanone, 410g/l THF, 610g/l dimethylformamide, <20g/l mineral oils
Octanol/water PC:
LD_{50}: rat orl 190 mg/kg; rat skn >2 mg/kg; mus orl 110 mg/kg; mus ivn 56 mg/kg; dog orl 100 mg/kg; rbt orl 260 mg/kg; rbt skn 3660 mg/kg; gpg orl 63 mg/kg; ckn orl 240 mg/kg; dck orl 200 mg/kg; mam unr 190 mg/kg; bwd orl 50 mg/kg.
40CFR180.324

CAS RN 106-96-7
C_3H_3Br

Propargyl bromide (ACN)
1-Propyne, 3-bromo- (9CI)
Propyne, 3-bromo- (8CI)
γ-Bromoallylene
Methylacetylene, 3-bromo-
Propynyl bromide
1-Bromo-2-propyne
2-Propynyl bromide
3-Bromo-1-propyne
3-Bromopropyne

Br
HC≡C-CH_2

Merck Index No.
MP:
BP: 88 - 90°; BP: $33^{\circ 130}$
Density: 1.58
Solubility: v. sol. alcohol, diethyl ether, benzene, sol. carbon tetrachloride, chloroform
Octanol/water PC:
LD_{50}: rat orl LDLo 53 mg/kg; gpg orl 29 mg/kg.
CFR: 40CFR180.199

CAS RN 106-93-4
$C_2H_4Br_2$

ethylene dibromide
Ethane, 1,2-dibromo- (8CI9CI)
α,β-Dibromoethane
E-D-Bee
s-Dibromoethane
s-Dibromoethylene
Aadibroom
Acetylene dibromide
Aethylenbromid (German)
Bromofume
Bromuro di etile (Italian)
Celmid
Celmide
Dibromoethane
Dibromure d'ethylene (French)
Dowfume
Dowfume edb
Dowfume EDB
Dowfume MC-2
Dowfume W-8

Br–CH_2–CH_2–Br

Merck Index No. 3753
MP: 9°
BP: 131-132°
Density: 2.172
Solubility: soluble in 250 parts water, misc. alcohol, ether
Octanol/water PC:
LD_{50}: wmn orl LDLo 90 mg/kg; rat orl 108 mg/kg; rat ihl LC_{50} 14300 mg/m^3/30M; rat skn 300 mg/kg; mus orl 250 mg/kg; mus ipr 220mg/kg; mus unr 146 mg/kg; rbt orl 55 mg/kg; rbt skn 300 mg/kg; rbt rec LDLo 2500 mg/kg; gpg orl 110 mg/kg; gp
CFR: 40CFR180.126; 40CFR180.146

CAS RN 7778-44-1
AsH_3O_4.3/2Ca

Calcium arsenate ($Ca_3(AsO_4)_2$)
Arsenic acid (H_3AsO_4), calcium salt (2:3) (8CI9CI)
Arseniate de calcium (French)
Arsenic acid, calcium salt
Arsenic acid, calcium salt (2:3)
Calcium arsenate, solid (DOT)
Calcium orthoarsenate
Chip-Cal
Chip-Cal Granular
Cucumber dust
Cucumber Dust
Fencal
FLAC
Kalziumarseniat (German)
Kilmag
KALO
Pencal
Pencel
Security
Spra-cal
Tricalcium arsenate
Tricalcium orthoarsenate
Tricalciumarsenat (German)

O=As(OH)–OH 3/2 Ca (HO–As(=O)(OH)–OH 3/2 Ca)

Merck Index No. 1646
MP:
BP:
Density: 1.455 SG
Solubility: 130 ppm water
Octanol/water PC:
LD_{50}: rat orl 298mg/kg; rat orl 20 mg/kg; rat skn 2400 mg/kg; mus orl 794 mg/kg; dog orl 38 mg/kg; rbt orl 50 mg/kg.
CFR: 40CFR180.192 (removed 4/3/91)

CAS RN 96-12-8
$C_3H_5Br_2Cl$

Nemagon
Propane, 1,2-dibromo-3-chloro- (8CI9CI)
BBC
BBC 12
Dibromchlorpropan (German)
Dibromochloropropane
DBCP
ENT-18445
Fumagon
Fumazone
Fumazone 86
Fumazone 86e
Fumazone 86E
Nemabrom
Nemafume
Nemagon Soil Fumigant
Nemagon 20
Nemagon 20G
Nemagon 206
Nemagon 90

Merck Index No. 3003
MP:
BP: 196°; BP $78^{\circ 16}$; BP $21^{\circ 0.8}$
Density: 2.093
Solubility: sl. sol. water, misc. oils, dichloropropane, isopropanol
Octanol/water PC:
LD_{50}: rat orl 0.17g/kg, mus orl 0.26g/kg
CFR: 40CFR180.197

CAS RN 18181-80-1
$C_{17}H_{16}Br_2O_3$

Bromopropylate (ACN)
Benzeneacetic acid, 4-bromo-α-(4-bromophenyl)-α-hydroxy-, 1-methylethyl ester (9CI)
Benzilic acid, 4,4'-dibromo-, isopropyl ester (8CI)
Acarol
ENT 27552
Geigy 19851
GS 19851
Isopropyl dibromobenzilate
Isopropyl-4,4'-dibromobenzilate
Neoron
Phenisobromolate
1-Methylethyl 4-bromo-α-(4-bromophenyl)-α-hydroxybenzeneacetat
1-Methylethyl 4-bromo-α-(4-bromophenyl)-α-hydroxybenzeneacetate

Merck Index No. 1422
MP: 77°
BP:
Density: 1.59
Solubility: <5mg/l water, 850g/kg acetone, 970g/kg dichloromethane, 870g/kg dioxane, 750g/kg benzene, 280g/kg methanol, 530g/kg xylene, 90g/kg isopropanol
Octanol/water PC:
LD_{50}: rat orl 5000mg/kg
CFR: 40CFR180. (Pending, 1981)

CAS RN 8065-36-9
$C_{13}H_{19}NO_2.C_{13}H_{19}NO_2$

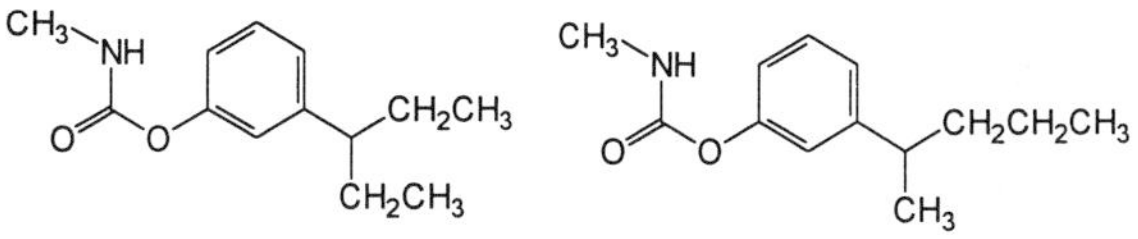

Bufencarb (ACN)
Bux 2 (8CI)
Bux
Bux insecticide
BUX-Ten
Carbamic acid, methyl-, *m*-(1-methyl)butyl)phenyl ester mixed with carbamic acid, methyl-, *m*-(1-ethylpropyl)phenyl ester (3:1)
Carbamic acid, methyl-, *m*-(1-methylbutyl)phenyl ester mixed with carbamic acid, methyl-, *m*-(1-ethylpropyl)phenyl ester (4:1)
Chevron re-5353
Compound 5353
Metalkamate
Mixture of *m*-(1-ethylpropyl)phenyl methylcarbamate & *m*-(1-methylbutyl)phenyl methylcarbamate (ratio of 1:3)
RE-5353
3-(1-Methylbutyl)phenyl methylcarbamate and 3-(1-Ethylpropyl)phenyl methylcarbamate (3:1) (ACN)

Merck Index No. 1459
MP: 26 - 39°
BP: 125° $^{0.04}$
Density: 1.024^{26}
Solubility: < 50 ppm water, v. sol. xylene, methanol,
Octanol/water PC:
LD_{50}: rat orl 97 mg/kg, rat orl 61 mg/kg.
CFR: 40CFR180.255

CAS RN 21564-17-0
$C_9H_6N_2S_3$

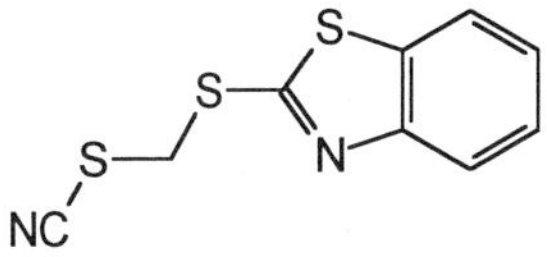

BUSAN 72A
Thiocyanic acid, (2-benzothiazolylthio)methyl ester (8CI9CI)
(2-Benzothiazolylthio)methyl thiocyanate
component of Busan 44
component of Busan 74
component of Halt
Benzothiazole, 2-(thiocyanomethylthio)-
Busan 30
Busan 44
Busan 72
Busan 72A
Halt
Ichiban
KVK 733059
Thiocyanic acid, 2-(benzothiazolylthio)methyl ester
TCMTB
2-(Benzothiazolylthio)methylthiocyanate
2-(Thiocyanomethylthio)benzothiazole (ACN)

Merck Index No.:
MP:
BP:
d:
Solubility:
Octanol/water PC:
LD_{50}: rat orl 2000 mg/kg; rat skn >5000 mg/kg; rat ipr 73 mg/kg; rat scu 1300 mg/kg; mus orl 445 mg/kg; mus ipr 143 mg/kg; mus scu 205 mg/kg; rbt skn 10000 mg/kg.
CFR: 40CFR180.288

CAS RN 23184-66-9
$C_{17}H_{26}ClNO_2$

Butachlor (ACN)
Acetamide, N-(butoxymethyl)-2-chloro-N-(2,6-diethylphenyl)- (9CI)
Acetanilide, N-(butoxymethyl)-2-chloro-2',6'-diethyl- (8CI)
Acetanilide, 2-chloro-2',6'-diethyl-N-(butoxymethyl)-
Butanex
CP 53619
Machete
Machette
N-(Butoxymethyl)-2-chloro-N-(2,6-diethylphenyl)acetamide
N-(Butoxymethyl)-2-chloro-2',6'-diethylacetanilide (ACN)
2-Chloro-2',6'-diethyl-N-(butoxymethyl)acetanilide
2-Chloro-2',6'-diethyl-N-(n-butoxymethyl)acetanilide
2-Chloro-2',6'-diethyl-N-(N-butoxymethyl)acetanilide

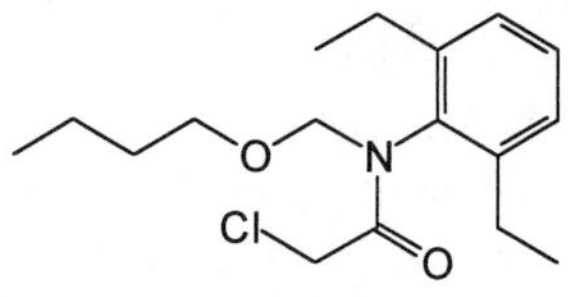

Merck Index No. 1498
MP: < -5°
BP: $156^{\circ 0.5}$
Density: 1.070^{25}
20mg/l water 20°, sol. in diethyl ether, acetone, benzene, ethanol, ethyl acetate, hexane
Octanol/water PC:
LD_{50}: rat orl 1740mg/kg
CFR: 40CFR180.288

CAS RN 33629-47-9
$C_{14}H_{21}N_3O_4$

Butralin (ACN)
Benzenamine, 4-(1,1-dimethylethyl)-N-(1-methylpropyl)-2,6-dinitro- (9CI)
Aniline, N-*sec*-butyl-4-*tert*-butyl-2,6-dinitro- (8CI)
A 820
Amchem 70-25
Amex
Amex 820
Amex. 220
Aniline,
4-(1,1-dimethylethyl)-N-(1-methylpropyl)-2,6-dinitro-
AMEX
Butalin
Dibutalin
N-*sec*-Butyl-4-*tert*-butyl-2,6-dinitroaniline
N-*sec*-Butyl-4-*tert*-butyl-2,6-dinitrobenzamine
No-Crab
Rutralin
Sector
Tamex
4-(1,1-Dimethylethyl)-N-(1-methylpropyl)-2,6-dinitro-
benzenamine (ACN)

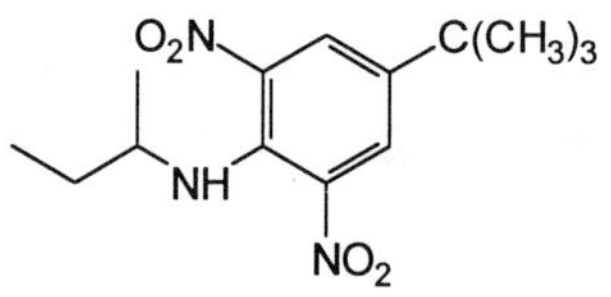

Merck Index No. 1532
MP: 60-61°
BP: $134\text{-}136^{\circ 0.5}$
Density:
Solubility: 1 mg/l water 24°, 4.48 kg/kg acetone, 2.7 kg/kg benzene, 9.55 kg/kg methyl ethyl ketone, 1.46 kg/kg carbon tetrachloride, 3.88 kg/kg xylene
Octanol/water PC:
LD_{50} rat orl 2500mg/kg; rat ihl LC_{50} 50 g/m^3/4H; rbt skn 200 mg/kg; mam orl 12600 mg/kg.
CFR: 40CFR180.358

CAS RN 563-83-7
C_4H_9NO

***sec*-Butylamide**
Propanamide, 2-methyl- (9CI)
Isobutyramide (8CI)
Isobutylamide
Isobutyrimidic acid
Isopropylformamide
2-Methylpropanamide
2-Methylpropionamide

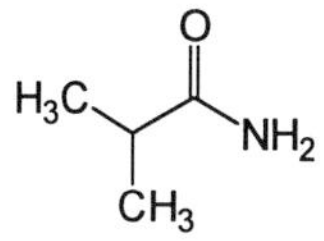

Merck Index No.
MP: 127 - 129°
BP: 216 - 220°
Density:
Solubility:sol. chloroform
Octanol/water PC:
CFR: 40CFR180.321

CAS RN 886-50-0
$C_{10}H_{19}N_5S$

Terbutryn (ACN)
1,3,5-Triazine-2,4-diamine, N-(1,1-dimethylethyl)-N'-ethyl-6-(methylthio)- (9CI)
s-Triazine, 2-(*tert*-butylamino)-4-(ethylamino)-6-(methylthio)- (8CI)
N-(1,1-Dimethylethyl)-N'-ethyl-6-(methylthio)-1,3,5-triazine-2,4-diamine
A 1866 (VAN)
Clarosan
Igran
Igran 50
Igran 500
Prebane
Shortstop
Shortstop E
Terbutrex
Terbutrin
Terbutryne
2-(*tert*-Butylamino)-4-(ethylamino)-6-(methylmercapto)-*s*-triazine
2-(*tert*-Butylamino)-4-(ethylamino)-6-(methylthio)-*s*-triazine (ACN)
2-(*tert*-Butylamino)-4-(ethylamino)-6-(methylthio)-1,3,5-triazine
4-Aethylamino-2-tert-butylamino-6-methylthio-*s*-triazin (German)

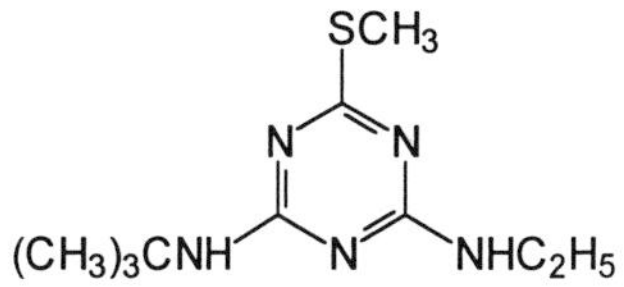

Merck Index No.
MP: 104-105°
BP: 154-160° $^{0.06}$
Density: 1.115^{20}
Sol: 25mg/l water 20, 280g/l acetone, 9 g/l hexane, 300g/l dichloromethane, 130g/l octanol, 280g/l methanol, 45g/l toluene
Octanol/water PC: 3070 (octanol/water)
LD_{50}: rat orl 2045mg/kg; mus orl 500 mg/kg; rat skn >2000 mg/kg; rat ipr 699 mg/kg; mus orl 3884 mg/kg; mus ipr 554 mg/kg; rbt skn >10200 mg/kg; mam unr 2900 mg/kg.
CFR: 40CFR180.265

CAS RN 23947-60-6
$C_{11}H_{19}N_3O$

Ethirimol
4(1H)-Pyrimidinone, 5-butyl-2-(ethylamino)-6-methyl- (9CI)
4(3H)-Pyrimidinone, 5-butyl-2-(ethylamino)-6-methyl- (8CI)
Milcurb super
Milgo
Milgo E
Milstem
New milstem
PP 149
PP149
4-(1H)Pyrimidinone, 5-butyl-2-(ethylamino)-6-methyl-
5-*n*-Butyl-2-(ethylamino)-4-hydroxy-6-methyl-pyrimidine
5-Butyl-2-(ethylamino)-4-hydroxy-6-methylpyrimidine (ACN)
5-Butyl-2-(ethylamino)-6-methyl-4(1H)-pyrimidinone
5-Butyl-2-(ethylamino)-6-methyl-4-pyrimidinol

$CH_2CH_2CH_2CH_3$
CH_3
OH
N
N
NHC_2H_5

Merck Index No. 3695
MP: 159-160° (phase change at 140°)
BP:
Density: 1.21^{25}
Sol: 200mg/l water 20, 150g/kg chloroform, 24g/kg ethanol, 5g/kg acetone
Octanol/water PC: 200 (20, pH7)
LD_{50}: rat orl 4000 mg/kg, rat orl 6340 mg/kg, mus orl 4000mg/kg
CFR: 40CFR180. (pending 1977)

CAS RN 94-82-6
$C_{10}H_{10}Cl_2O_3$

Butyrac 118
Butanoic acid, 4-(2,4-dichlorophenoxy)- (9CI)
Butyric acid, 4-(2,4-dichlorophenoxy)- (8CI)
γ-(2,4-Dichlorophenoxy)butyric acid
(2,4-Dichlorophenoxy)butyric acid
Butoxon
Butoxone
Butyrac
Butyrac 200
Embutox
Embutox klean-up
Legumex
2-4-DB
2,4-D Butyric
2,4-D Butyric acid
2,4-DB (VAN)
2,4-DB acid
4-(2,4-Dichlorophenoxy)butanoic acid
4-(2,4-Dichlorophenoxy)butyric acid (ACN)
4-(2,4-DB)

O
COOH
Cl
Cl

Merck Index No. 2828
MP: 117-119°
BP:
Density:
Solubility: 46mg/l water 25°, readily sol. acetone, ethanol, diethyl ether, slightly sol. benzene, toluene, kerosene
Octanol/water PC:
LD_{50}: rat orl 370-700mg/kg, sodium salt: rat orl 1500mg/kg, mus orl 400mg/kg; rat skn 800 mg/kg; rbt skn >10000mg/kg.
CFR: 40CFR180.331

CAS RN 13071-79-9
$C_9H_{21}O_2PS_3$

Terbufos (ACN)
Phosphorodithioic acid, S-[[(1,1-dimethylethyl)thio]methyl] O,O-diethyl ester (9CI)
Phosphorodithioic acid, S-[(*tert*-butylthio)methyl] O,O-diethyl ester (8CI)
AC 92100
Counter
Counter 15 G
Counter 15g soil insecticide
Counter 15G
O,O-Diethyl S-(*tert*-butylthio)methyl phosphorodithioate
Phosphorodithioic acid, O,O-diethyl S-[[(1,1-dimethylethyl)thio] methyl] ester
S-[(tert-Butylthio)methyl] O,O-diethyl phosphorodithioate
S-[(1,1-Dimethyl)thio]methyl O,O-diethyl phosphorodithioate
S-[(1,1-Dimethylethylthio)methyl] O,O-diethyl phosphorodithioate
S-[[(1,1-Dimethylethyl)thio]methyl] O,O-diethyl phosphorodithioate (ACN)
ST-100

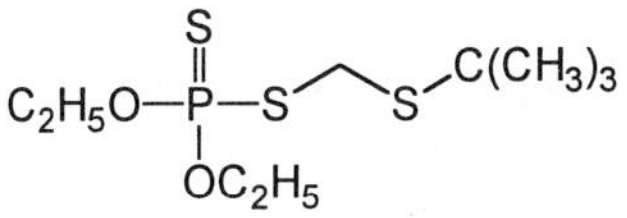

$C_2H_5O-P(=S)(OC_2H_5)-S-CH_2-S-C(CH_3)_3$

Merck Index No. 9088
MP: -29.2°
BP: $69^{\circ\ 0.01}$
Density: 1.105^{24}
Solubility: 4.5 mg/ml water, ca. 300 g/l aromatic hydrocarbons, chlorinated hydrocarbons, alcohols, acetone.
Octanol/water PC: 33,000
LD_{50}: rat orl 1.6mg/kg, rat orl 4.5 mg/kg; rat skn 7.4 mg/kg; mus orl 5.0 mg/kg; mus orl 3.5 mg/kg; rbt skn 1.1 mg/kg.
CFR: 40CFR180.352

CAS RN 592-01-8
C_2CaN_2

Calcium cyanide (ACN)(8CI)
Calcium cyanide ($Ca(CN)_2$) (9CI)
Calcid
Calcium cyanide (mixture)
Calcium cyanide, solid (DOT)
Calcyan
Calcyanide
Cyanogas
Cyanure de calcium (French)

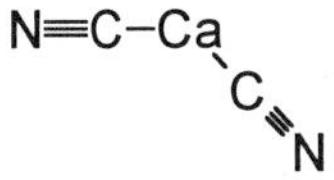

$N\equiv C-Ca-C\equiv N$

Merck Index No. 1664
MP: decomposes at 35°
BP: decomposes at 35°
Density:
Solubility: Sol. water (reacts with water; liberates HCN)
Octanol/water PC:
LD_{50}: rat orl 39mg/kg.
CFR: 40CFR180.125

CAS RN 8001-35-2
W99

Structure information not available for this compound

Toxaphene (ACN)(DOT)(8CI9CI)
Agricide maggot killer (f)
Alltex
Alltox
Camphechlor
Camphene, chlorinated
Camphochlor
Camphoclor
Camphofene huileux
Chem-Phene
Chlorinated camphene
Chlorinated camphene, technical
Chlorinated camphenes
Chlorocamphene
Clor chem t-590
Compound 3956
Compound-3956
Coopertox
Crestoxo
Cristoxo
Cristoxo 90
$C_{10}H_{10}Cl_6$-Technical chlorinated camphene, 67-69% chlorine
$C_{10}H_{10}Cl_8$-Technical chlorinated camphene, 67-69% chlorine
Estonox
ENT 9,735
ENT-9735
Fasco-Terpene
Geniphene
Gy-Phene
Hercules toxaphene

Merck Index No. 9478
MP: 35° (dec); 65-90°
BP: (dec)
Density:
Solubility: Insol. water, sol. aromatic hydrocarbons
Octanol/water PC:
LD_{50}: rat orl 90, 80mg/kg; rat orl 50 mg/kg; hum orl LDLo 28 mg/kg; man orl 29 mg/kg; hum skn TDLo 657 mg/kg; man unr LDLo 44 mg/kg; rat skn 600 mg/kg; rat ipr LDLo 70 mg/kg; rat par 90 mg/kg; rat unr 240 mg/kg; mus orl 112 mg/kg; mus ihl LCLo 2000 mg/m3/2H; mus ipr 42 mg/kg; mus unr 45 mg/kg; dog orl 15 mg/kg; rbt orl 75 mg/kg; rbt skn 1025 mg/kg; gpg orl 250 mg/kg; ham orl 200 mg/kg; dck orl 31 mg/kg.
CFR: 40CFR180.138

CAS RN 2425-06-1 (2939-80-2)
$C_{10}H_9Cl_4NO_2S$

Captafol
1H-Isoindole-1,3(2H)-dione, 3a,4,7,7a-tetrahydro-2-[(1,1,2,2-tetrachloroethyl)thio]- (9CI)
4-Cyclohexene-1,2-dicarboximide, N-[(1,1,2,2-tetrachloroethyl)thio]- (8CI)
(Tetrachloroethylthio)tetrahydrophthalimidecis-N-[(1,1,2,2-Tetrachloroethyl)thio]-4-cyclo-hexene-1,2-dicarboximide
Captatol
Captofol
CS 5623
Difolatan
Difolatan 4F1
Difolatan-R
Difosan
Folcid
N-(Tetrachloroethylthio)tetrahydrophthalimide
N-(1,1,2,2-Tetrachloraethylthio)-cyclohex-4-en-1,4-diacarboximid (German)
N-[(1,1,2,2-Tetrachloroethyl)sulfenyl]-*cis*-4-cyclohexene-1,2-dicarboximide
N-[(1,1,2,2-Tetrachloroethyl)thio]-4-cyclohexene-1,2-dicarboximide
N-1,1,2,2-Tetrachloroethylmercapto-4-cyclohexene-1,2-carboximide
Sanspor
Sulfonimide
3a,4,7,7a-Tetrahydro-2-[(1,1,2,2-tetrachloroethyl)thio]-1H-isoindole-1,3-(2H)-dione
4-Cyclohexene-1,2-dicarboximide, *cis*-N-[(1,1,2,2-tetrachloroethyl)thio]-

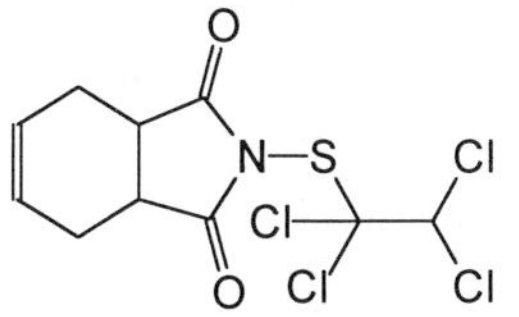

Merck Index No. 1770
MP: 160-161°
BP:
Density:
Solubility: 1.4 mg/l water 20, 13 g/kg isopropanol, 25 g/kg benzene, 17 g/kg toluene, 100 g/kg xylene, 43 g/kg acetone, 44 g/kg methyl ethyl ketone, 170 g/kg dimethylsulfoxide
Octanol/water PC:
LD_{50}: rat orl 6200mg/kg, 4200 mg/kg; rat orl 2500 mg/kg; rat unr >6000 mg/kg; mus ipr LDLo 3 mg/kg; rbt skn 15400 mg/kg.
CFR: 40CFR180.267

CAS RN 133-06-2
$C_9H_8Cl_3NO_2S$

Captan (ACN)
1H-Isoindole-1,3(2H)-dione, 3a,4,7,7a-tetrahydro-2-[(trichloromethyl)thio]- (9CI)
4-Cyclohexene-1,2-dicarboximide, N-[(trichloromethyl)thio]- (8CI)
[(Trichloromethyl)thio]tetrahydrophthalamide
Aacaptan
Amercide
Captaf
Captaf 85W
Captan 50W
Captancapteneet 26,538
Captex
Esso fungicide 406
Essofungicide 406
Flit 406
Fungicide contg. captan
Fungus Ban Type II
Glyodex 37-22
Kaptan
Malipur
Merpan
Merpan 90

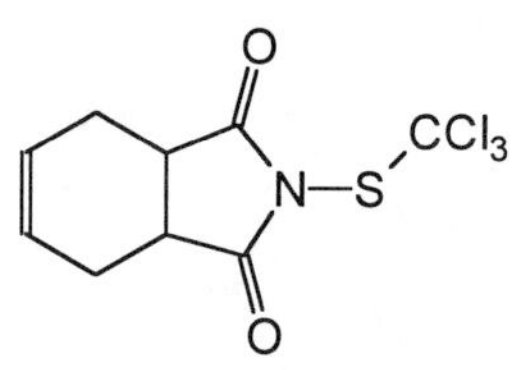

Merck Index No. 1771
MP: 178°
BP:
Density: 1.74 SG
Solubility: Insol water, 7.78g/100ml chloroform, 8.15g/100ml tetrachoroethane, 4.96g/100ml cyclohexanone, 4.70g/100ml dioxane, 2.13g/100ml benzene, 0.69g/100ml toluene, 0.04g/100ml C7H16, 0.29g/100ml ethanol, 0.25g/100ml diethyl ether
Octanol/water PC: 610
LD_{50}: hum orl LDLo 1071 mg/kg; rat orl 9000mg/kg; rat ihl LC_{50} >5700 mg/m^3/2H; rat skn >5000 mg/kg; rat ipr LDLo >25 mg/kg; rat unr 2650 mg/kg; mus orl 7000 mg/kg; mus ihl 4500 mg/m^3/2H; mus irp 30 mg/kg; mus unr 138 mg/kg; rbt unr 740 mg/kg; brd unr LDLo >100 mg/kg.
CFR: 40CFR180.103

CAS RN 63-25-2
$C_{12}H_{11}NO_2$

Carbaryl (ACN)
1-Naphthalenol, methylcarbamate (9CI)
Carbamic acid, methyl-, 1-naphthyl ester (8CI)
α-Naftyl-N-methylkarbamat (Czech)
α-Naphthalenyl methylcarbamate
α-Naphthyl methylcarbamate
α-Naphthyl N-methylcarbamate
Atoxan
Bercema NMC50
Caprolin
Carbamic acid, N-methyl-1-naphthyl-
Carbamic acid, N-methyl, 1-naphthyl ester
Carbamine
Carbaril
Carbarilo
Carbarilum
Carbatox
Carbatox 75
Carbatox-60
Carbavur
Carpolin
Compound 7744
Denapon
Dicarbam
Dyna-carbyl

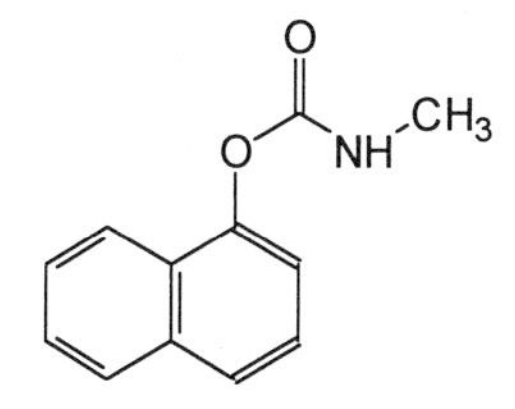

Merck Index No. 1789
MP: 142°
BP:
Density: 1.232 SG
Solubility: 1000 ppm water; 40mg/l water 30, 400-450g/kg dimethylformamide, 400-450g/l dimethylsulfoxide, 200-300g/l acetone, 200-250g/l cyclohexanone, 100g/l isopropanol, 100g/l xylene
Octanol/water PC:
LD_{50}: man orl TDLo 500 mg/kg; wmn orl LDLo 5000 mg/kg; rat orl 230 mg/kg; rat orl 850, 500 mg/kg; rat skn 4000 mg/kg; rat ipr 64 mg/kg; rat scu 1400 mg/kg; rat ivn 18mg/kg; rat unr 500 mg/kg; mus orl 128 mg/kg; mus ipr 25 mg/kg; mus scu 900 mg/kg; dog orl 759 mg/kg; cat orl 150 mg/kg; rbt orl 710 mg/kg; rbt skn 2000 mg/kg; rbt ipr 223 mg/kg; rbt scu LD >2000 mg/kg; gpg orl 250 mg/kg; ham orl LDLo 250 mg/kg; ham ipr 640 mg/kg; ckn scu LDLo 2000 mg/kg; grb orl 491 mg/kg; mam orl 310 mg/kg; bwd orl 56 mg/kg.
CFR: 40CFR180.169

CAS RN 10605-21-7
$C_9H_9N_3O_2$

Carbendazim
Carbamic acid, 1H-benzimidazol-2-yl-, methyl ester (9CI)
2-Benzimidazolecarbamic acid, methyl ester (8CI)
Bavistin
Bavistin 3460
Benzimidazole-2-carbamic acid, methyl ester
BAS 3460
BAS 3460F
BAS 67054F
BCM
BMK (VAN)
BMK (fungicide)
Carbendazime
Carbendazol
Carbendazole
Carbendazym
G 665
Kemdazin
Mecarzole
Methyl benzimidazol-2-ylcarbamate
Methyl benzimidazolylcarbamate
Methyl 1H-benzimidazol-2-ylcarbamate
Methyl 1H-benzimidazole-2-carbamate
Methyl 2-benzimidazolecarbamate
Methyl 2-benzimidazolylcarbamate

NH
NHCOOCH$_3$
N

Merck Index No. 1794
MP: 302-307° (dec)
BP:
Density: 1.45
Solubility: 8 mg/l water pH 7, 29 mg/l water pH 4, 0.5 mg/l hexane, 36 mg/l benzene, 68 mg/l dichloromethane, 300 mg/l ethanol
Octanol/water PC: 36
LD_{50}: rat orl 6400 mg/kg, rat orl >15000 mg/kg; rat skn 2000 mg/kg; rat ipr 1720 mg/kg; mus orl 11000 mg/kg; mus ipr 1225 mg/kg; dog orl >2500 mg/kg; rbt orl 81160 mg/kg; rbt skn >10000 mg/kg; gpg orl 4150 mg/kg; qal orl >10000 mg/kg.
CFR: 40CFR180. Pend (1988)

CAS RN 75-15-0
CS_2

Carbon disulfide (ACN)(DOT)(8CI9CI)
Carbon bisulfide (DOT)
Carbon bisulphide
Carbone (sufure de) (French)
Carbonio (solfuro di) (Italian)
Dithiocarbonic anhydride
ENT 8935
Kohlendisulfid (schwefelkohlenstoff) (German)
Koolstofdisulfide (zwavelkoolstof) (Dutch)
NCI-C04591
Schwefelkohlenstoff (German)
Wegla dwusiarczek (Polish)

Merck Index No. 1818
MP: -111.6°
BP: $-73.8^{1.0}$, -44.7^{10}, -5.1^{100}, 28.0^{400}, 46.5^{760}
Density: 1.29272^{0}, 1.27055^{15}, 1.2632^{20}, 1.24817^{30}

S=C=S

Solubility: 0.294% water 20°, misc. methanol, ethanol, diethyl ether, benzene, chloroform, carbon tetrachloride, oils
Octanol/water PC:
LD_{50}: hmn ihl LCLo 4000 ppm/30M; hmn ihl LCLo 2000 ppm/5H; man unr LDLo 186 mg/kg; rat orl 3188 mg/kg; rat ihl 25000 mg/m^3/2H; mus orl 2780 mg/kg; mus ihl LC_{50} 10000 mg/m^3/2H; rbt orl 2550 mg/kg; gpg orl 2125 mg/kg; gpg ipr LDLo 400 mg/kg; mam ihl LCLo 2000 ppm/5H.
CFR: 40CFR180. (temp. 1986)

CAS RN 1563-66-2
$C_{12}H_{15}NO_3$

Carbofuran (ACN)
7-Benzofuranol, 2,3-dihydro-2,2-dimethyl-, methylcarbamate (9CI)
Carbamic acid, methyl-, 2,3-dihydro-2,2-dimethyl-7-benzofuranyl ester (8CI)
BAY 70143
BAY 78537
Carbamic acid, methyl-, 2,2-dimethyl-2,3-dihydro-7-benzofuranyl ester
Carbamic acid, methyl-, 2,2-dimethyl-2,3-dihydrobenzofuran-7-yl ester
Carbamic acid, methyl-2,3-dihydro-2,2-dimethyl-7-benzofuranyl ester
Curaterr
Furadan
Furadan 3G
Niagaral 242
NIA 10242
OMS 864
Yaltox
2,2-Dimethyl-2,2-dihydrobenzofuranyl-7 N-methylcarbamate
2,2-Dimethyl-2,3-dihydro-7-benzofuranyl-N-methylcarbamate
2,2-Dimethyl-2,3-dihydrobenzofuranyl 7-methylcarbamate
2,2-Dimethyl-7-coumaranyl N-methylcarbamate
2,3-Dihydro-2,2-dimethyl-7-benzofuranyl methylcarbamate (ACN)
2,3-Dihydro-2,2-dimethyl-7-benzofuranyl N-methylcarbamate
2,3-Dihydro-2,2-dimethylbenzofuran-7-yl methylcarbamate
2,3-Dihydro-2,2-dimethylbenzofuranyl-7-N-methylcarbamate

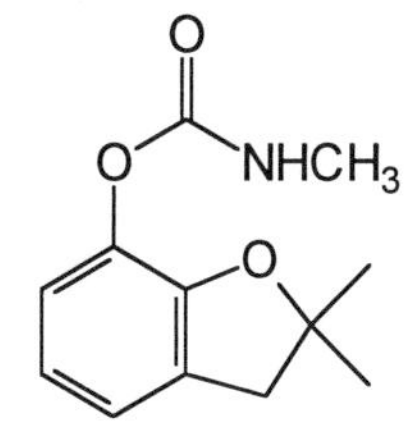

Merck Index No. 1810
MP: 150-153°
BP:
Density: 1.180 SG
Solubility: 700 ppm water, 320 mg/l water 25, 150 g/kg acetone, 140 g/kg acetonitrile, 120 g/kg dichloromethane, 90 g/kg cyclohexanone, 40 g/kg benzene, 40 g/kg ethanol, 250 g/kg dimethylsulfoxide, 270 g/kg dimethylformamide, 300 g/kg N-methylpyrrolidone, sp. sol. xylene, petroleum ether
Octanol/water PC: 17-26
LD_{50}: rat orl 5 mg/kg; rat orl 8.2-14.1 mg/kg; rat per >3000mg/kg; rat ihl LC_{50} 85 mg/m^3; rat skn 120 mg/kg; rat unr 19 mg/kg; mus orl 2 mg/kg; mus ivn 0.45 mg/kg; mus unr 5 mg/kg; dog orl 19 mg/kg; dog ihl LC_{50} 52 mg/m^3; rbt skn 885 mg/kg; rbt unr 885 mg/kg; gpg ihl LC_{50} 43 mg/m^3/4H; pgn orl 1.33 mg/kg; ckn orl 6.3 mg/kg; qal orl 3.16 mg/kg; dck orl 0.415 mg/kg; mam skn 837 mg/kg; bwd orl 0.420 mg/kg; bwd skn 100 mg/kg.
CFR: 40CFR180.254

CAS RN 786-19-6
$C_{11}H_{16}ClO_2PS_3$

Carbophenothion (ACN)
Phosphorodithioic acid, S-[[(4-chlorophenyl)thio]methyl] O,O-diethyl ester (9CI)
Phosphorodithioic acid, S-[[(p-chlorophenyl)thio]methyl] O,O-diethyl ester (8CI)
Acarithion
Akarithion
Carbofenothion
Carbofenotion
Carbofenotionum
Carbofenthion
Dagadip
Dithiophosphate de O,o-diethyle et de (4-chloro-phenyl) thiomethyle (French)
Endyl
Ent 23,708
Ethyl carbophenothion
ENT 23,708
Hexathion
Lethox
Nephocarp
O,O-Diethyl dithiophosphoric acid p-chlorophenylthiomethyl ester
O,O-Diethyl *p*-chlorophenylmercaptomethyl dithiophosphate
O,O-Diethyl [[(*p*-chlorophenyl)mercapto]methyl] dithiophosphate
O,O-Diethyl S-(*p*-chlorophenylthio)methyl phosphorodithioate
O,O-Diethyl S-[[(*p*-chlorophenyl)thio]methyl] dithiophosphate
O,O-Diethyl S-[[(*p*-chlorophenyl)thio]methyl] phosphorodithioate

$C_2H_5O-P(=S)(OC_2H_5)-S-CH_2-S-C_6H_4-Cl$

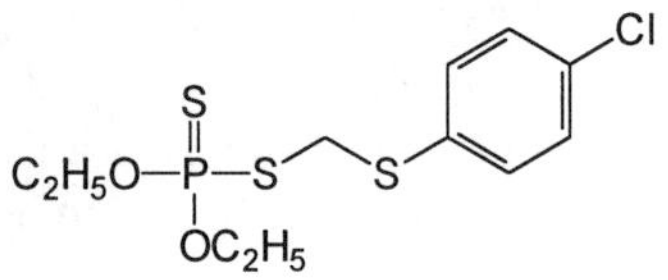

Merck Index No. 1827
MP:
BP: $82^{\circ 0.01}$
Density: 1.271^{20}
Solubility: 40 ppm water 25°; <1mg/l water, sol. most organics
Octanol/water PC:
LD_{50}: rat orl 6.8 mg/kg; rat orl 10, 30 mg/kg; rat skn 27 mg/kg; rat der 27, 54 mg/kg; mus orl 218 mg/kg; mus ipr 27 mg/kg; rbt orl 1250 mg/kg; rbt skn 1270 mg/kg; prn orl 34.8 mg/kg; ckn orl 57.2 mg/kg; ckn scu 640 mg/kg; qal orl 56.8 mg/kg; bwd orl 5.6 mg/kg.
CFR: 40CFR180.156

CAS RN 55285-14-8
$C_{20}H_{32}N_2O_3S$

Carbosulfan (ACN)
FMC 35001
2,3-Dihydro-2,2-dimethyl-7-benzofuranyl (dibutylamino)thio methylcarbamate

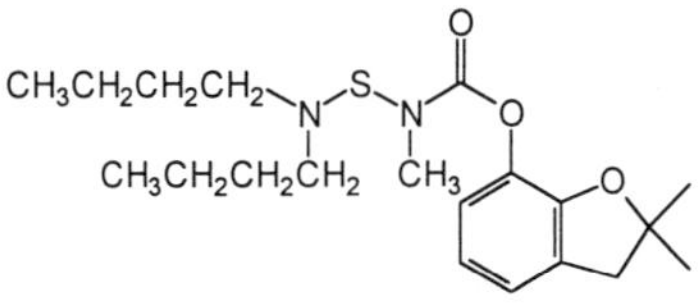

Merck Index No.
MP:
BP: 124 - 128°
Density: 1.056^{20}
Solubility: 0.03mg/l water 25°, misc. org. solvents
Octanol/water PC: 157
LD_{50}: rat orl 250, 185 mg/kg; rat orl 51 mg/kg; rat ihl 1530 mg/m^3/1H; rat skn >2000 mg/kg; mus orl 74 mg/kg; rbt skn >2000 mg/kg; qal orl 82 mg/kg; dck orl 8.1 mg/kg; bwd orl 26 mg/kg.
40CFR180. (pend. 1982)

CAS RN 5234-68-4
$C_{12}H_{13}NO_2S$

Carboxin (ACN)
1,4-Oxathiin-3-carboxamide, 5,6-dihydro-2-methyl-N-phenyl- (9CI)
1,4-Oxathiin-3-carboxanilide, 5,6-dihydro-2-methyl- (8CI)
Carbathiin
Carboxine
D 735
DCMO
DMOC (VAN)
F 735
Vitavax
Vitavax 100
Vitavax 735D
Vitavax 75W
1,4-Oxathiin, 2,3-dihydro-5-carboxanilido-6-methyl-
2,3-Dihydro-5-carboxanilido-6-methyl-1,4-oxathiin
2,3-Dihydro-6-methyl-1,4-oxathiin-5-carboxanilide
5-Carboxanilido-2,3-dihydro-6-methyl-1,4-oxathiin
5,6-Dihydro-2-methyl-N-phenyl-1,4-oxathiin-3-carboxamide
5,6-Dihydro-2-methyl-1,4-oxathiin-3-carboxanilide (ACN)

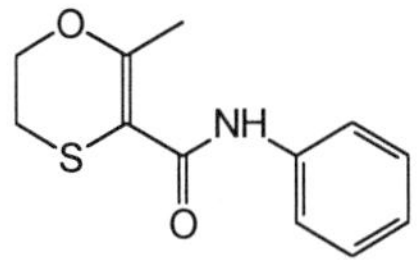

Merck Index No. 1832
MP: 93-95°
BP:
Density:
Solubility: 170mg/l water 25°, 1500g/kg dimethylsulfoxide, 600g/kg acetone, 210g/kg methanol, 150g/kg benzene, 110g/kg ethanol,
Octanol/water PC: 148
LD_{50}: rat orl 430 mg/kg; rat ihl LC50 >20000 mg/kg; rat skn 1050 mg/kg; mus orl 3200 mg/kg; rbt skn 8000 mg/kg; ckn orl 24 mg/kg; mam unr 3200 mg/kg; bwd orl 42.2 mg/kg.
CFR: 40CFR180.301

CAS RN 101-27-9
$C_{11}H_9Cl_2NO_2$

Barban (ACN)
Carbamic acid, (3-chlorophenyl)-, 4-chloro-2-butynyl ester (9CI)
Carbanilic acid, *m*-chloro-, 4-chloro-2-butynyl ester (8CI)
(4-Chloor-but-2-yn-yl)-N-(3-chloor-fenyl)-carbamaat (Dutch)
(4-Chlor-but-2-in-yl)-N-(3-chlor-phenyl)-carbamat (German)
(4-Cloro-but-2-in-il)-N-(3-cloro-fenil)-carbammato (Italian)
A980
Barbamate
Barbane
C-847
Carbanilic acid, 4-chloro-, 4-chloro-2-butynyl ester
Carbin
Carbine
Carbyne
Carbyne (herbicide)
Caryne
Chlorinat
Chlorobutynyl Chlorocarbanilate
CBN
ENT 28201
Fisons B25
N-(3-Chloro phenyl)carbamate de 4-chloro 2-butynyle (French)

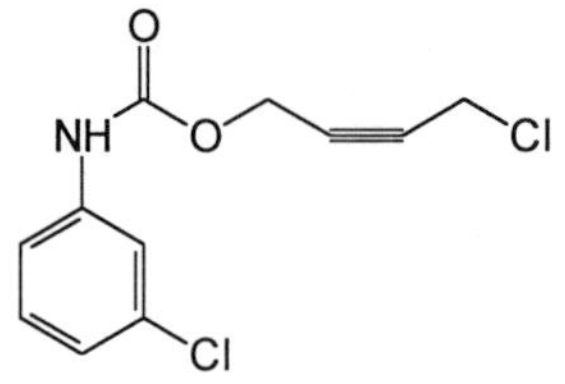

Merck Index No. 969
MP: 75-76°
BP:
Density:
Solubility: 11ppm water 25°, sl. sol. hexane, sol. benzene, ethylene dichloride
Octanol/water PC:
LD_{50}: rat orl 600mg/kg; rat orl 527 mg/kg; rat ihl 27400 mg/m^3/4H; rat skn >1600 mg/kg; rat unr 527 mg/kg; mus orl 322 mg/kg; rbt orl 600 mg/kg; rbt skn 23000 mg/kg; gpg orl 240 mg/kg; mam unr 240 mg/kg.
CFR: 40CFR180.268

CAS RN 1194-65-6
$C_7H_3Cl_2N$

Dichlobenil (ACN)
Benzonitrile, 2,6-dichloro- (8CI9CI)
Carsoron
Casaron
Casoron
Casoron G
Casoron 133
Code H 133
Dbn (the herbicide)
Dichlorobenil
Du-Sprex
DBN
DCB
ENT-26665
H 1313
H 133
Nia 5996
Niagara 5,996
NIA 5996
2,6 dBN

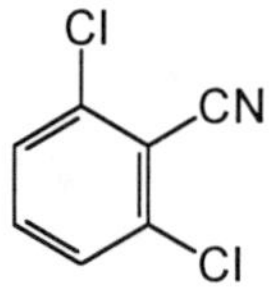

Merck Index No. 3029
MP: 144 - 145°, 145 - 146°, 139 - 145° (technical)
BP: 270°
Density: > 1 SG
Solubility: 18 ppm water 20°, 25 ppm water 25, 50 g/l acetone, 50 g/l dioxane, 50 g/l xylene, 100 g/k dichloromethane, sl. sol. non-polar solvents .
Octanol/water PC:
LD_{50}: rat orl 2710 mg/kg; rat orl >3160mg/kg; rat unr 500 mg/kg; mus orl 6800 mg/kg; mus orl 2126, 2056 mg/kg; mus ipr 360 mg/kg; rbt orl 270 mg/kg; rbt skn 1350 mg/kg; gpg orl 681 mg/kg; mam unr 1000 mg/kg.
CFR: 40CFR180.231

CAS RN 93-71-0
$C_8H_{12}ClNO$

CDAA 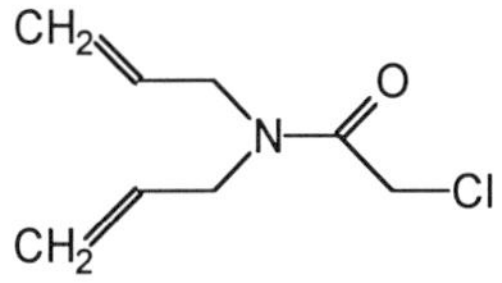
Acetamide, 2-chloro-N,N-di-2-propenyl- (9CI)
Acetamide, N,N-diallyl-2-chloro- (8CI)
α-Chloro-N,N-diallylacetamide
component of Limit
Acetamide, 2-chloro-N,N-diallyl-
Alidochlor
Allidochlor
Chloroacetamide, N,N-diallyl-
CDAAT
CP 6,343
Diallylchloroacetamide
Limit
N-Diallyl-2-chloroacetamide
N,N-Diallyl-α-chloroacetamide
N,N-Diallyl-2-chloroacetamide (ACN)
N,N-Diallylchloroacetamide
NCI-C04035
Radox
Rantox T
2-Chloro-N,N-di(2-propenyl)acetamide
2-Chloro-N,N-diallylacetamide

Merck Index No. 250
MP:
BP: $74^{o0.3}$, $115\text{-}117^{o1}$, 92^{o2}
Density:
Solubility: 2% water, sol. alcohol, hexane, xylene
LD_{50}: rat orl 700mg/kg; rat skn 360 mg/kg.
Octanol/water PC:
CFR: 40CFR180.282

CAS RN 95-06-7
$C_8H_{14}ClNS_2$

CDEC 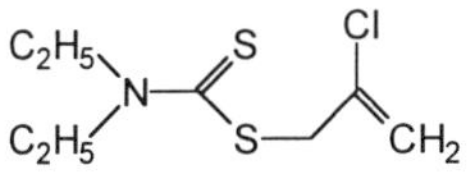
Carbamodithioic acid, diethyl-, 2-chloro-2-propenyl ester (9CI)
Carbamic acid, diethyldithio-, 2-chloroallyl ester (8CI)
Chlorallyl diethyldithiocarbamate
CP 4,742
CP 4572
Diethyldithiocarbamic acid 2-chloroallyl ester
Diethyldithiocarbamic acid, 2-chloroallyl ester
NCI-C00453
Sulfallate
Thioallate
Vegadex
Vegadex super
Vegedex
2-Chlorallyl diethyldithiocarbamate
2-Chloro-2-propenyl diethylcarbamodithioate
2-Chloroallyl diethyldithiocarbamate (ACN)
2-Chloroallyl N,N-diethyldithiocarbamate
2-Propene-1-thiol, 2-chloro-, diethyldithiocarbamate

Merck Index No. 8882
MP:
BP: $128\text{-}130^{o1}$
Density: 1.088
Solubility: 100ppm water 25°, sol. most organics
Octanol/water PC:
LD_{50}; rat orl 850 mg/kg; rbt skn 2200 mg/kg.
CFR: 40CFR180.247

CAS RN 133-90-4
$C_7H_5Cl_2NO_2$

Chloramben (ACN)
Benzoic acid, 3-amino-2,5-dichloro- (8CI9CI)
Acp-m-728
Ambiben
Amiben
Amibin
Amoben
Chlorambed
Chlorambene
NCI-C00055
Ornamental weeder
Vegaben
Vegiben
2,5-Dichloro-3-aminobenzoic acid
3-Amino-2,5-dichlorobenzoic acid (ACN)

COOH
Cl
Cl
NH_2

Merck Index No. 2063
MP: 200-201°
BP:
Density:
Solubility: 700 mg/l water 25°, 23.3 g/100ml acetone, 22.3 g/100ml methanol, 17.3 g/100ml ethanol, 0.09 g/100ml chloroform, 0.02 g/100ml benzene, insol carbon tetrachloride, 120.6 g/100g dimethylformamide,
Octanol/water PC:
LD_{50}: rat orl 6520 mg/kg, rat orl 3500 mg/kg; rat perc >3160 mg/kg; rat skn >2200 mg/kg; mus orl 3725 mg/kg; rbt skn 3136mg/kg.
CFR: 40CFR180.266

CAS RN 103-17-3
$C_{13}H_{10}Cl_2S$

Chlorbenside (ACN)
Benzene, 1-chloro-4-[[(4-chlorophenyl)methyl]thio]- (9CI)
Sulfide, *p*-chlorobenzyl *p*-chlorophenyl (8CI)
(4-Chloor-benzyl)-(4-chloor-fenyl)-sulfide (Dutch)
(4-Chlor-benzyl)-(4-chlor-phenyl)-sulfid (German)
(4-Cloro-benzil)-(4-cloro-fenil)-solfuro (Italian)
p-Chlorobenzyl *p*-chlorophenyl sulfide (ACN)
p-Chlorobenzyl *p*-chlorophenyl sulphide
p-Chlorophenyl *p*-chlorobenzyl sulfide
p,p'-Dichlorodiphenyl sulfide
Chloorbenzide (Dutch)
Chloracid (VAN)
Chlorbensid (German)
Chlorbenxide
Chlorbenzide
Chlorocid
Chlorocide
Chloroparacide
Chlorosulfacide
Chlorparacide
Chlorsulphacide
CP 20

Cl
S
Cl

Merck Index No. 2074
MP: 75-76°
BP:
Density: 1.4210^{20}
Solubility: <1:5000 water 25°, 2.9% ethanol, 5-7.5% kerosene, sol. acetone, benzene, toluene, petroleum ether
Octanol/water PC:
LD_{50}: rat orl 2000 mg/kg; mus unr 3000 mg/kg.
CFR: 40CFR180.168 (revoked 5/4/88)

CAS RN 13360-45-7
$C_9H_{10}BrClN_2O_2$

Chlorbromuron (ACN)
Urea, N'-(4-bromo-3-chlorophenyl)-N-methoxy-N-methyl- (9CI)
Urea, 3-(4-bromo-3-chlorophenyl)-1-methoxy-1-methyl- (8CI)
Bromex (VAN)
C 6313
Chlorobromuron
Ciba 6313
Maloran
N-(4-Bromo-3-chlorophenyl)-N'-methoxy-N'-methylurea
N'-(4-Bromo-3-chlorophenyl)-N-methoxy-N-methylurea
Urea, 1-(4-bromo-3-chlorophenyl)-3-methoxy-3-methyl-
1-(3-Chloro-4-bromophenyl)-3-methyl-3-methoxyurea
3-(4-Bromo-3-chlorophenyl)-1-methoxy-1-methylurea
(ACN)

Merck Index No.
MP: 95 - 97°
BP:
Density: 1.69^{20}
Solubility: 35 mg/l water 20°, 460 g/kg acetone, 170 g/kg dichloromethane, 89 g/kg hexane, 72 g/kg benzene, 12 g/kg isopropanol
Octanol/water PC:
LD_{50}: rat orl >5000 mg/kg; rat orl 2150 mg/kg; rat ihl >1050 mg/m^3/6H; rat skn >2000 mg/kg; rbt skn >10000 mg/kg.
CFR: 40CFR180.279 (revoked 5/4/88)

CAS RN 57-74-9
$C_{10}H_6Cl_8$

Chlordane
4,7-Methano-1H-indene, 1,2,4,5,6,7,8,8-octachloro-2,3,3a,4,7,7a-hexahydro- (9CI)
4,7-Methanoindan, 1,2,4,5,6,7,8,8-octachloro-3a,4,7,7a-tetrahydro- (8CI)
γ-Chlordane
Aspon-Chlordane
AG Chlordane
Belt
Chlor-Kill
Chlordan
Chlordane, γ-
Chlordane, liquid (DOT)
Chlorindan
Chlorodane
Clordan (Italian)
Compound 1068
Cortilan-neu
Cortilan-Neu
CD 68
CD-68
Dichlorochlordene
Dowchlor

Merck Index No. 2079

MP: 106 - 107° (*cis*), 104 - 105° (*trans*)
BP: $175°^{1}$
Density: $1.59 - 1.63^{25}$
Solubility: 0.1 mg/l water 25°, misc. hydrocarbons
Octanol/water PC:
LD_{50}: man orl TDLo 3.071 mg/kg; hmn orl 29 mg/kg; wmn orl 0.12 mg/kg; hmn skn 428 mg/kg; man unr LDLo 1118 mg/kg; rat orl 200 mg/kg; rat orl 457-590 mg/kg; rat ipr 343 mg/kg; rat skn 690 mg/kg; mus orl 145 mg/kg; mus orl 430 mg/kg; mus ipr LDLo 240 mg/kg; mus ivn 100 mg/kg; cat ihl 100 mg/m^3/4H; rbt orl 100 mg/kg; rbt skn 780 mg/kg; rbt ivn LDLo 10 mg/kg; ham orl 1720 mg/kg; ckn orl 220 mg/kg; dck orl 1200 mg/kg; dom orl 50 mg/kg; mam orl 180 mg/kg; mam ihl LC_{50} >20000 mg/m^3/4H.
CFR: 40CFR180.122 (removed 12/24/86)

CAS RN 143-50-0
$C_{10}Cl_{10}O$

Kepone
1,3,4-Metheno-2H-cyclobuta[cd]pentalen-2-one,1,1a,3,3a,4,5,5,5a,5b,6-decachloro-octahydro- (8CI9CI)
Chlordecone
Ciba 8514
Clordecone
Compound 1189
Decachloro-1,3,4-metheno-2H-cyclobuta[cd]pentalen-2-one
Decachloroketone
Decachlorooctahydro-1-3-4-metheno-2H-cyclobuta[cd]pentalen-2-one
Decachlorooctahydro-1,3,4-metheno-2H-cyclobuta[cd]pentalin-2-one
Decachlorooctahydro-1,3,4-metheno-2H-cyclobuta[6d]pentalen-2-one
Decachloropentacyclo[5.2.1.$0^{2,6}$.$0^{3,9}$.$0^{5,8}$]decan-4-one
Decachloropentacyclo[5.2.1.$0^{2,6}$.$0^{3,9}$.$0^{5,8}$]decan-4-one
Decachloropentacyclo[5.3.0.0.0.$0^{2,6,4,10,5}$,9]decane-3-ones
Decachloropentacyclo[5.3.0.$0^{2,6}$.$0^{4,10}$.$0^{5,9}$]decan-3-one
Decachlorotetracyclodecanone
Decachlorotetrahydro-4,7-methanoindeneone
Kepone-2-one
Kepone-2-one, decachlorooctahydro-
Merex
NCI-C00191
1,1a,3,3a,4,5,5,5a,5b,6-Decachlorooctahydro-1,3,4-metheno-2H-cyclobuta[cd]pentalen-2-one

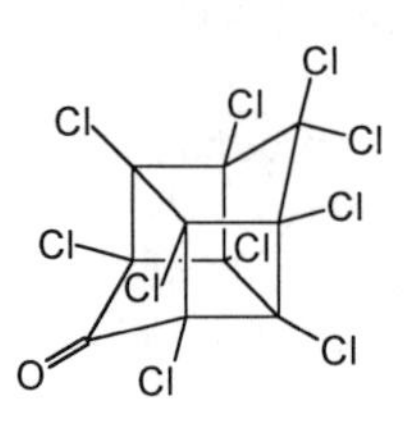

Merck Index No. 2081
MP: 350°(dec)
BP:
Density: 1.59 - 1.63^{25}
Solubility: 2 ppm water; sl. sol. water, hydrocarbons
Octanol/water PC:
LD_{50}: rat orl 95 mg/kg; rat ipr 343 mg/kg; rat skn >2000 mg/kg; dog orl 250 mg/kg; rbt orl 65 mg/kg; rbt skn 345 mg/kg; qal orl 237 mg/kg; mam unr 126mg/kg.
CFR: 40CFR180. (Temp. 1978; Admin. Guidelines)

CAS RN 6164-98-3 (19750-95-9; 61886-56-4)
$C_{10}H_{13}ClN_2$

Chlordimeform (ACN)
Methanimidamide, N'-(4-chloro-2-methylphenyl)-N,N-dimethyl- (9CI)
Formamidine, N'-(4-chloro-*o*-tolyl)-N,N-dimethyl- (8CI)
Acaron
Bermat
C 8514
Carzol
Chlorfenamidine
Chlorodimeform
Chlorophedine
Chlorophenamide
Chlorphenamidine
Ciba 8514
Ciba-C8514
CDM
Galecron
N^2-(4-Chloro-*o*-tolyl)-N^1,N^1-dimethylformamidine

Merck Index No. 2083
MP: 35°
BP: 156 - 157° $^{0.4}$
Density: 1.105^{25} SG
Solubility: 250 ppm water 20°; sl. sol. water, easily sol. organics
Octanol/water PC:
LD_{50}; rat orl 160 mg/kg; rat skn 263 mg/kg; rat ipr 90 mg/kg; rat ipr 238 mg/kg; mus orl 160 mg/kg; mus skn 225 mg/kg; mus ipr 71 mg/kg; rbt orl 625 mg/kg; rbt skn 640 mg/kg.
CFR: 40CFR180.285

CAS RN 470-90-6
$C_{12}H_{14}Cl_3O_4P$

Chlorfenvinphos
Phosphoric acid, 2-chloro-1-(2,4-dichlorophenyl)ethenyl diethyl ester (9CI)
Phosphoric acid, 2-chloro-1-(2,4-dichlorophenyl)vinyl diethyl ester (8CI)
β-2-Chloro-1-(2',4'-dichlorophenyl)vinyl diethyl phosphate
O-2-Chloor-1-(2,4-dichloor-fenyl)-vinyl-O,O-diethylfosfaat (Dutch)
O-2-Chloor-1-(2,4-dichloor-fenyl)-vinyl-O,O-diethylfosfaat (Dutch)
O-2-Chlor-1-(2,4-dichlor-phenyl)-vinyl-O,O-diaethylphosphat (German)
O-2-Cloro-1-(2,4-dicloro-fenil)-vinil-O,O-dietilfosfato (Italian)
Benzyl alcohol, 2,4-dichloro-α-(chloromethylene)-, diethyl phosphate
Birlan
Birlane
Birlane 10 G
C 8949
C-10015
Chlofenvinphos
Chlofenvinphoso-2-chloor-1-(2,4-dichloor-fenyl)-vinyl-O,O-diethylfosfaat (Dutch)
Chlorfenvinfos
Chlorofeninphos
Chlorofenvinphos
Chlorphenvinfos
Chlorphenvinphos
Clofenvineosum
Clofenvinfos

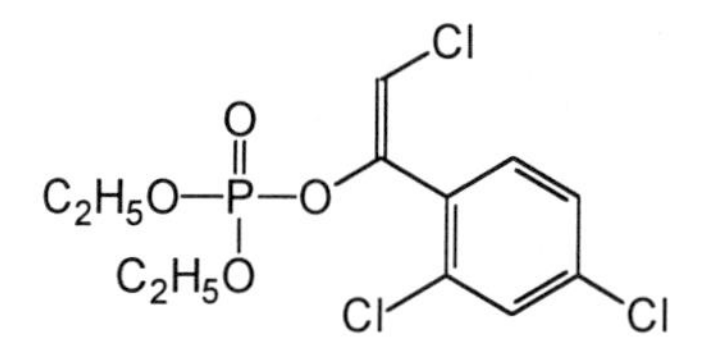

Merck Index No. 2087
MP: -22° - -16°
BP: $120^{\circ 0.001}$, $167\text{-}170^{\circ 0.05}$
Density:
Solubility: 145 ppm water, miscible acetone, ethanol, propylene glycol
Octanol/water PC:
LD_{50}: hmn skn TDLo 10 mg/kg; rat orl 9.6 mg/kg; rat orl 10 mg/kg; rat ihl 50 mg/m^3/4H; rat skn 26400 mg/kg; rat ipr 8.5 mg/kg; rat iv 6.6 mg/kg;rat scu 7 mg/kg; mus orl 65 mg/kg; mus skn 336 mg/kg; mus ipr 87 mg/kg; mus scu 339 mg/kg; mus ivn 87 mg/kg; dog orl 1200 mg/kg; dog ivn 51 mg/kg; rbt orl 300 mg/kg; rbt skn 400 mg/kg; gpg orl 125 mg/kg; pgn orl 13.3 mg/kg; ckn orl 29.1 mg/kg; ckn ipr 23.1 mg/kg; qal orl 148 mg/kg; dom orl 71.25 mg/kg; ctl orl 20 mg/kg; mam orl 10 mg/kg; bwd orl 13 mg/kg.
CFR: 40CFR180.322

CAS RN 2464-37-1
$C_{14}H_9ClO_3$

Chlorflurecol
9H-Fluorene-9-carboxylic acid, 2-chloro-9-hydroxy- (9CI)
Fluorene-9-carboxylic acid, 2-chloro-9-hydroxy- (8CI)
Chloflurecol
Chlorflurenol
CF 125
CME 73170P
IT 3299
2-Chloro-9-hydroxy-9H-fluorene-9-carboxylic acid
2-Chloro-9-hydroxyfluorene-9-carboxylic acid
(ACN)

Merck Index No.
MP: (methyl ester) 152°, 136 - 142° (technical)
BP:
Density: 1.496^{20}
Solubility: 18 mg/l water 20°, 260 g/kg acetone, 150 g/kg methanol, 80 g/kg ethanol, 70 g/kg benzene, 24 g/kg isopropanol, 24 g/kg carbon tetrachloride, 2.4 g/kg cyclohexane
Octanol/water PC:
LD_{50}: rat orl >12,800 mg/kg, rat skn >1,000 mg/kg; qal orl >10000 mg/kg.
CFR: 40CFR180. (Temp. 1979)

CAS RN 38727-55-8
$C_{16}H_{22}ClNO_3$

Diethatyl-ethyl (ACN)
Glycine, N-(chloroacetyl)-N-(2,6-diethylphenyl)-, ethyl ester (9CI)
Antor
AC 22234
Bay nnt 6867
Diethatyl ethyl
H 22234
Hercules 22234
N-(Chloroacetyl)-N-(2,6-diethylphenyl)glycine ethyl ester
N-(Chloroacetyl)-N-(2,6-diethylphenyl)glycine, ethyl ester (ACN)

Merck Index No.
MP: 49 - 50° (ethyl ester)
BP:
Density: 1.38^{25} (ethyl ester)
Solubility: (ethyl ester) 105 mg/l water 25°, 810 g/kg methanol, 810 g/kg ethanol, 750 g/kg isopropanol, 820 g/kg acetone, 820 g/kg xylene, 830 g/kg methyl isobutyl ketone, 770 g/kg chloroform, 200 g/kg kerosene
Octanol/water PC: (ethyl ester) 3970
LD_{50}: rat orl 2318, 3720 mg/kg, mus orl 1653 mg/kg; mus orl 4118 mg/kg; rbt skn 4000 mg/kg.
CFR: 40CFR180.402

CAS RN 510-15-6
$C_{16}H_{14}Cl_2O_3$

Chlorobenzilate
Benzeneacetic acid, 4-chloro-α-(4-chlorophenyl)-α-hydroxy-, ethyl ester (9CI)
Benzilic acid, 4,4'-dichloro-, ethyl ester (8CI)
Acar
Acaraben
Akar
Akar 338
Benzilan
Chlorbenzilat
Chlorbenzylate
Chlorobenzylate
Compound 338
Ethyl di(*p*-chlorophenyl)glycollate
Ethyl ester of 4,4'-dichlorobenzilic acid
Ethyl *p,p*'-dichlorobenzilate
Ethyl 4,4'-dichlorobenzilate (ACN)
Ethyl 4,4'-dichlorodiphenyl glycollate
Ethyl 4,4'-dichlorophenyl glycollate
Ethyl-2-hydroxy-2,2-bis(4-chlorophenyl)acetate
ENT 18,596
Folbex
Folbex Smoke-Strips
G 23992
G 338

Cl Cl HO $COOC_2H_5$

Merck Index No. 2123
MP: 36 - 37.5°; 35 - 37°
BP: 146-148°$^{0.04}$, 156 - 158°$^{0.07}$
Density: 1.2816^{20}
Solubility: sl. sol. water, sol. most organics, 10 mg/l water 20, 1000 g/kg acetone, 1000 g/kg dichloromethane, 1000 g/kg methanol, 1000 g/kg toluene, 600 g/kg hexane
Octanol/water PC:
LD_{50}: rat orl 700 mg/kg; rat orl 1040 mg/kg; rat orl 1220 mg/kg; rat orl 2784-3880 mg/kg; rat skn >10200 mg/kg; mus orl 729 mg/kg; mus unr 729 mg/kg; rbt skn >1000 mg/kg; ham orl 700 mg/kg.
CFR: 40CFR180.109

CAS RN 7003-89-6 (chloride 999-81-5)
$C_5H_{13}ClN$

Chlormequat
Ethanaminium, 2-chloro-N,N,N-trimethyl- (9CI)
(2-Chloroethyl)trimethylammonium
Chlorocholine
2-Chloro-N,N,N-trimethylethanaminium

Cl $N^{\oplus}$ CH_3 CH_3 CH_3 $Cl^{\ominus}$

Merck Index No. 2103 (chloride)
MP: 245° (dec)(chloride)
BP:
Density:
Solubility: (chloride) >1kg/kg water 20°, 320 g/kg ethanol, 0.3 g/kg acetone, 0.3 g/kg chloroform
Octanol/water PC:
LD_{50}: (chloride) rat orl 883 mg/kg; mus orl 589 mg/kg; mus ivr 7 mg/kg; mus orl 54 mg/kg.
CFR: 40CFR180. (Temp. 1974)

CAS RN 22936-86-3
$C_9H_{14}ClN_5$

Cyprazine (ACN)
1,3,5-Triazine-2,4-diamine, 6-chloro-N-cyclopropyl-N'-(1-methylethyl)- (9CI)
s-Triazine, 2-chloro-4-(cyclopropylamino)-6-(isopropylamino)- (8CI)
component of Prefox
Cypraznie
Cyprozine
K 6295
Outfox
S 6115
S-6115
S-9115
S9115
2-Chloro-4-(cyclopropylamino)-6-(isopropylamino)-*s*-triazine (ACN)
2-Chloro-4-(cyclopropylamino)-6-(isopropylamino)-1,3,5-triazine

NH N NHCH(CH3)2 N N Cl

Merck Index No.
MP: 167 - 168°
BP:
Density:
Solubility:
Octanol/water PC:
LD50: rat orl 1200 mg/kg; rbt skn 7500 mg/kg.
CFR: 40CFR180.306

CAS RN 42874-03-3
$C_{15}H_{11}ClF_3NO_4$

Oxyfluorfen (ACN)
Benzene, 2-chloro-1-(3-ethoxy-4-nitrophenoxy)-4-(trifluoromethyl)- (9CI)
Ether, 2-chloro-α,α,α-trifluoro-*p*-tolyl 3-ethoxy-4-nitrophenyl
Goal
RH-2915
RH-32,915
RH-32915
RH-915
2-Chloro-α,α,α-trifluoro-*p*-tolyl 3-ethoxy-4-nitrophenyl ether
2-Chloro-α,α,α,-trifluoro-*p*-tolyl-3-ethoxy-4-nitrophenyl ether
2-Chloro-1-(3-ethoxy-4-nitrophenoxy)-4-(trifluoromethyl)benzene (ACN)
2-Chloro-4-trifluoromethyl-3'-ethoxy-4'-nitrodiphenyl ether
2-Chloro-4-trifluoromethylphenyl 3-ethoxy-4-nitrophenyl ether
2-ENBFT

OC_2H_5 CF_3 NO_2 O Cl

Merck Index No. 6916
MP: 83 - 84°, 84 - 85°
BP: 358.2° (dec)
Density: 1.35^{73}
Solubility: 0.1 mg/ml water, 72.5 g/100g acetone, 61.5 g/100g cyclohexanone, 61.5 g/100g isophorone, >50 g/100g dimethylformamide, 50-55 g/100g chloroform, 40-50 g/100g mesityl oxide,
Octanol/water PC: 29,400
LD_{50}: rat orl >5000 mg/kg; dog orl >5000 mg/kg; rbt skn >10000 mg/kg.
CFR: 40CFR180.381

CAS RN 21725-46-2
$C_9H_{13}ClN_6$

Cyanazine 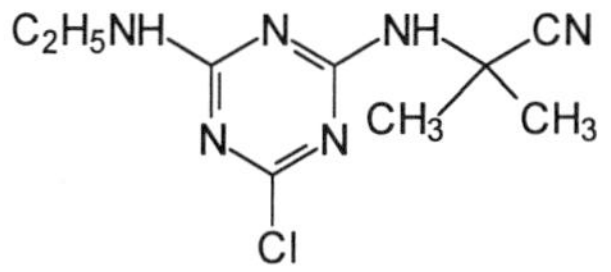
Propanenitrile, 2-[[4-chloro-6-(ethylamino)-1,3,5-triazin-2-yl]amino]-2-methyl- (9CI)
Propionitrile, 2-[[4-chloro-6-(ethylamino)-s-triazin-2-yl]amino]-2-methyl- (8CI)
s-Triazine, 2-chloro-4-(ethylamino)-6-(1-cyano-1-methyl)(ethylamino)-
Bladex
Bladex 80WP
Cyanazine SD 15418
DW 3418
Fortol
Fortrol
Payze
SD 15418
WL 19805
2-[[4-Chloro-6-(ethylamino)-*s*-triazin-2-yl]amino]-2-methylpropionitrile (ACN)

Merck Index No. 2692
MP: 167.5 - 169°
BP:
Density:
Solubility: 171 mg/l water 25°, 15g/l benzene, 210 g/l chloroform, 45 g/l ethanol, 15 g/l hexane, 210 g/l methylcyclohexane, 210 g/l chloroform, 195 g/l acetone, 45 g/l ethanol, 15 g/l benzene, 15 g/l hexane, <10 g/l carbon tetrachloride
Octanol/water PC:
LD_{50}: rat orl 149 mg/kg; rat orl 182 mg/kg; rat orl 182-334 mg/kg; rat ihl LCL0 >4900 mg/m^3; rat skn 1200 mg/kg; rat ipr 112 mg/kg; rat scu 1738 mg/kg; mus orl 380 mg/kg; mus ihl 2470 mg/m^3/4H; mus skn >6590 mg/kg; mus ipr 174 mg/kg; mus scu 3715 mg/kg; rbt orl 141 mg/kg; rbt skn >2000 mg/kg; ckn orl 750 mg/kg; qal orl 400 mg/kg; dck orl >2000 mg/kg.
CFR: 40CFR180.307

CAS RN 33245-39-5
$C_{12}H_{13}ClF_3N_3O_4$

Fluchloralin (ACN) 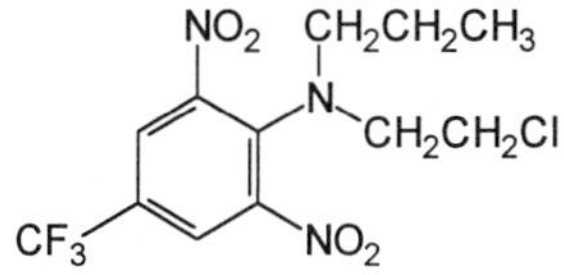
Benzenamine, N-(2-chloroethyl)-2,6-dinitro-N-propyl-4-(trifluoromethyl)- (9CI)
p-Toluidine, N-(2-chloroethyl)-α,α,α-trifluoro-2,6-dinitro-N-propyl- (8CI)
p-Toluidine, N-(2-chloroethyl)-2,6-dinitro-N-propyl-α,α,α-trifluoro-
Basalin
BAS 392 H
BAS 392-H
BAS 392H
BAS 3921
N-(2-Chloroethyl)-α,α,
α-trifluoro-2,6-dinitro-N-propyl-*p*-toluidine (ACN)
N-(2-Chloroethyl)-α,α,α-trifluoro-2,6-dinitro-N-propyl-*p*-toluidine
N-Propyl-N-(2-chloroethyl)-α,α,α-trifluoro-2,6-dinitro-*p*-toluidine

Merck Index No. 4052
MP: 42-43°
BP:
Density:
Solubility: 10 ppm water, <1 mg/ml water 20°, >1000 g/kg acetone, >1000 g/kg benzene, >1000 g/kg chloroform, >1000 g/kg diethyl ether, >1000 g/kg ethyl acetate, 251 g/kg cyclohexane, 177 g/kg ethanol, 260 g/kg olive oil
Octanol/water PC:
LD_{50}: rat orl 2940 mg/kg; rat orl 6400 mg/kg; rat skn >4000 mg/kg; mus orl 730 mg/kg; mus unr 730 mg/kg; rbt skn >10000 mg/kg; rbt unr 8000 mg/kg; qal orl 7000 mg/kg; dck orl 1300 mg/kg.
CFR: 40CFR180.363

CAS RN 6814-58-0
$C_4H_4ClN_3O_2$

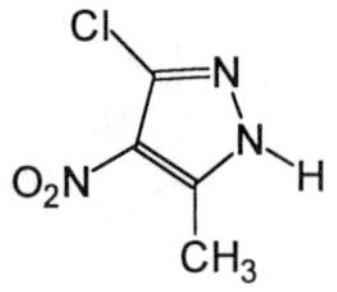

Pyrazachlor
5-Chloro-3-methyl-4-nitro-1H-pyrazole (ACN)
1H-Pyrazole, 3-chloro-5-methyl-4-nitro- (9CI)
A-23703
ABG 3030
Release
5-Chloro-3-methyl-4-nitropyrazole

Merck Index No.
MP: 112°, 108 - 110°
BP:
Density:
Solubility:
Octanol/water PC:
LD_{50}: rat orl 350 mg/kg; rat ihl LC50 6200 mg/m3/4H; rbt skn >2000 mg/kg.
40CFR180. temp(1975)

CAS RN 2675-77-6
$C_8H_8Cl_2O_2$

Chloroneb (ACN)
Benzene, 1,4-dichloro-2,5-dimethoxy- (8CI9CI)
Chlorneb
Chloronebe (French)
Demosan
Demosan 65W
Soil Fungicide 1823
Tersan sp
Tersan-SP
1,4-Dichloro-2,5-dimethoxybenzene (ACN)

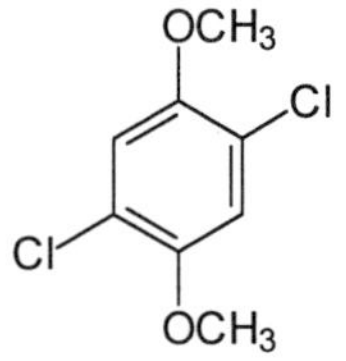

Merck Index No.
MP: 133 - 135°
BP: 268°
Density:
Solubility: 8 mg/l water 25°, 115 g/kg acetone, 118 g/kg dimethylformamide, 89 g/kg xylene, 133 g/kg dichloromethane
Octanol/water PC:
LD_{50}: rat orl 11000 mg/kg; rat orl >11000 mg/kg; rbt skn >5000 mg/kg; qal orl >5000 mg/kg; dck orl >5000 mg/kg; bwd orl >5000 mg/kg.
40CFR180.257

CAS RN 42874-01-1
$C_{13}H_7ClF_3NO_3$

Nitrofluorfen (ACN)
Benzene, 2-chloro-1-(4-nitrophenoxy)-4-(trifluoromethyl)- (9CI)
Ether, 2-chloro-α,α,α-trifluoro-*p*-tolyl *p*-nitrophenyl
RH 2512
RH-2512
RH-512
2-Chloro-α,α,α-trifluoro-*p*-tolyl *p*-nitrophenyl ether
2-Chloro-1-(4-nitrophenoxy)-4-(trifluoromethyl)benzene (ACN)
2-Chloro-1-(4-nitrophenoxy)-4-trifluoromethylbenzene
2-Chloro-4-(trifluoromethyl) 4'-nitrodiphenyl ether
2-Chloro-4-(trifluoromethyl)phenyl 4-nitrophenyl ether

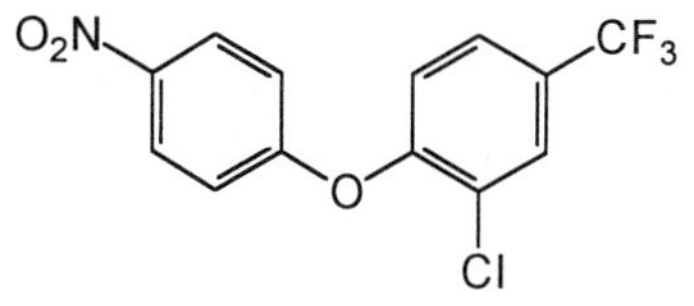

O_2N CF_3 O Cl

Merck Index No.
MP: 67 - 70°
BP:
Density:
Solubility:
Octanol/water PC:
LD_{50}: rat orl 1000 mg/kg.
40CFR180. temp 1975

CAS RN 122-88-3
$C_8H_7ClO_3$

(*p*-Chlorophenoxy)acetic acid (ACN)
Acetic acid, (4-chlorophenoxy)- (9CI)
Acetic acid, (*p*-chlorophenoxy)- (8CI)
(4-Chlorophenoxy)acetic acid (ACN)
p-Chlorphenoxyessigsaeure (German)
CPA
Marks 4-CPA
Phenoxyacetic acid, 4-chloro-
PCPA
Sure-Set
Tomato fix
Tomatotone
4-Chlorophenoxyacetic acid
4-CP
4-CPA

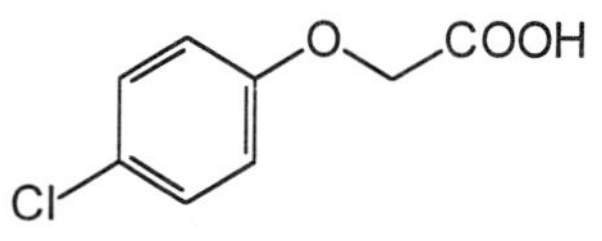

O COOH Cl

Merck Index No.
MP: 157 - 158°
BP:
Density:
Solubility: v. sol. water
Octanol/water PC:
LD_{50}: rat orl 850 mg/kg; mus ipr 680 mg/kg;
40CFR180.202

CAS RN 101-10-0
$C_9H_9ClO_3$

Cloprop 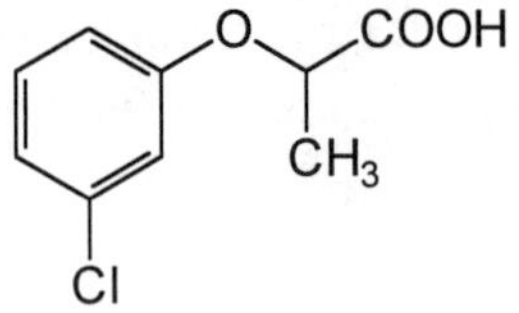
Propanoic acid, 2-(3-chlorophenoxy)- (9CI)
Propionic acid, 2-(*m*-chlorophenoxy)- (8CI)
α-(3-Chlorophenoxy)propionic acid
Amchem 3-cp
Bidisin Forte
CPA
Metachlorphenprop
Propanoic acid, 2-(*m*-chlorophenoxy)-
2-(*m*-Chlorophenoxy)propanoic acid
2-(*m*-Chlorophenoxy)propionic acid (ACN)
2-(3-Chlorophenoxy)propanoic acid (ACN)
2-(3-Chlorophenoxy)propionic acid
3-Chlorophenoxypropionic acid
3-CPA
3CP

Merck Index No.
MP:
BP:
Density:
Solubility: 1.2 g/l water 22°, 790.9 g/l acetone, 2685 g/l dimethylsulfoxide, 710.8 g/l ethanol, 716.5 g/l methanol, 247.3 g/l isooctanol, 24.2 g/l benzene, 17.1 g/l C6H5Cl, 17.6 g.l toluene, 390.6 g/l ethylene glycol, 2354.5 g/l dimethylformamide, 789.2 g/l dioxane
Octanol/water PC:
LD_{50}: rat orl >750 mg/kg; rat orl 3360 mg/kg; rat orl 2140 mg/kg; rbt skn >2000 mg/kg.
40CFR180.325

CAS RN 7778-42-9
ClH_2NO_2S

Chlorosulfamic acid (ACN) 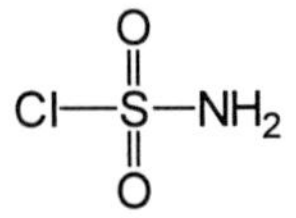
Sulfamoyl chloride (8CI9CI)
Amidosulfonyl chloride
Aminosulfonyl chloride
Chloroimidosulfuric acid (VAN)

Merck Index No.
MP:
BP:
Density:
Solubility:
Octanol/water PC:
LD_{50}:
40CFR180.201

CAS RN 1982-47-4
$C_{15}H_{15}ClN_2O_2$

Chloroxuron (ACN)
Urea, N'-[4-(4-chlorophenoxy)phenyl]-N,N-dimethyl- (9CI)
Urea, 3-[*p*-(*p*-chlorophenoxy)phenyl]-1,1-dimethyl- (8CI)
C 1983
Chloroxifenidim
Ciba 1983
C1983
N-[4-(*p*-chlorophenoxy)phenyl]-N',N'-dimethylurea
Norex
Tenoran
1-(4-(4-Chloro-phenoxy)phenyl)-3,3-D'methyluree (French)
3-(4-(4-Chloor-fenoxy)-fenoxy)-fenyl)-1,1-dimethylureum (Dutch)
3-(4-(4-Chlor-phenoxy)-phenyl)-1,1-dimethylharnstoff (German)
3-(4-(4-Chloro-fenossil)-1,1-dimetil-urea (Italian)
3-[*p*-(*p*-Chlorophenoxy)phenyl]-1,1-dimethylurea (ACN)

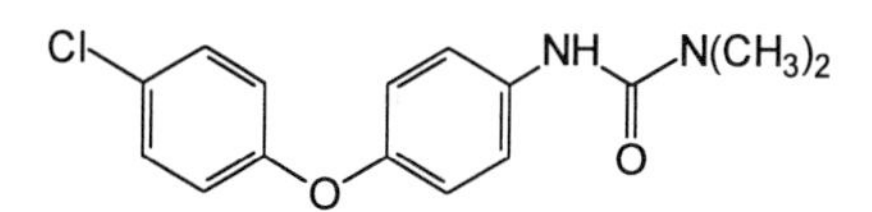

Merck Index No.
MP: 151 - 152°
BP:
Density:
Solubility: 3.7 mg/l water 20, 106 g/kg dichloromethane, 44 g/kg acetone, 35 g/kg methanol, 4 g/kg toluene, sol. dimethylformamide, chloroform, sl. sol. benzene, diethyl ether
Octanol/water PC: 1585
LD_{50}: rat orl 3700 mg/kg; rat orl 5400 mg/kg; rat ihl LC_{50} >1350 mg/m^3/6H; rat skn >3000 mg/kg; rat unr 1000 mg/kg; dog orl 10000 mg/kg; mus orl >1000 mg/kg; rbt skn >10000 mg/kg.
CFR: 40CFR180.216

CAS RN 2921-88-2
$C_9H_{11}Cl_3NO_3PS$

Chlorpyrifos (ACN)
Phosphorothioic acid, O,O-diethyl O-(3,5,6-trichloro-2-pyridinyl) ester (9CI)
Phosphorothioic acid, O,O-diethyl O-(3,5,6-trichloro-2-pyridyl) ester (8CI)
Brodan
Chloropyrifos
Chloropyriphos
Chloropyripos
Chlorpyriphos
Chlorpyrofos
Chlorpyrophos
Detmol U.A.
Dowco 179
Dursban
Dursban F
Dursban 4E
DOWCO
Eradex
ENT 27,311
Killmaster
Lorsban
O,O-diethyl O-(3,5,6-trichloro-2-pyridinyl) phosphorothioate
O,O-Diaethyl-O-3,5,6-trichlor-2-pyridylmonothiophosphat (German)
O,O-Diethyl O-(3,5,6-trichloro-2-pyridyl) phosphorothioate (ACN)
O,O-Diethyl O-(3,5,6-trichloro-2-pyridyl) thiophosphate
OMS 971
Phosphorothioic acid O,O-diethyl O-(3,5,6-trichloro-2-pyridinyl) ester
Trichlorpyrphos
2-Pyridinol, 3,5,6-trichloro-, O-ester with O,O-diethyl phosphorothioate

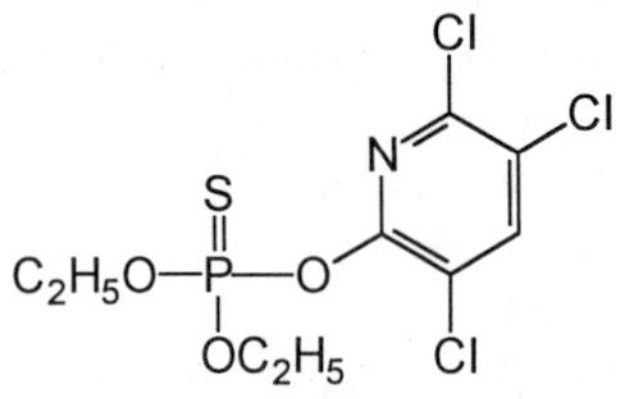

Cl
Cl
N
S
C_2H_5O—P—O
OC_2H_5 Cl

Merck Index No. 2190
MP: 41-42°, 42 - 43.5°
BP:
Density:
Solubility: 2 ppm water, 79% isooctone, 43% methanol, read. sol. other organics, 2 mg/l water 25, 7900 g/kg benzene, 6500 g/kg acetone, 6300 g/kg chloroform, 5900 g/kg CS2, 5100 g/kg diethyl ether, 4000 g/kg xylene, 4000 g/kg dichloromethane, 790 g/kg isooctane, 450 g/kg methanol
Octanol/water PC: 50,000
LD_{50}: man orl TDLo 300 mg/kg; rat orl 82 mg/kg; rat orl 145 mg/kg; rat orl 135-163 mg/kg; rat ihl LC_{50} >200 $mg/m^3/4H$; rat skn 202 mg/kg; rat unr 150 mg/kg; mus orl 60 mg/kg; mus ipr 192 mg/kg; rbt orl 1000 mg/kg; rbt skn 2000 mg/kg; gpg orl 504 mg/kg; gpg scu LDLo 100 mg/kg; prn orl 10 mg/kg; ckn orl 25.4 mg/kg; qal orl 13.3 mg/kg; dcl orl 76 mg/kg; mam unr 163 mg/kg; bwd orl 5 mg/kg.
CFR: 40CFR180.342

CAS RN 5598-13-0
$C_7H_7Cl_3NO_3PS$

Chlorpyrifos-methyl (ACN)
Phosphorothioic acid, O,O-dimethyl O-(3,5,6-trichloro-2-pyridinyl) ester (9CI)
Phosphorothioic acid, O,O-dimethyl O-(3,5,6-trichloro-2-pyridyl) ester (8CI)
Chlormethylfos
Chloropyrifos-methyl
Chloropyriphosmethyl
Chlorpyrifos, methyl analog
Dowco 214
ENT 27520
Fospirate
M 3196
Methyl chlorpyrifos
Methyl Dursban
Noltran
O,O-Dimethyl O-(3,5,6-trichloro-2-pyridinyl)phosphorothioate
O,O-Dimethyl O-(3,5,6-trichloro-2-pyridyl) phosphorothioate (ACN)
O,O-Dimethyl O-(3,5,6-trichloro-2-pyridyl)phosphorothioate
O,O-Dimethyl-O-(3,5,6-trichloro-2-pyridyl)phosphorothioate
OMS 1155
Reldan
Reldan F
Trichlormethylfos
Zertell

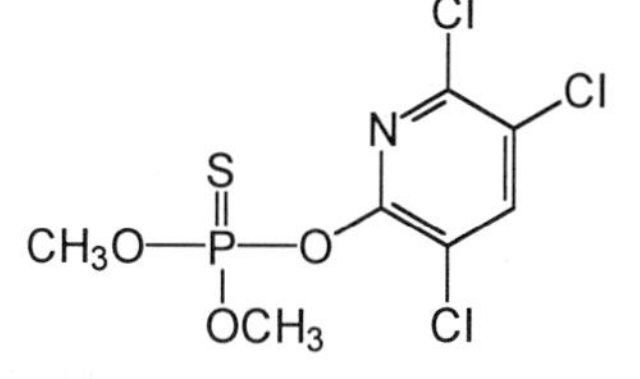

Merck Index No.
MP: 42 - 43.5°
BP:
Density:
Solubility: 4 mg/l water 24°, 5200 g/kg benzene, 6400 g/kg acetone, 3500 g/kg chloroform, 4800 g/kg diethyl ether, 390 g/kg methanol, 230 g/kg hexane
Octanol/water PC: 50,000
LD_{50}: rat orl 1828 mg/kg; rat orl 1630-2140 mg/kg; rat ihl LC_{50} >670 mg/m^3/4H; rat skn 3713 mg/kg; rat scu 6900 mg/kg; mus orl 2032 mg/kg; mus orl 1100-2250 mg/kg; mus ipr 2325 mg/kg; mus scu 23800 mg/kg; rbt orl 2000 mg/kg; rbt skn >2000 mg/kg; gpg orl 2250 mg/kg; ckn orl >7950 mg/kg; dom orl LDLo 1200 mg/kg; bwd orl 13 mg/kg;
CFR: 40CFR180.419

CAS RN 21923-23-9
$C_{11}H_{15}Cl_2O_3PS_2$

Chlorthiophos (ACN)
Phosphorothioic acid, O-[2,5-dichloro-4-(methylthio)phenyl] O,O-diethyl ester (8CI9CI)
Celathion 50
Cm S 2957
CELA S 2957
ENT 27635
O-[Dichloro(methylthio)phenyl] O,O-diethyl phosphorothioate
O-[2,5-Dichloro-4-(methylthio)phenyl] O,O-diethyl phosphorothioate & 2,4,5 and 4,5,2 isomers (73:14:13) (ACN)
O,O-Diethyl O-[2,5-dichloro-4-(methylthio)phenyl] thionophosphate
OMS 1342
Phosphorothioic acid, O,O-diethyl O-[(2,5-dichloro-4-methylthio)phenyl] ester

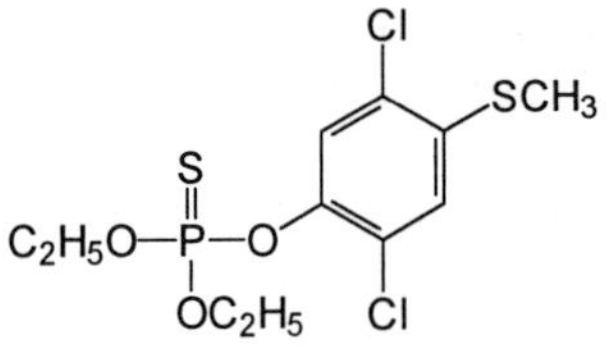

Merck Index No.
MP:
BP: 150 - 151$^{o0.001}$
Density:
Solubility:
Octanol/water PC:
LD_{50}: rat orl 13 mg/kg; rat skn 58 mg/kg; mus orl 141 mg/kg; rbt orl 20 mg/kg; rbt skn 31 mg/kg; gpg orl 58 mg/kg; ckn orl 45 mg/kg; qal orl 45 mg/kg.
CFR: 40CFR180.398

CAS RN 7700-17-6
$C_{14}H_{19}O_6P$

Ciodrin
2-Butenoic acid, 3-[(dimethoxyphosphinyl)oxy]-, 1-phenylethyl ester, (E)- (9CI)
Crotonic acid, 3-hydroxy-, α-methylbenzyl ester, dimethyl phosphate, (E)-
α-Methylbenzyl (E)-3-hydroxycrotonate dimethyl phosphate
(E)-Dimethyl 1-methyl-2-(1-phenylethoxycarbonyl)vinyl phosphate
(E)-3-[(Dimethoxyphosphinyl)oxy]-2-butenoic acid 1-phenylethyl ester
cis-Crotonic acid, 3-hydroxy-, α-methylbenzyl ester, dimethyl phosphate of
cis-2-(1-Phenylethoxy)carbonyl-1-methylvinyl dimethylphosphate 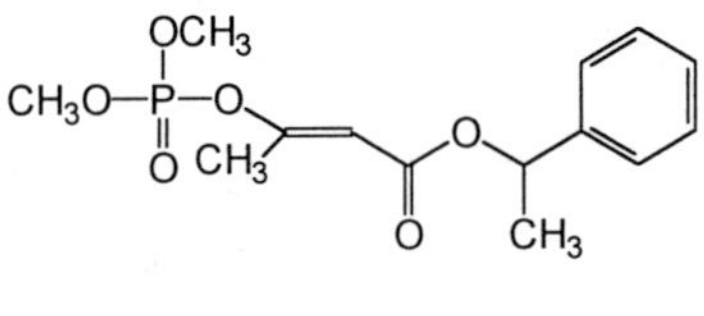
Ciovap
Crotonic acid, 3-hydroxy-,α-methylbenzyl ester dimethyl phosphate,(E)-
Crotoxyphos
Cyodrin
Dimethyl phosphate of α-methylbenzyl 3-hydroxy-*cis*-crotonate (ACN)
ENT 24,717
Pantozol 1

Merck Index No.2603
MP:
BP: 135$^{o0.03}$
Density: 1.19^{25}
Solubility: 0.1% water, sol. acetone, chloroform, ethanol, chlorinated hydrocarbons, sl. sol. hydrocarbons
Octanol/water PC:
LD_{50}: rat orl 38.4 mg/kg; rat orl 74 mg/kg; rat orl 110 mg/kg; rat skn 202 mg/kg; rat scu 47 mg/kg; rat unr 64.8 mg/kg; mus orl 39.8 mg/kg; mus ipr 71 mg/kg; mus scu 15 mg/kg; mus ivn 4.5 mg/kg; mus unr 40 mg/kg; cat orl 802 mg/kg; rbt skn 385 mg/kg; ckn orl 111 mg/kg; mam ihl LC_{50} 88 mg/m^3; mam unr 125 mg/kg; bwd orl 56.2 mg/kg.
CFR: 40CFR180.280

CAS RN 101-21-3
$C_{10}H_{12}ClNO_2$

CIPC
Carbamic acid, (3-chlorophenyl)-, 1-methylethyl ester (9CI)
Carbanilic acid, *m*-chloro-, isopropyl ester (8CI)
(3-Chlorophenyl)carbamic acid, 1-methylethyl ester
m-Chlorocarbanilic acid isopropyl ester
O-Isopropyl N-(3-chlorophenyl)carbamate
Beet-Kleen
Carbanilic acid, 4-chloro-, isopropyl ester
Chloro IPC
Chloro-IFK
Chloro-IPC
Chloropropham
Chlorpropham
Chlorprophame (French)
ChlorIPC
CI-IPC
Elbanil
ENT 18,060
Fasco WY-HOE
Furloe
Furloe 3 EC
Isopropyl chlorocarbanilate
Isopropyl *m*-chlorocarbanilate (ACN)
Isopropyl *m*-chlorophenylcarbamate
Isopropyl N-(*m*-chlorophenyl)carbamate
Isopropyl N-(3-chlorophenyl)carbamate (ACN)
Isopropyl N-chlorophenylcarbamate
Isopropyl 3-chlorocarbanilate
Isopropyl 3-chlorophenylcarbamate
IPC

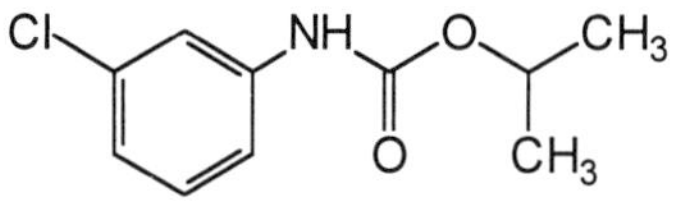

Cl NH O CH3 O CH3

Merck Index No. 2188
MP: 40.7 - 41.1°, 41.4°, 38.5 - 40° (technical)
BP: $149^{\circ 2}$
Density: 1.180^{30}
Solubility: Sl. sol. water, misc. most oils, org, solvents, 89 mg/l water 25, 100 g/kg kerosene
Octanol/water PC:
LD_{50}: rat orl 1200 mg/kg; rat orl 5000-7500 mg/kg; rat ipr 700 mg/kg; rat unr 3350 mg/kg; mus ipr 2600 mg/kg; rbt orl 5000 mg/kg; dck unr >2000 mg/kg; mam unr 3000 mg/kg.
CFR: 40CFR180.181

CAS RN 56-72-4
$C_{14}H_{16}ClO_5PS$

Coumaphos
Phosphorothioic acid, O-(3-chloro-4-methyl-2-oxo-2H-1-benzopyran-7-yl) O,O-diethyl ester
Coumarin, 3-chloro-7-hydroxy-4-methyl-, O-ester with O,O-diethylphosphorothioate (8CI)
Agridip
Agridip-Summer Sheep Dip
Asunthol
Asuntol
Azunthol
Bayer 21/199
Bayer 21199
Baymix
Baymix 50
BAY 21/199
Co-ral
Co-Ral
Coral
Coumafos
Coumafosum
Coumarin
Coumarin, 3-chloro-7-hydroxy-4-methyl, O-ester with O,O-diethylphosphorothioate
Coumophos
Cumafos (Dutch)
Cumafosum
Diethyl thiophosphoric acid ester of 3-chloro-4-methyl-7-hydroxycoumarin
Diethyl 3-chloro-4-methylumbelliferyl thionophosphate
ENT 17,957
Meldane
Meldone
Muscatox
NCI-C08662

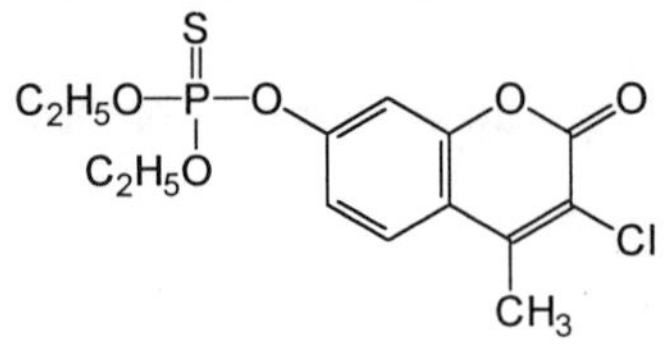

Merck Index No. 2559
MP: 91°, 95°, 90 - 92° (technical)
BP:
Density: 1.474
Solubility: Insol. water, sl. sol. acetone, chloroform, oils, 1.5 mg/l water 20°
Octanol/water PC:
LD_{50}: rat orl 13 mg/kg; rat orl 16 mg/kg; rat orl 41 mg/kg; rat orl 41 mg/kg; rat orl 15.5 mg/kg; rat skn 860 mg/kg; rat ihl LC_{50} 303 mg/m^3; rat skn 850 mg/kg; rat ipr 7.5 mg/kg; mus orl 28 mg/kg; mus ipr 50 mg/kg; rbt orl 80 mg/kg; rbt skn 500 mg/kg; rbt unr 2.6 mg/kg; gpg orl 58 mg/kg; gpg ipr 140 mg/kg; pgn orl 5.62mg/kg; qal orl 13.3 mg/kg; dck orl 29.8 mg/kg; dom orl 22.6 mg/kg; mam unr 55 mg/kg; bwd orl 1.78 mg/kg; bwd skn 7.5 mg/kg.
CFR: 40CFR180.189

CAS RN 2164-17-2
$C_{10}H_{11}F_3N_2O$

Fluometuron (ACN)
Urea, N,N-dimethyl-N'-[3-(trifluoromethyl)phenyl]- (9CI)
Urea, 1,1-dimethyl-3-(α,α,α-trifluoro-*m*-tolyl)- (8CI)
C 2059
Ciba 2059
Cotoran
Cotoron
Cottonex
Fluometureon
Herbicide C-2059
Kotoran
Lanex
N-[3-(Trifluoromethyl)phenyl]-1,1-dimethylurea
N,N-Dimethyl-N'-[3-(trifluoromethyl)phenyl]urea
NCI-C08695
Pakhtaran
1,1-Dimethyl-3-(α,α,α-trifluoro-*m*-tolyl)urea
3-(*m*-Trifluoromethylphenyl)-1,1-dimethylurea
3-(3-Trifluoromethylphenyl)-1,1-dimethylurea

NH N(CH3)2 O CF3

Merck Index No. 4080
MP: 163-164.5°
BP:
Density: 1.39^{20}
Solubility: 80 ppm water, sol. acetone, ethanol, isopropanol, dimethylformamide, organic solvents, 105 mg/l water 20, 110 g/l methanol, 105 g/l acetone, 23 g/l dichloromethane, 22 g/l n-octanol, 0.17 g/l hexane
Octanol/water PC: 171
LD_{50}: rat orl 1450 mg/kg; rat orl 89 mg/kg; rat skn >2000 mg/kg; rat ipr 685 mg/kg; mus orl 900 mg/kg; mus ipr 552 mg/kg; mus unr 850 mg/kg; dog orl >10000 mg/kg; rbt orl 2500 mg/kg; rbt skn >10000 mg/kg; gpg orl 810 mg/kg.
CFR: 40CFR180.229

CAS RN 15096-52-3 (61105-08-6)
$AlF_6.3Na$

Cryolite (ACN)(8CI)
Cryolite ($Na_3(AlF_6)$) (9CI)
Aluminum sodium fluoride
Cryolite syn
ENT 24,984
Fluoaluminate (sodium)
Greenland spar, Icetone
Koyoside
Kriolit
Kryocide
Kryolith (German)
Natriumaluminiumfluorid (German)
Natriumhexafluoroaluminate (German)
Sodium aluminofluoride
Sodium aluminum fluoride
Sodium cryolite
Sodium fluoaluminate (ACN)
Sodium fluoanateVola
Sodium hexafluoroaluminate
Villiaumite

F F F Al F F F 3⊖ 3Na⊕

Merck Index No. 2609
MP:
BP:
Density: 2.95
Solubility:
Octanol/water PC:
LD_{50}: rat orl 200 mg/kg; rbt orl LDLo 9000 mg/kg.
CFR: 40CFR180.145

CAS RN 66-81-9
$C_{15}H_{23}NO_4$

Cycloheximide (ACN)(USAN)
2,6-Piperidinedione, 4-[2-(3,5-dimethyl-2-oxocyclohexyl)-2-hydroxyethyl]-, [1S-[1α(S*),3α,5β]]- (9CI)
Glutarimide, 3-[2-(3,5-dimethyl-2-oxocyclohexyl)-2-hydroxyethyl] -
β-[2-(3,5-Dimethyl-2-oxocyclohexyl)-2-hydroxyethyl]glutarimide
[1S-[1(S*),3α,5β]]-4-[2-(3,5-Dimethyl-2-oxocyclohexyl)-2-hydroxyethyl]- 2,6-piperidinedione
Acti-dione BR
Acti-Aid
Actidion
Actidione
Actidione PM
Actidone
Actispray
Cicloheximide
Cyclohemimide
Cyclohexamide
Cyloheximide
ENT-15541
EU-4527
Glutaramide, 3-[2-(3,5-dimethyl-2-oxocyclohexyl)-2-hydroxyethyl]
Glutarimide, 3-[2-(3,5-dimethyl-2-oxocyclyheyl)-2-hydroxethyl]-
Hizarocin
Isocycloheximide
Kaken
Naramycin
Naramycin A
Neocycloheximide
NSC-185

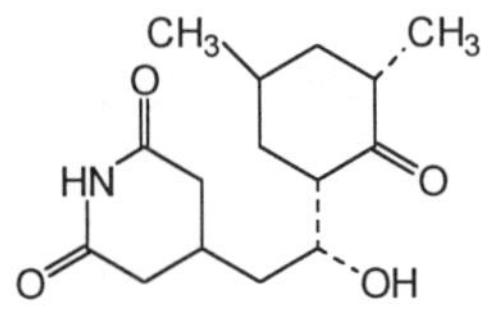

CH3 CH3 O HN O O OH

Merck Index No. 2734
MP: 119.5 - 121°, 115 - 116°
BP:
Density:
Solubility: 2.1 g/100ml water, 7 g/100ml amyl acetate
Octanol/water PC:
LD_{50}: rat orl 2 mg/kg; rat ipr 3.7 mg/kg; rat scu 2.5 mg/kg; rat ivn 2 mg/kg; rat unr 133 mg/kg; mus orl 133 mg/kg; mus iv 150 mg/kg; mus ipr 100 mg/kg; mus scu 160 mg/kg; dog orl 65 mg/kg; dog ivn 1 mg/kg; mky orl 60 mg/kg; cat ipr 4 mg/kg; rbt ivn 17 mg/kg; gpg orl 65 mg/kg; gpg ipr 60 mg/kg; gpg scu 60 mg/kg; ham ipr LDLo 40 mg/kg; ckn scu 2mg/kg.
CFR: 40CFR180.336

CAS RN 52315-07-8
$C_{22}H_{19}Cl_2NO_3$

Cypermethrin
Cyclopropanecarboxylic acid, 3-(2,2-dichloroethenyl)-2,2-dimethyl-,cyano (3-phenoxy phenyl) methyl ester (9CI)
α-Cyano-3-phenoxybenzyl 2,2-dimethyl-3-(2,2-dichlorovinyl)cyclopropane carboxylate
α-Cyano-3-phenoxybenzyl 3-(2,2-dichlorovinyl)-2,2-dimethylcyclopropanecarboxylate
Barricade
Cymbush
FMC 30980
FMC 45806
Imperator
Kafil
PP 383
Ripcord
WL 43467

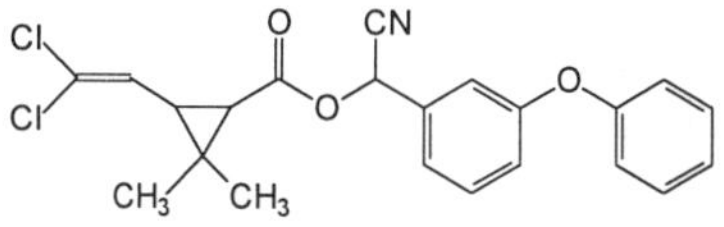

Merck Index No. 2775
MP: 60 - 80° (technical)
BP:
Density: 1.25^{20}
Solubility: Insol. water, sol. methanol, acetone, xylene, dichloromethane, 0.01 mg/l water 20, >450 g/l acetone, chloroform, cyclohexanone, xylene, 337 g/l ethanol, 103 g/l hexane
Octanol/water PC: 4000000
LD_{50}: rat orl 70 mg/kg; rat orl 250-4150 mg/kg; rat ihl LC_{50} 7889 mg/m^3/4H; rat ivn LDLo 6 mg/kg; rat unr 400 mg/kg; rat skn >1600 mg/kg; mus orl 138 mg/kg; rbt orl 3000 mg/kg; rbt skn >2400 mg/kg; gpg orl 500 mg/kg; ham orl 400 mg/kg; ckn orl 7000 mg/kg; ckn unr 2000 mg/kg; brd unr 3000 mg/kg.
CFR: 40CFR180.418

CAS RN 51338-27-3
$C_{16}H_{14}Cl_2O_4$

Diclofop-methyl (ACN)
Propanoic acid, 2-[4-(2,4-dichlorophenoxy)phenoxy]-, methyl ester
Dichlofop-methyl
Dichlordiphenprop
Hoegrass
Illoxan
Iloxan
Methyl ester of 2-[4-(2,4-dichlorophenoxy)phenoxy]propanoic acid
Propionic acid, 2-[4-(2,4-dichlorophenoxy) phenoxy]-, methyl ester
2-[4-(2,4-Dichlorophenoxy)phenoxy]methylpropionate
2-[4-(2,4-Dichlorophenoxy)phenoxy]propanoic acid, methyl ester (ACN)

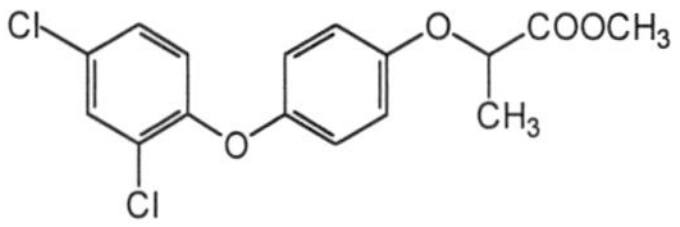

Merck Index No. 3072
MP: 39 - 41°
BP: 175 - 177°$^{0.1}$
Density:
Solubility: 0.3mg/100ml water, 249 g/100ml acetone, 11 g/100ml ethanol, 253 g/100ml xylene
Octanol/water PC:
LD_{50}: rat orl 512 mg/kg; rat skn >5000 mg/kg; mus orl 586 mg/kg; dog orl 1600 mg/kg; qal orl >10000 mg/kg.
CFR: 40CFR180.385

CAS RN 62-73-7
$C_4H_7Cl_2O_4P$

Dichlorvos (USAN)

Phosphoric acid, 2,2-dichloroethenyl dimethyl ester (9CI)
Phosphoric acid, 2,2-dichlorovinyl dimethyl ester (8CI)
(2,2-Dichloor-vinyl)-dimethyl-fosfaat (Dutch)
(2,2-Dichlor-vinyl)-dimethyl-phosphat (German)
(2,2-Dichloro-vinil)dimetil-fosfato (Italian)
O-(2,2-Dichlorvinyl)-O,O-dimethylphosphat (German)
Atgard
Atgard C
Atgard V
Bibesol
Brevinyl
Brevinyl E-50
Brevinyl E50
Canogard
Celcusan
Chlorvinphos
Dedevap
Deriban
Derribante
Dichlorman
Dichlorophos
Dichlorovos
Dichlorphos
Dichlorvosum
Diclorvos
Dimethyl dichlorovinyl phosphate
Dimethyl 2,2-dichloroethenyl phosphate
Dimethyl 2,2-dichlorovinyl phosphate
Dimethyl 2,2-dichlorovinyl phosphoric acid ester

$CH_3O-P(=O)(OCH_3)-O-CH=CCl_2$

Merck Index No. 3069
MP:
BP: 140°^{20}, 84°^{1}, 72°$^{0.5}$,30°$^{0.01}$
Density: 1.415^{25}
Solubility: Misc. alcohol, non-polar solvents, 10 g/l water 20°
Octanol/water PC:
LD_{50}: rat orl 17 mg/kg; rat orl 80 mg/kg; rat orl 56 mg/kg; rat ihl LC_{50} 15 mg/m^3/4H; rat skn 70.4 mg/kg; rat ipr 23.3 mg/kg; rat scu 10.8 mg/kg; rat unr 57 mg/kg; mus orl 61 mg/kg; mus orl 140-275 mg/kg; mus ihl LC_{50} 13 mg/m^3/4H; mus skn 206 mg/kg; mus ipr 22 mg/kg; mus scu 24 mg/kg; mus ivn 18 mg/kg; mus unr 75 mg/kg; dog orl 100 mg/kg; dog ivn 2.2 mg/kg; rbt orl 10 mg/kg; rbt skn 107 mg/kg; rbt unr 23 mg/kg; pig orl 157 mg/kg; ham ipr 30 mg/kg; pgn orl 23.7 mg/kg; ckn orl 6.45 mg/kg; qal orl 22.010 mg/kg; dck orl 7.80 mg/kg; mam unr 125 mg/kg; bwd orl 12 mg/kg.
CFR: 40CFR180.235

CAS RN 115-32-2
$C_{14}H_9Cl_5O$

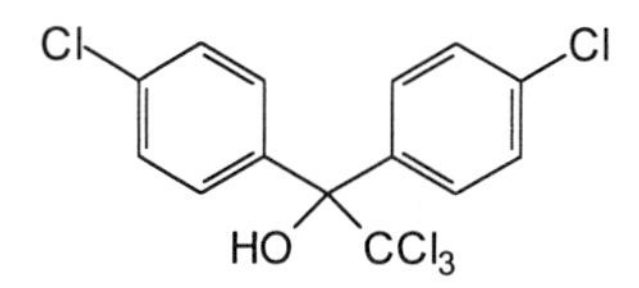

Dicofol
Benzenemethanol, 4-chloro-α-(4-chlorophenyl)-α-(trichloromethyl)- (9CI)
Benzhydrol, 4,4'-dichloro-α-(trichloromethyl)- (8CI)
p,p'-Kelthane
Acarin
Carbax
CPCA
Decofol
Di(*p*-chlorophenyl)trichloromethylcarbinol
Dichlorokelthane
DTMC
Ethanol, 1,1-bis(*p*-chlorophenyl)-2,2,2-trichloro-
Ethanol, 2,2,2-trichloro-1,1-bis(*p*-chlorophenyl)-
Ethanol, 2,2,2-trichloro-1,1-bis(4-chlorophenyl)-
ENT 23,648
FW 293
Keltane
Kelthane
Kelthane A
Kelthanethanol
Kelthone
Milbol
Mitigan
NCI-C00486
1,1-Bis(chlorophenyl)-2,2,2-trichloroethanol (ACN)
1,1-Bis(*p*-chlorophenyl)-2,2,2-trichloroethanol
1,1-Bis(4-chlorophenyl)-2,2,2-trichloroethanol
1,1'-Bis(*p*-chlorophenyl)-2,2,2-trichlorethanol

Merck Index No. 3075
MP: 77 - 78°
BP:
Density:
Solubility: Insol. water, sol. most org, solvents
Octanol/water PC:
LD_{50}: rat orl 1495 mg/kg; rat orl 575 mg/kg; rat ihl LC_{50} 5000 mg/m^3/4H; rat ipr 1150 mg/kg; rat skn 100 mg/kg; mus orl 420 mg/kg; dog orl 4000 mg/kg; rbt orl 1810 mg/kg; rbt skn 1870 mg/kg; gpg orl 1810 mg/kg; ckn orl 4365 mg/kg; mam unr 433 mg/kg.
CFR: 40CFR180.163

CAS RN 141-66-2
$C_8H_{16}NO_5P$

Dicrotophos
Phosphoric acid, 3-(dimethylamino)-1-methyl-3-oxo-1-propenyl dimethyl ester, (E)- (9CI)
Phosphoric acid, dimethyl ester, ester with 3-hydroxy-N,N-dimethylcrotonamide, (E)- (E)-Phosphoric acid, 3-(dimethylamino)-1-methyl-3-oxo-1-propenyl dimethyl ester
cis-2-Dimethylcarbamoyl-1-methylvinyl dimethylphosphate
trans-Bidrin
Bidrin
Bidrin-R
C 709 (VAN)
Carbicron
Carbomicron
Ciba 709
Crotonamide, 3-hydroxy-N,N-dimethyl-, *cis*-, dimethyl phosphate
Crotonamide, 3-hydroxy-N,N-dimethyl-, dimethyl phosphate, (E)-
Crotonamide, 3-hydroxy-N,N-dimethyl-, dimethyl phosphate, cis-
Crotonamide, 3-hydroxy-N,N-dimethyl, *cis*-dimethyl phosphate
Dicrotofos (Dutch)
Dimethyl *cis*-2-dimethylcarbamoyl-1-methylvinyl phosphate
Dimethyl phosphate ester of (E)-3-hydroxy-N,N-dimethylcrotonamide
Dimethyl phosphate ester of 3-hydroxy-N,N-dimethyl-*cis*-crotonamide
Dimethyl phosphate ester with 3-hydroxy-N,N-dimethyl-*cis*-crotonamide (ACN)
Dimethyl phosphate ester with 3-hydroxy-N,N-dimethylcrotonamide
Dimethyl phosphate of 3-hydroxy-N,N-dimethyl-*cis*-crotonamide
Dimethyl 1-dimethylcarbamoyl-1-propen-2-yl phosphate
Dimethyl 2-dimethylcarbamoyl-1-methylvinyl phosphate
Ektafos
Ektofos
ENT 24,482
Karbicron
O,O-Dimethyl O-(N,N-dimethylcarbamoyl-1-methylvinyl) phosphate 5 -

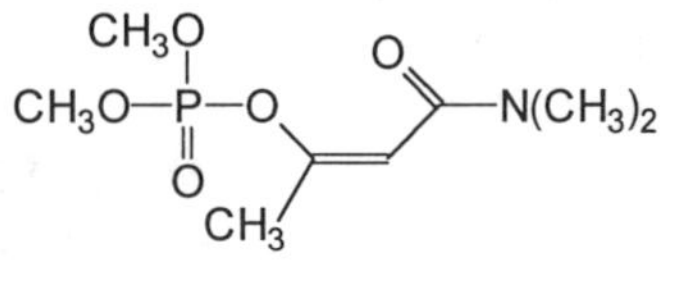

CH_3O — CH_3O—P(=O)—O— C(CH_3)=CH—C(=O)—$N(CH_3)_2$

Merck Index No. 3077
MP:
BP: 400°, $130^{o0.1}$
Density: 1.216^{15}
Solubility: Misc. water, ethanol, xylene, acetone, acetonitrile, CHCl2, carbon tetrachloride
Octanol/water PC:
LD_{50}: rat orl 13 mg/kg; rat orl 16 mg/kg; rat orl 21 mg/kg; rat skn 42 mg/kg; rat ihl LC_{50} 90 mg/m^3/4H; rat scu 8.137 mg/kg; rat unr 22 mg/kg; mus orl 11 mg/kg; mus orl 15 mg/kg; mus ipr 9.5 mg/kg; mus scu 11.5 mg/kg; mus ivn 9.9 mg/kg; rbt skn 168 mg/kg; pgn orl 2 mg/kg; ckn orl 7.97 mg/kg; qal orl 4 mg/kg; dck orl 4.14 mg/kg; bwd orl 1 mg/kg; bwd skn 1.3 mg/kg.
CFR: 40CFR180.299

CAS RN 60-57-1
$C_{12}H_8Cl_6O$

Dieldrin
2,7:3,6-Dimethanonaphth[2,3-b]oxirene,3,4,5,6,9,9-hexachloro-1a,2,2a,3,6,6a,7,7a-octahydro-, (1aα,2β,2aα,3β,6β,6aα,7β,7aα)-1,4:5,8-Dimethanonaphthalene,1,2,3,4,10,10-hexachloro-6,7-epoxy-1,4,4a,5,6,7,8,8a-octahydro-, *endo,exo-*
(1aα,2β,2aα,3β,6β,6aα,7β,7aα)-3,4,5,6,9,9-Hexachloro- 1a,2,2a,3,6,6a,7,7a-octahydro-2,7:3,6- dimethanonaphth [2,3-b] oxirene
exo-Dieldrin
Aldrin epoxide
Alvit 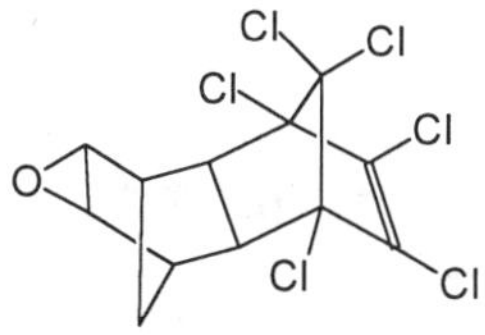
Alvit 55
Compound 497
Compound-497
Dieldrex
Dieldrin-R
Dieldrina
Dieldrine (French)
Dieldrinum
Dieldrite
Dielmoth
Dildrin
Dorytox
ENT-16225
Hexachloroepoxyoctahydro-*endo,exo*-dimethanonaphthalene
HEOD
Illoxol
Insectlack
Kombi-Albertan
Moth Snub D
NCI-C00124
Octalox
OMS 18
Panoram D-31

Merck Index No. 3093
MP: 176 - 177°
BP:
Density:
Solubility: Insol. water, sol. organic sovents
Octanol/water PC:
LD_{50}: man orl LDLo 65 mg/kg; hmn unr LDLo 28 mg/kg; rat orl 38.3 mg/kg; rat orl 46 mg/kg; rat ihl LC_{50} 13 mg/m^3/4H; rat skn 56 mg/kg; rat ipr 35 mg/kg; rat scu 49 mg/kg; rat ivn 9 mg/kg; mus orl 38 mg/kg; mus ipr LDLo 26 mg/kg; mus ivn 10.5 mg/kg; dog orl 65 mg/kg; dog unr LDLo 65 mg/kg; mky orl 3 mg/kg; cat orl LDLo 500 mg/kg; cat ihl LC_{50} 80 mg/m^3/4H; cat skn LDLo 750 mg/kg; rbt orl 45 mg/kg; rbt skn 250 mg/kg; rbt scu LDLo 150 mg/kg; pig orl 38 mg/kg; gpg orl 49 mg/kg; ham orl 60 mg/kg; pgn orl 23.7 mg/kg; pgn ivn 1200 mg/kg; ckn orl 20 mg/kg; qal orl 10.78 mg/kg; dck orl 381 mg/kg; mam unr 25 mg/kg; bwd orl 13.3 mg/kg.
CFR: 40CFR180.137 (removed)

CAS RN 29232-93-7
$C_{11}H_{20}N_3O_3PS$

Pirimiphos-methyl (ACN)
Phosphorothioic acid, O-[2-(diethylamino)-6-methyl-4-pyrimidinyl] O,O-dimethyl ester
Actellic
Actellifog
BLEX
ENT 27699Gc
Methylpirimiphos
Methylpyrimiphos
O-(2-Diethylamino-6-methylpyrimidin-4-yl) O,O-dimethyl phosphorothioate
O-[2-(Diethylamino)-6-methyl-4-pyrimidinyl] O,O-dimethyl phosphorothioate
O-[2-(Diethylamino)-6-methylpyrimidin-4-yl] O,O-diethyl phosphorothioate
O-[2-(Diethylamino)-6-methylpyrimidin-4-yl] O,O-dimethyl phosphorothioate
O,O-Dimethyl O-[2-(diethylamino)-6-methyl-4-pyrimidinyl] phosphorothioate
O,O-Dimethyl O-[2-(diethylamino)-6-methylpyrimidin-4-yl]phosphorothioate
Phosphorothioic acid, O-[2-(diethylamino)-6-methyl-4-pyridinyl] O,O-diethyl ester
Pirimifosmethyl
Pirimiphos methyl
Pirimiphos Me
Plant Protection PP511
Pyridimine phosphate
PP 511
R-33986
Silosan
2-(Diethylamino)-6-methylpyrimidin-4-yl dimethyl phosphorothionate
2-Diethylamino-6-methylpyrimidin-4-yl dimethyl phosphorothionate

Merck Index No.
MP: ca. 15°
BP: (dec)
Density: 1.17^{20}
Solubility: 5 mg/l water 30, misc. most org. solvents
Octanol/water PC: 16000
LD_{50}: rat orl 1250 mg/kg; rat orl 2050 mg/kg; rat skn >4592 mg/kg; mus orl 1180 mg/kg; mus unr 1180 mg/kg; rbt orl 1150 mg/kg; rbt skn >2000 mg/kg; gpg orl 1000 mg/kg; ckn orl 30 mg/kg; qal orl 140 mg/kg; mam orl 2000 mg/kg.
40CFR180.409

CAS RN 23505-41-1
$C_{13}H_{24}N_3O_3PS$

Pirimiphos-ethyl (ACN)
Phosphorothioic acid, O-[2-(diethylamino)-6-methyl-4-pyrimidinyl] O,O-diethyl ester

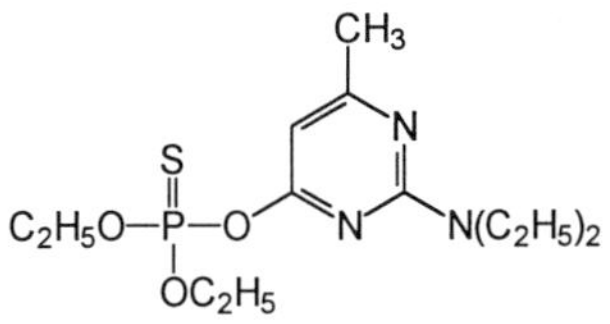

O-[2-(Diethylamino)-6-methyl-4-pyrimidinyl] O,O-diethyl phosphorothioate
O,O-Diethyl O-[2-(diethylamino)-6-methylpyrimidin-4-yl]phosphorothioate
Phosphorothioic acid, O,O-diethyl O-[2-(diethylamino)-6-methyl-4-pyrimidinyl] ester
Fernex
Pirimifosethyl
Pirimiphos ethyl
Primicid
Primotec
Prinicid
PP 211
R 42211

Merck Index No. 7469
MP:
BP: dec >130°

Density: 1.14^{20}
Solubility: 1 mg/l 30°, misc. most org. solvents
Octanol/water PC: 28,800
LD_{50}: rat orl 140mg/kg; rat orl 192 mg/kg; rat skn 1000-2000 mg/kg; mus orl 105 mg/kg; cat orl 25 mg/kg; gpg orl 50 mg/kg; pgn orl 6 mg/kg; qal orl 6 mg/kg; dcl orl 2.5 mg/kg; bwd orl 3 mg/kg.
CFR: 40CFR180.308

CAS RN 35367-38-5
$C_{14}H_9ClF_2N_2O_2$

Diflubenzuron (ACN)
Benzamide, N-[[(4-chlorophenyl)amino]carbonyl]-2,6-difluoro- (9CI)
Benzamide, 2,6-difluoro-N-[(4-chlorophenyl)aminocarbonyl]- 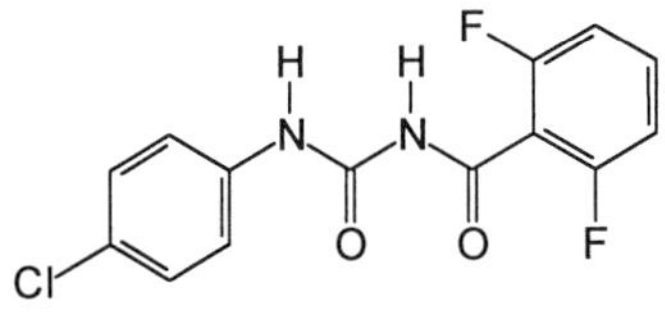
Deflubenzon
Diflubenuron
Diflubenzon
Difluron
Dimilin
Duphar PH 60-40
DU 112307
ENT 29054
ENT-29054
Largon
N-[[(4-Chlorophenyl)amino]carbonyl]-2,6-difluorobenzamide
N-[[[4-Chlorophenyl)amino]carbonyl]-2,6-difluorobenzamide
OMS 1804
Ph 6040
PDD 6040I
PH 60-40
Thompson-Hayward 6040
TH 60 40
TH 6040
Urea, 1-(*p*-chlorophenyl)-3-(2,6-difluorobenzoyl)-
1-(4-Chlorophenyl)-3-(2,6-difluorobenzoyl)urea

Merck Index No. 3128
MP: 239°, 230 - 232°
BP:
Density:
Solubility: 0.3 ppm water, 0.14 mg/l water 20, 6.5 g/l acetone, 104 g/l dimethylformamide, 20 g/l dioxane
Octanol/water PC:
LD_{50}: rat orl >4640 mg/kg; rat skn >10000 mg/kg; mus orl 4640 mg/kg; mus ipr <2150 mg/kg; mus unr 4600 mg/kg; rbt skn 2000 mg/kg.
CFR: 40CFR180.377

CAS RN 29091-05-2
$C_{11}H_{13}F_3N_4O_4$

Dinitramine (ACN)
1,3-Benzenediamine, N3,N3-diethyl-2,4-dinitro-6-(trifluoromethyl)- (9CI)
Toluene-2,4-diamine, N4,N4-diethyl-α,α,α-trifluoro-3,5-dinitro- (8CI)
Cobex
Cobex (herbicide)
Cobexo
Diethamine
Dinitroamine
N^3,N^3-diethyl-2,4-dinitro-6-(trifluoromethyl)-*m*-phenylenediamine
N^3,N^3-Diethyl-2,4-dinitro-6-(trifluoromethyl)-*m*-phenylenediamine
N^3,N^3-Diethyl-2,4-dinitro-6-(trifluoromethyl)-1,3-benzenediamine
N^3,N^3-Diethyl-2,4-dinitro-6-(trifluoromethyl)-1,3-phenylenediamine
N^4,N^4-Diethyl-α,α,α-trifluoro-3,5-dinitrotoluene-2,4-diamine
N^4,N^4-Diethyl-α,α,α-trifluoro-3,5-dinitrotoluene, 2,4-diamine
N',N'-Diethyl-2,6-dinitro-4-trifluoromethyl-m-phenylenediamine
N4,N4-Diethyl-α,α,α-trifluoro-3,5-dinitro-toluene-2,4-diamine
Toluene-2,4-diamine, N^4,N^4-diethyl-α,α,α-trifluoro-3,5-dinitro-
Toluene-2,4-diamine, N^4,N^4-Diethyl-3,5-dinitro-α,α,α-trifluoro-
USB 3584
USB-3584
1,3-Benzenediamine, N^3,N^3-diethyl-2,4-dinitro-6-(trifluoromethyl)-

NH_2
CF_3
NO_2
$N(C_2H_5)_2$
NO_2

Merck Index No.
MP: 98 - 99°
BP:
Density:
Solubility: 1 mg/l water 20, 1040 g/l acetone, 670 g/l chloroform, 473 g/l benzene, 227 g/l xylene, 107 g/l ethanol, 14 g/l hexane
Octanol/water PC: 20000
LD_{50}: rat orl 3000 mg/kg; rbt skn 2000 mg/kg; rbt skn >6800 mg/kg; qal orl >1200 mg/kg; dck orl >10000 mg/kg.
40CFR180.327

CAS RN 92-52-4
$C_{12}H_{10}$

Biphenyl (ACN)(8CI)
1,1'-Biphenyl (9CI)
Bibenzene
Diphenyl
Lemonene
Phenador-X
Phenylbenzene
PHPH
Xenene
1,1'-Diphenyl

Merck Index No. 3314
MP: 69 - 71°
BP: 254 - 255°
Density: 1.041
Solubility: insol. water, sol. alcohol, phenol,
Octanol/water PC:
LD_{50}: hmn ihl TCLo 4.4 mg/m^3; rat orl 2400 mg/kg; rat orl 3280 mg/kg; rat unr 4500 mg/kg; mus orl 1900 mg/kg; mus ivn 56 mg/kg; rbt orl 2400 mg/kg; rbt skn >5010 mg/kg.
CFR: 40CFR180.141

CAS RN 534-52-1
$C_7H_6N_2O_5$

***o*-Dinitrocresol** 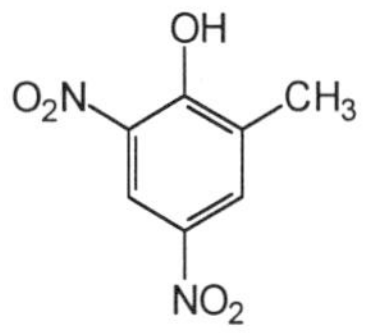
Phenol, 2-methyl-4,6-dinitro- (9CI)
o-Cresol, 4,6-dinitro- (8CI)
Antinonin
Antinonnin
Arborol
Capsine
Chemsect
Chemsect DNOC
Degrassan
Dekrysil
Detal
Dillex
Dinitro
Dinitro-*o*-cresol (VAN)
Dinitrocresol
Dinitrodendtroxal
Dinitrol
Dinitromethyl cyclohexyltrienol
Dinitromethylcyclohexytrienol
Ditrosol
Dn-dry mix no. 2
Dnok (Czech)
Dwunitro-*o*-krezol (Polish)
DN
DN Dry Mix No. 2
DNC
DNOC

Merck Index No. 3272
MP: 87.5°, 86°, 83 - 85° (95-98%)
BP:
Density:
Solubility: sp. sol. water, sol. diethyl ether, acetone, ethanol, sp. sol. Petroleum ether, 130 mg/l water 15, 372 g/l chloroform, 43 g/l ethanol. Salts sol. water
Octanol/water PC:
LD_{50}: man orl TDLo 7.5 mg/kg; hmn ihl TCLo 1 mg/m^3; chd skn LDLo 500 mg/kg; man unr LDLo 29 mg/kg; rat orl 7 mg/kg; rat orl 30 mg/kg, rat orl 25-40 mg/kg, rat skn 200 mg/kg; rat ipr LDLo 28 mg/kg; rat scu 25.6 mg/kg; rat unr 85 mg/kg; mus orl 24 mg/kg; mus orl 21 mg/kg; mus ipr 19 mg/kg; mus unr 40 mg/kg; dog ivn LDLo 15 mg/kg; cat orl 50 mg/kg; cat ihl LCLo 40 mg/m^3/4H; rbt orl 24.6 mg/kg; rbt skn 1000 mg/kg; rbt ipr 23.5 mg/kg; gpg orl 24.6 mg/kg; gpg skn LDLo 500 mg/kg; gpg ipr 22.5 mg/kg; pgn ivn LDLo 7 mg/kg; dom orl 100 mg/kg; mam ihl LC_{30} 100 mg/m^3.
CFR: 40CFR180.344

CAS RN 6119-92-2
$C_{18}H_{24}N_2O_6$

Dinocap
2-Butenoic acid, 2-(1-methylheptyl)-4,6-dinitrophenyl ester (9CI)
Crotonic acid, 2-(1-methylheptyl)-4,6-dinitrophenyl ester (8CI)
[6-(1-Methyl-heptyl)-2,3-dinitro-phenyl]-crotonat (German)
[6-(1-Methyl-heptyl)-2,4-dinitro-fenyl]-crotonaat (Dutch)
[6-(1-Metil-epitl)-2,4-dinitro-fenil]-crotonato (Italian)
Arathane
Caprane
Capryl
Capryldinitrophenyl crotonate
Cr 1639
Crothane
Crotonate de 2,4-dinitro 6-(1-methyl-heptyl)-phenyle (French)
Crotonates of nitrophenols and derivatives, chiefly 2-(1-methylheptyl)-4,6-dinitrophenol
Crotothane
CR 1639
Dinitro(1-methylheptyl)phenyl crotonate
Dinokap
Dnocp
Dnopc
DPC
Ent 24727
ENT 24727
Iscothan
Isothan
Karathane
Karathane wd
Karathene
Mildex
Nitrooctylphenols(principally dinitro) (ACN)

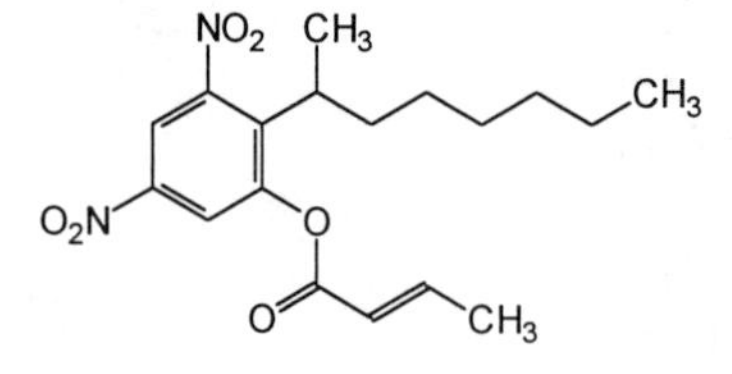

Merck Index No.
MP:
BP: 134 - 138$^{\circ 0.01}$
Density:
Solubility:
Octanol/water PC:
LD_{50}:
40CFR180.341

CAS RN 78-34-2
$C_{12}H_{26}O_6P_2S_4$

Dioxathion (ACN)
Phosphorodithioic acid, S,S'-1,4-dioxane-2,3-diyl O,O,O',O'-tetraethyl ester
Phosphorodithioic acid, S,S'-p-dioxane-2,3-diyl O,O,O',O'-tetraethyl ester (8CI)
p-Dioxane-2,3-dithiol, S,S-diester with O,O-diethyl phosphorodithioate
p-Dioxane-2,3-diyl ethyl phosphorodithioate
AC 528
AC-528
AENT 22897
Bis(dithiophosphate de O,O-diethyle) de S,S'-(1,4-dioxanne-2,3-diyle) (French)
Delnatex
Delnav
Dioxation
Dioxationum
Dioxothion
ENT 22,897
Hercules AC528
Hercules 528
Kavadel
Navadel
NCI-C00395
Phosphorodithioic acid, O,O-diethyl ester, S,S-diester with p-dioxane-2,3-dithiol
Phosphorodithioic acid, S,S'-1H-dioxane-2,3-diyl O,O,O',O'-tetraethyl ester
Ruphos
S,S-p-Dioxane-2,3-diyl O,O,O',O'-tetraethyl di(phosphorodithioate)
S,S-1,4-Dioxan-2,3-ylidene bis-O,O-diethyl phosphorothiolothionate, 70%, & related cpds.
S,S'-(1,4-Dioxane-2,3-diyl) O,O,O',O'-tetraethyldi(phosphorothioate)
S,S'-p-Dioxane-2,3-diyl-O,O,O',O'-tetraethyl phosphorodithioate
S,S'-1,4-Dioxane-2,3-diyl O,O,O',O'-tetraethyl di(phosphorodithioate)
1,4-Diossan-2,3-diyl-bis(O,O-dietil-ditiofosfato) (Italian)

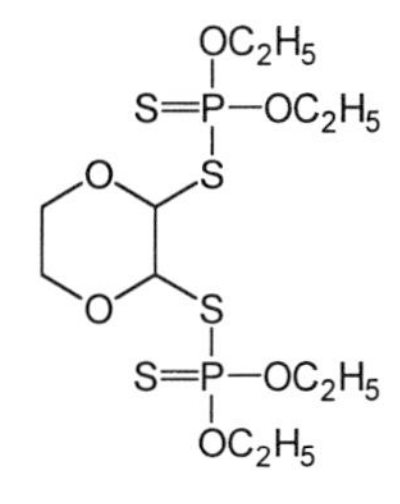

Merck Index No. 3296
MP: -20°
BP:
Density: 1.257^{26}
Solubility: Insol. water, partly sol. hexane
Octanol/water PC:
LD_{50}: hmn orl TDLo 9 mg/kg/60D; rat orl 20 mg/kg; rat orl 23 mg/kg; rat orl 43 mg/kg; rat skn 63 mg/kg; rat skn 235 mg/kg; rat ihl LC_{50} 1398 mg/m^3/1H; rat ipr 30 mg/kg; mus orl 176 mg/kg; mus ihl LC_{50} 340 mg/m^3/1H; mus ipr 33 mg/kg; dog orl 10 mg/kg; rbt skn 85 mg/kg; ckn orl >316 mg/kg; ckn scu >200 mg/kg.
CFR: 40CFR180.171

CAS RN 88-85-7
$C_{10}H_{12}N_2O_5$

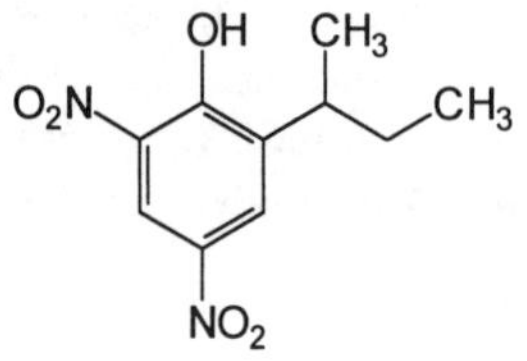

Dinoseb (ACN)
Phenol, 2-(1-methylpropyl)-4,6-dinitro- (9CI)
Phenol, 2-*sec*-butyl-4,6-dinitro- (8CI)
Aatox
Aretit
Basanite
Butaphene
BNP 20
BNP 30
Caldon
Chemox general
Chemox P.E.
Chemox PE
Dibutox
Dinitrall
Dinitro Weed Killer
Dinitro-*o*-butyl phenol
Dinitro-*o*-*sec*-butyl phenol
Dinitrobutylphenol
Dinosch
Dinosebe (French)
Dnosbp
Dnsbp
Dow general weed killer
Dow selective weed killer
Dow General
Dyanap
Dytop
DBNF
DN 289 (VAN)

OH CH3
O2N CH3
NO2

Merck Index No. 3282
MP: 38 - 42°, 30 - 40° (95-98% technical)
BP:
Density: 1.265^{45}
Solubility: 52 mg/l water 20, 480 g/kg ethanol, 270 g/kg C7H17, sol. most org. solvents & petr. ether
Octanol/water PC:
LD_{50}: rat orl 25 mg/kg; rat orl 27 mg/kg; rat orl 28 mg/kg; rat orl 58 mg/kg; rat skn 80 mg/kg; rat scu 20.37 mg/kg; mus orl 16 mg/kg; mus skn 40 mg/kg; mus ipr 10 mg/kg; cat ihl LCLo 45 mg/m^3/3H; rbt skn 80 mg/kg; gpg orl 20 mg/kg; gpg skn LDLo 500 mg/kg; ckn orl 26 mg/kg; dck orl 9 mg/kg; mam unr 60 mg/kg; bwd orl 7.1 mg/kg.
CFR: 40CFR180.281

CAS RN 957-51-7
$C_{16}H_{17}NO$

Diphenamid (ACN)
Benzeneacetamide, N,N-dimethyl-α-phenyl- (9CI)
Acetamide, N,N-dimethyl-2,2-diphenyl- (8CI)
Dif 4
Dimid
Diphenamide
Dymid
DIF 4
Enide
Enide 50
Enide 50W
ENT-28567
Fenam
L-34314
Lilly 34,314
N,N-Dimethyl-α-phenylbenzeneacetamide
N,N-Dimethyl-α,α-diphenylacetamide
N,N-Dimethyl-2,2-diphenylacetamide (ACN)
N,N-Dimethyldiphenylacetamide
U 4513
2,2-Diphenyl-N,N-dimethylacetamide

O N(CH3)2

Merck Index No.
3305
MP: 134.5 - 135.5°, 132 - 134° (technical)
BP:
Density: $1.17^{23.3}$
Solubility: sol. water, acetone, dimethylformamide, xylene, phenyl cellsosolve, 260 mg/l water 27, 189 g/l acetone, 165 g/l dimethylformamide, 50 g/l xylene
Octanol/water PC:
LD_{50}: rat orl 685 mg/kg; rat orl 700 mg/kg; rat orl 1050 mg/kg; rat skn >6320 mg/kg; mus orl 600 mg/kg; mus ipr 500 mg/kg; mus scu 800 mg/kg; mus unr 700 mg/kg; dog orl 1000 mg/kg; mky orl 1000 mg/kg; rbt orl 1500 mg/kg; mam unr 700 mg/kg.
CFR: 40CFR180.230

CAS RN 122-39-4
$C_{12}H_{11}N$

Diphenylamine (ACN)(8CI)
Benzenamine, N-phenyl- (9CI)
Aniline, N-phenyl-
Anilinobenzene
Benzene, (phenylamino)-
Benzene, anilino-
Big Dipper
C.I. 10355
DFA
DPA (VAN)
ENT-781
N-Petro ULF phenyl aniline
N-Phenylaniline
N-Phenylbenzenamine
N-Phenylbenzeneamine
N,N-Diphenylamine
No scald
Phenylaniline
Scaldip

NH

Merck Index No. 3317
MP: 53 - 54°
BP: 302°
Density: 1.16
Solubility: 1 gm/2.2ml ethanol, 1 gm/4.5ml isopropanol, feely sol. benzene, diethyl ether, AcOH, CS2
Octanol/water PC:
LD_{50}: rat orl 2 mg/kg; mus orl 1750 mg/kg; gpg orl 300 g/kg; mam orl 3200 mg/kg.
CFR: 40CFR180.190

CAS RN 4147-51-7
$C_{11}H_{21}N_5S$

Dipropetryn (ACN)
1,3,5-Triazine-2,4-diamine, 6-(ethylthio)-N,N'-bis(1-methylethyl)- (9CI)
s-Triazine, 2-(ethylthio)-4,6-bis(isopropylamino)- (8CI)
s-Triazine, 2,4-bis(isopropylamino)-6-ethylthio-
Cotofer
Cotofor
Dipropetryne
GS 16068
Sancap
2-(Ethylthio)-4,6-bis(isopropylamino)-*s*-triazine (ACN)
2-(Ethylthio)-4,6-bis(isopropylamino)-1,3,5-triazine
2-(Ethylthio)-4,6-bis[isopropylamino]-*s*-triazine
2-Ethylthio-4,6-bis(isopropylamino)-*s*-triazine
6-Ethylthio-N,N'-bis(1-methylethyl)-1,3,5-triazine-2,4-diamine

(CH3)2CHNH, N, NHCH(CH3)2, N, N, SC2H5

Merck Index No. 3349
MP: 104 - 106°
BP:
Density:
Solubility: 16 mg/l water, 270 g/l acetone, 300 g/l dichloromethane, 220 g/l toluene, 190 g/l methanol, 130 g/l n-octanol, 9 g/l C7H16, sol. org. solvents
Octanol/water PC:
LD_{50}: rat orl 3900-4200 mg/kg; rbt skn 10000 mg/kg.
CFR: 40CFR180.329

CAS RN 85-00-7
$C_{12}H_{12}N_2.2Br$

Diquat dibromide (ACN)
Dipyrido[1,2-a:2',1'-c]pyrazinediium, 6,7-dihydro-, dibromide (8CI9CI)
Aquacide
Deiquat
Diquat (VAN)
Diquat bromide
DDB
Ethylene dipyridylium dibromide
Preeglone
Reglone
Reglox
Weedtrine D
1,1'-Ethylene-2,2'-dipyridylium dibromide
5,6-Dihydrodipyrido[1,2a,2,1c]pyrazinium dibromide
6,7-Dihydrodipyrido(1,2-a:2',1'-c)pyrazidinium dibromide
6,7-Dihydrodipyrido[1,2-a:2',1'-c]pyrazinediium dibromide

N, N, $2Br^{\ominus}$

Merck Index No. 3359
MP: 335 - 340° (<320)
BP:
Density: 1.22 - 1.27^{20}
Solubility: 70% water 20°
Octanol/water PC: 0.000025
LD_{50}: rat orl 120 mg/kg; rat orl 231 mg/kg; rat skn 433 mg/kg; rat ipr LDLo 500 mg/kg; rat scu 20 mg/kg; rat ivn LDLo 14 mg/kg; mus orl 233 mg/kg; mus orl 125 mg/kg; dog orl 187 mg/kg; rbt orl 101 mg/kg; rbt orl 187 mg/kg; rbt skn >500 mg/kg; gpg orl 100 mg/kg; gpg orl 187 mg/kg; ckn orl 373 mg/kg; dck orl 564 mg/kg; ctl orl 56 mg/kg; dom orl 30 mg/kg; mam unr 400 mg/kg.
CFR: 40CFR180.226

CAS RN 298-04-4
$C_8H_{19}O_2PS_3$

Disulfoton
Phosphorodithioic acid, O,O-diethyl S-[2-(ethylthio)ethyl] ester (8CI9CI) 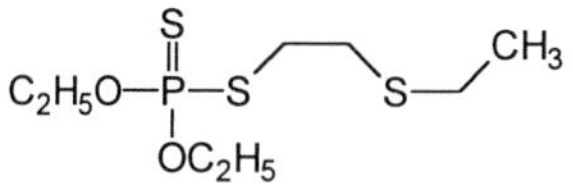
Bay 19639
Bayer 19639
Di-syston
Di-Syston
Di-Syston G
Dimaz
Disipton
Disulfaton
Disulfton
Disystox
Dithiodemeton
Dithiosystox
Ekatin TD
Ethylthiometon B
ENT 23,437
ENT 23437
ENT-23427
ENT-23437
ENT-23437O
Frumin
Frumin AL
Frumin G
Glebofos
M 74 (VAN)

Merck Index No. 3371
MP: , -25°
BP: $108^{o0.01}$, 128^{o1}, 132 - $133^{o1.5}$
Density: 1.144^{20}
Solubility: 12 mg/l water 22°, readily. misc. most org. solvents
Octanol/water PC:
LD_{50}: rat orl 4 mg/kg; rat orl 2.3 mg/kg; rat orl 6.8 mg/kg; rat orl 2-12 mg;/kg, rat ihl LC50 200 mg/m^3; rat skn 6 mg/kg; rat ipr 2 mg/kg; rat ivn 5.5 mg/kg; rat unr 2.5 mg/kg; mus orl 5.5-6.5 mg/kg; mus orl 4.8 mg/kg; mus ipr 5.5 mg/kg; dog orl 5 mg/kg; cat ihl LCLo 10 mg/m^3/4H; gpg orl 10.8 mg/kg; gpg ipr 7 mg/kg; ckn unr 28 mg/kg; qal orl 12 mg/kg; dck orl 6.5 mg/kg; bwd orl 2.4 mg/kg.
CFR: 40CFR180.183

CAS RN 330-54-1
$C_9H_{10}Cl_2N_2O$

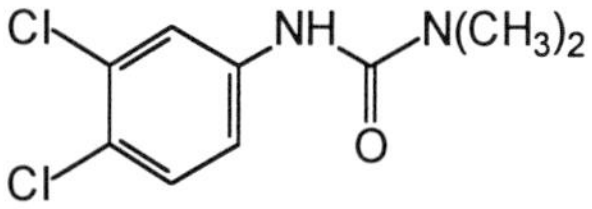

Diuron (ACN)
Urea, N'-(3,4-dichlorophenyl)-N,N-dimethyl- (9CI)
Urea, 3-(3,4-dichlorophenyl)-1,1-dimethyl- (8CI)
AF 101
Crisuron
Dailon
Diater
Dichlorfenidim
Dichlorophenyldimethylurea
Diurex
Diurol
Diuron 4l
Drexel
Duran
Dynex
DMU (VAN)

Merck Index No. 3388
MP: 158 - 159°
BP:
Density:
Solubility: 42 ppm water 25°, 53 g/kg acetone, 1.4 g/kg butyl stearate, 1.2 g/kg benzene, insol. hydrocarbons
Octanol/water PC:
LD_{50}: rat orl 1017 mg/kg; rat orl 437 mg/kg; rat orl 3400 mg/kg; rat skn >5000 mg/kg; rat unr 3400 mg/kg; mus ipr 500 mg/kg; dck unr >2000 mg/kg.
CFR: 40CFR180.106

CAS RN 2439-10-3
$C_{13}H_{29}N_3.C_2H_4O_2$

Dodine (ACN)
Guanidine, dodecyl-, monoacetate (8CI9CI)
n-Dodecylguanidine acetate
Aadodin
Acetic acid, dodecylguanidine salt
American Cyanamid 5,223
American Cyanamid 5223
AC 5223
Carpene
Curitan
Cyprex
Cyprex 65W
Dodecylguanidine acetate (ACN)
Dodecylguanidine, monoacetate
Dodin
Dodine acetate
Dodine, mixture with glyodin
Doguadine
Doquadine
Experimental Fungicide 5223
ENT 16,436
Guanidine, dodecyl-, acetate
Karbine
Karpen
Laurylguanidine acetate
Melprex
Melprex 65
Melprex 65-W
Milprex
N-Dodecylguanidine acetate

NH NH2 · CH3—COOH
NH

Merck Index No. 3406
MP: 136°
BP:
Density:
Solubility: 630 mg/l water 25°, ethanol, hot water, sl. Sol. others
Octanol/water PC:
LD_{50}: rat orl 660 mg/kg; rat orl 566 mg/kg; rat orl ca. 1000 mg/kg; rat skn >6000 mg/kg; rat unr 1000 mg/kg; mus orl 266 mg/kg; mus ihl LC_{50} 129 mg/m^3/2H; rbt orl 535 mg/kg; rbt skn 2100 mg/kg; gpg orl 176 mg/kg; gpg skn LDLo 2000 mg/kg; qal orl 788 mg/kg; dcl orl 1142 mg/kg.
CFR: 40CFR180.172

CAS RN 944-22-9
$C_{10}H_{15}OPS_2$

Dyfonate
Fonofos (ISO,BSI)
Phosphonodithioic acid, ethyl-, O-ethyl S-phenyl ester (8CI9CI)
Difonate
Difonatul
Dyphonate
Ethylphosphonodithioic acid, O-ethyl S-phenyl ester
ENT 25,796
Fonophos
N 2790
N-2788
O-Aethyl S-phenyl-aethyl-dithiophosphonat (German)
Stauffer N-2790

C_2H_5—P(=S)(—S—C_6H_5)—OC_2H_5

Merck Index No. 4147
MP:
BP: $130^{\circ 0.1}$
Density: 1.16^{25}
Solubility: 13 mg/l water, misc. org. solvents
Octanol/water PC: 8000
LD_{50}: rat orl 3 mg/kg; rat orl 24.5 mg/kg; rat orl 10.8 mg/kg; rat ihl LC_{50} 1900 mg/m^3/1H mg/kg; rat skn 147 mg/kg; dog orl 3 mg/kg; rbt skn 25 mg/kg; gpg skn 278 mg/kg; pgn orl 13.3 mg/kg; qal orl 12 mg/kg; dom orl 1.3 mg/kg; mam unr 17 mg/kg; bwd orl 10 mg/kg.
CFR: 40CFR180.221

CAS RN 26225-79-6
$C_{13}H_{18}O_5S$

Ethofumesate (ACN)
5-Benzofuranol, 2-ethoxy-2,3-dihydro-3,3-dimethyl-, methanesulfonate, (±)-
Nortran
Nortron (VAN)
NC 8438
Tramat
2-Ethoxy-2,3-dihydro-3,3-dimethyl-5-benzofuranyl methanesulfonate (ACN)
2-Ethoxy-2,3-dihydro-3,3-dimethylbenzofuran-5-yl methanesulfonate
2,3-Dihydro-3,3-dimethyl-2-ethoxy-5-benzofuranyl methanesulfonate
5-Benzofuranol, 2-ethoxy-2,3-dihydro-3,3-dimethyl-, methanesulfonate

CH_3SO_2—O; O; OC_2H_5

Merck Index No. 3697
MP: 70 - 72°
BP:
Density: 1.14
Solubility: 110 mg/l water, 400 g/l chloroform, 400 g/l dioxane, 400 g/l benzene, 400 g/l acetone, 100 g/l ethanol, 4 g/kl hexane, 100 g/kg ethanol, 400 g/kg acetone, 4 g/kg hexane
Octanol/water PC: 501 (octanol/water, 25)
LD_{50}: rat orl 1130 mg/kg; rat orl 6400 mg/kg; rat skn 1440 mg/kg; mus orl >1600 mg/kg, rbt orl >1000 mg/kg; qal unr 6000 mg/kg.
40CFR180.345

CAS RN 115-29-7
$C_9H_6Cl_6O_3S$

Endosulfan (ACN)
5,9-Methano-2,4,3-benzodioxathiepin,6,7,8,9,10,10-hexachloro-1,5,5a,6,9,9a-hexahydro-,3-oxide (9CI)
5-Norbornene-2,3-dimethanol, 1,4,5,6,7,7-hexachloro-,cyclic sulfite (8CI)
α,β-1,2,3,4,7,7-Hexachlorobicyclo[2.2.1]-2-heptene-5,6-bisoxymethylene sulfite
α,β-1,2,3,4,7,7-Hexachlorobicyclo[2.2.1]hepten-2-bis(oxymethylon-5,6-)sulfite
β-6,7,8,9,10,10-Hexachloro-1,5,5a,6,9,9a-hexahydro-*endo*-6,9-methano-2,4,3-benzo-dioxathiepin3-oxide
(Hexachloro)hexahydromethano 2,4,3-benzodioxathiepin-3-oxide
Benzoepin
Beosit
BIO 5,462
BIO 5462
Chlorthiepin
Crisulfan
Cyclodan
Endocel
Endosol
Endosulfan 35EC
Endosulpham
Endosulphan
ENT 23,979
ENT 23979
FMC 5462
Hexachlorohexahydromethano-2,4,3-benzodioxathiepin oxide
Hexachlorohexahydromethano-2,4,3-benzodioxathiepin 3-oxide (ACN)
Hexachlorohexahydromethanobenzodioxathiepin 3-oxide
Hildan
HOE 2,671
HOE 2671
Insectophene
Kop-Thiodan
Malix
Niagara 5,462
Niagara 5462
NCI-C00566
NIA 5462
OMS 570
Sulfurous acid, cyclic ester with 1,4,5,6,7,7-hexachloro-5-norbornene-2,3-dimethanol
SD 4314
Thifor
Thimul

Cl Cl CCl2 S=O O O Cl Cl

Merck Index No. 3529
MP: 106°, 109.2°, 70 - 100° (technical))
BP: $106^{\circ 0.7}$
Density: 1.745^{20}
Solubility: 0.32 mg/l water 22°, 200 g/l ethyl acetate, 200 g/l dichloromethane, 200 g/l toluene, 65 g/l ethanol, 24 g/l hexane, sol. most org. solvents
Octanol/water PC:
LD_{50}: man orl TDLo 86 mg/kg; rat orl 43 mg/kg; rat orl 18 mg/kg; rat orl 70 mg/kg (aq), 100 mg/kg (oil); rat ihl LC_{50} 80 mg/m^3/4H; rat skn 34 mg/kg; rat ipr 8 mg/kg; rat unr 40 mg/kg; mus orl 7.36 mg/kg; mus ipr 7 mg/kg; mus unr 32 mg/kg; dog orl 76.7 mg/kg; cat orl 2 mg/kg; cat ihl LC_{50} 90 mg/m^3/4H; rbt orl 28 mg/kg; rbt skn 359 mg/kg (oil); ham orl 118 mg/kg; ham ipr 80 mg/kg; dcl orl 33 mg/kg; mam skn 147 mg/kg; bwd orl 35 mg/kg.
40CFR180.182

CAS RN 145-73-3
$C_8H_{10}O_5.2Na$

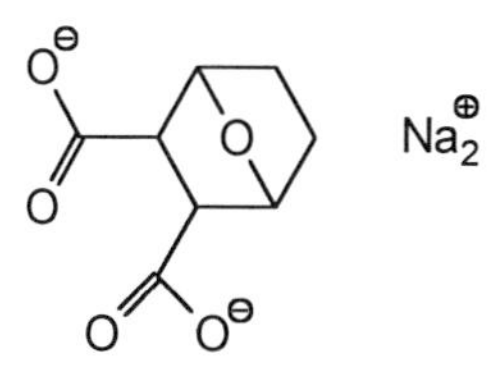

Endothall, disodium salt (ACN)
7-Oxabicyclo[2.2.1]heptane-2,3-dicarboxylic acid, disodium salt (ACN)(8CI9CI)
(3,6-Epossi-cicloesan-1,2-dicarbossilato)disodico (Italian)
Accelerate
Aguathol
Aquathol
Des-i-cate
Dinatrium-(3,6-epoxy-cyclohexaan-1,2-dicarboxylaat) (Dutch)
Dinatrium-(3,6-epoxy-cyclohexan-1,2-dicarboxylat) (German)
Disodium endothall
Disodium salt of endothall
Disodium salt of 7-oxabicyclo[2.2.1]heptane-2,3-dicarboxylic acid
Disodium 3,6-endohexahydrophthalate
Disodium 3,6-endoxohexahydrophthalate
Disodium 3,6-epoxycyclohexane-1,2-dicarboxylate
Disodium 7-oxabicyclo[2.2.1]heptane-2,3-dicarboxylate
Endotal
Endothal
Endothal disodium salt
Endothal weed killer
Endothal-natrium (Dutch)
Endothal-sodium
Endothall
Endothall, sodium salt
Herbicide 273
Hydout
Niagrathal
Ripenthol
Sodium endothall
Tri-Endothal

CAS RN 759-94-4
$C_9H_{19}NOS$

EPTC
Carbamothioic acid, dipropyl-, S-ethyl ester (9CI)
Carbamic acid, dipropylthio-, S-ethyl ester (8CI)
Eptam
Eptam 6E
Eradicane
Ethyl di-*n*-propylthiolcarbamate
Ethyl N,N-di-*n*-propylthiolcarbamate
Ethyl N,N-dipropylthiolcarbamate
FDA 1541
N,N-Dipropylthiocarbamic acid S-ethyl ester
R 1608
S-Aethyl-N,N-dipropylthiolcarbamat (German)
S-Ethyl di-N,N-propylthiocarbamate
S-Ethyl dipropylcarbamothioate
S-Ethyl dipropylthiocarbamate (ACN)
S-Ethyl N,N-di-*n*-propylthiocarbamate
S-Ethyl N,N-dipropylthiocarbamate
Stauffer R 1608
Torbin

O, CH2CH2CH3, N, CH3CH2S, CH2CH2CH3

Merck Index No. 3580
MP:
BP: $127^{\circ 20}$
Density: 0.9546^{30}
Sol: 365 mg/l water 20°, 375 mg/l water 25°, misc. benzene, alcohol, toluene, xylene
Octanol/water PC: 1600
LD_{50}: hmn ihl TCLo 135 mg/m³/90M; rat orl 916 mg/kg; rat orl 1631 mg/kg; rat orl 1630 mg/kg; rat ihl LCLo 200 mg/m³/4H; rat skn 3200 mg/kg; rat unr 1660 mg/kg; mus orl 750 mg/kg; mus orl 3160 mg/kg; mus ipr 58 mg/kg; mus ivn 320 mg/kg; cat orl 112 mg/kg; cat ihl LCLo 400 mg/m³/4H; rbt orl 2640 mg/kg; rbt skn 1460 mg/kg; mam unr 1660 mg/kg; bwd orl 100 mg/kg.
40CFR180.117

CAS RN 55283-68-6
$C_{13}H_{14}F_3N_3O_4$

Ethalfluralin (ACN)
Benzenamine, N-ethyl-N-(2-methyl-2-propenyl)-2,6-dinitro-4-(trifluoromethyl)-
p-Toluidine, 2,6-dinitro-N-ethyl-N-(2-methyl-2-propenyl)-α,α,α-trifluoro-
Compound 94961
Ethalflurlin
EL 161
EL-161
N-Ethyl-α,α,α-trifluoro-N-(2-methylallyl)-2,6-dinitro-*p*-toluidine
N-Ethyl-N-(2-methyl-2-propenyl)-2,6-dinitro-4-(trifluoromethyl)-
benzenamine (ACN)
Somilan
Sonalan
Sonalen

NO2, CH2, F3C, N, CH3, C2H5, NO2

Merck Index No. 3671
MP: 54 - 57°, 57 - 59°
BP: 256° (dec)
Density:
Solubility: 0.3 ppm water 25°, readily sol. (>500 g/lacetone, acetonitrile, benzene, chloroform, hexane, xylene, methanol
Octanol/water PC: 130000 pH 7, 25
LD_{50}: rat orl 10000 mg/kg; mus orl >10000 mg/kg; dog orl >200 mg/kg; cat orl >200 mg/kg; rbt skn >2000 mg/kg; qal orl >200 mg/kg; dck orl >200 mg/kg.
40CFR180.416

CAS RN 16672-87-0
$C_2H_6ClO_3P$

Ethephon (ACN)
Phosphonic acid, (2-chloroethyl)- (8CI9CI)
(2-Chloroethyl)phosphonic acid (ACN)
Amchem 68-250
Bromeflor
Bromoflor
Camposan
Cepha
Chloroethylphosphonic acid
CEP
CEPA
Ethefon
Ethel
Ethrel
Flordimex
Florel
G 996
Kamposan
2-Chloraethyl-phosphonsaeure (German)
2-Chloroacetyl phosphorate
2-Chloroethanephosphonic acid
2-Chloroethylenephosphonic acid
2-Chloroethylphosphonic acid

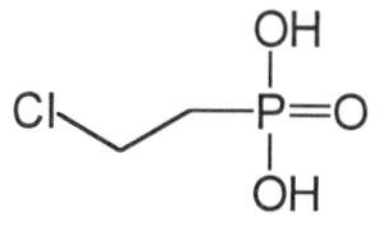

Merck Index No. 3686
MP: 74 - 75°
BP:
Density: 1.2 - 1.3
Solubility: Freely sol. (1 kg/l) water, methanol, acetone, ethylene glycol, propylene glycol, spar. sol. benzene, insol. petroleum ether
Octanol/water PC:
LD_{50}: rat orl 3400 mg/kg; rat orl 4229 mg/kg (24% in propylene glycol); rat ihl LC_{50} 90 mg/m^3/4H; mus orl 2850mg/kg; rbt orl 5000 mg/kg; rbt skn 5730 mg/kg; gpg orl 4200 mg/kg; qal orl 1072 mg/kg; mam unr 4200 mg/kg.
40CFR180.300

CAS RN 2941-55-1
$C_7H_{15}NOS$

Ethiolate (ACN)
Carbamothioic acid, diethyl-, S-ethyl ester (9CI)
Carbamic acid, diethylthio-, S-ethyl ester (8CI)
component of Prefox
Prefox
S-Ethyl diethylcarbamothioate
S-Ethyl diethylthiocarbamate (ACN)
S-Ethyl N,N-diethylthiocarbamate
S-15076
S6176

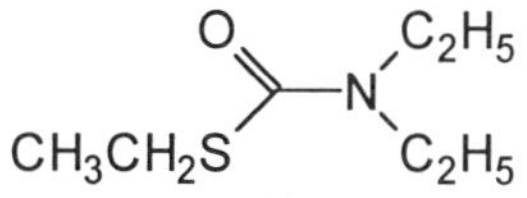

Merck Index No.
MP:
BP: 51 - 53°$^{0.5}$, 141.5 - 142°^{87}
Density: 0.9791^{30}
Solubility:
Octanol/water PC:
LD_{50}: rat orl 400 mg/kg.
40CFR180.343 (revoked 5-4-88)

CAS RN 563-12-2
$C_9H_{22}O_4P_2S_4$

Ethion (ACN)
Phosphorodithioic acid, S,S'-methylene O,O,O',O'-tetraethyl ester (8CI9CI)
AC 3422
Bis (dithiophosphate de O,o-diethyle) de S,s'-methylene (French)
Bis(diethoxyphosphinothioylthio)methane
Bis[S-(diethoxyphosphinethioyl)mercapto]methane
Bis[S-(diethoxyphosphinothioyl)mercapto]methane
Bladan
Diethion
Embathion
Ethodan
Ethopaz
Ethyl methylene phosphorodithioate
Ethyl methylene phosphorodithioate ($[(EtO)_2P(S)S]_2CH_2$)
ENT 24,105
Fosfatox E
Fosfono 50
FMC-1240
Hylemax
Hylemox
Itopaz
KWIT
Methanedithiol, S,S-diester with O,O-diethyl phosphorodithioate
Methyleen-S,S'-bis(O,O-diethyl-dithiofosfaat) (Dutch)
Methylene-S,s'-bis(O,o-diaethyl-dithiophosphat) (German)
Metilen-S,S'-bis(O,O-dietil-ditiofosfato) (Italian)
Niagara 1240
Nialate
NIA 1240
O,O,O,O'-Tetraethyl-S,S'-methylene bis-phosphorodithioate

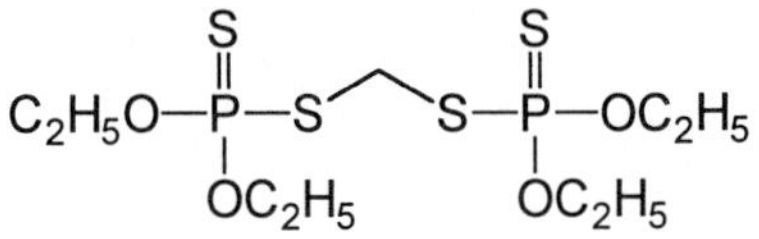

Merck Index No. 3691
MP: -13° - -12°, -15° - -12°,
BP: 164 - $165^{\circ 0.3}$
Density: 1.220^{20}, 1.215 - 1.230^{20} (technical)
Solubility: sl. sol. water, sol. xylene, chloroform, acetone, kerosene
Octanol/water PC:
LD_{50}: inf orl TDLo 1.57 mg/kg; hmn orl TDLo 100 rat orl 13 mg/kg; rat orl 27 mg/kg; rat orl 65 mg/kg; rat skn 62 mg/kg; rat skn 245 mg/kg; rat orl 208 mg/kg (96 mg/kg technical); rat ihl 964 mg/m^3; rat irp 26 mg/kg; rat unr 55 mg/kg; mus orl 40-45 mg/kg; mus ipr 35 mg/kg; gpg orl 40 mg/kg; gpg skn 915 mg/kg; bwd orl 45 mg/kg.
40CFR180.173

CAS RN 13194-48-4
$C_8H_{19}O_2PS_2$

Ethoprop (ACN)
Phosphorodithioic acid, O-ethyl S,S-dipropyl ester (8CI9CI)
Erthroprop
Ethoprophos
ENT 27,318
JOLT
Mocap
Mocap 10G
O-Ethyl S,S-dipropyl phosphorodithioate (ACN)
O-Ethyl S,S-dipropylphosphorodithioate
Phophos
Phosethoprop
Prophos
Rovokil
S,S-Dipropyl O-ethyl phosphorodithioate
V-C Chemical V-C 9-104
V-C 9-104
VC 9-104
VC9-104

$CH_3CH_2CH_2S-P(=O)(-OEt)-SCH_2CH_2CH_3$

Merck Index No. 3702
MP:
BP: 86 - 91$^{\circ 0.2}$
Density: 1.094^{20}
Solubility: 750 mg/l water, 700 mg/l water 20°, >300 g/kg acetone, ethanol, xylene, diethyl ether, cyclohexane
Octanol/water PC: 3890
LD_{50}: rat orl 34 mg/kg; rat orl 62 mg/kg; rat skn 60 mg/kg; rbt orl 55 mg/kg; rbt skn 2.4 mg/kg; pgn orl 13.3 mg/kg; ckn orl 5.5 mg/kg; qal orl 7.5 mg/kg; dck orl 1.26 mg/kg; dck skn 11 mg/kg; bwd orl 4.21 mg/kg.
40CFR180.262

CAS RN 91-53-2
$C_{14}H_{19}NO$

Ethoxyquin
Quinoline, 6-ethoxy-1,2-dihydro-2,2,4-trimethyl- (8CI9CI)
Alterungsschutzmittel EC
Amea 100
Antioxidant EC
Dawe's nutrigard
Ethoxyquine
EMQ
EQ
Niflex
Nix-Scald
Nocrack AW
Permanax 103
Quinol ED
Quinoline, 1,2-dihydro-2,2,4-trimethyl-
Quinoline, 1,2-dihydro-6-ethoxy-2,2,4-trimethyl-
Santoflex A
Santoflex AW
Santoquin
Santoquine (VAN)
Stop-Scald
USAF B-24
Vulkanox EC

H
N
CH_3
CH_3
CH_3CH_2O
CH_3

1,2-Dihydro-2,2,4-trimethyl-6-ethoxyquinoline
1,2-Dihydro-6-ethoxy-2,2,4-trimethylquinoline
2,2,4-Trimethyl-6-ethoxy-1,2-dihydroquinoline
6-Ethoxy-1,2-dihydro-2,2,4-trimethylquinoline (ACN)
6-Ethoxy-2,2,4-trimethyl-1,2-dihydroquinoline

Merck Index No. 3710
MP:
BP: 123 - 125$^{\circ 2}$, 123 - 125$^{\circ 2}$
Density: 1.029 - 1.031^{25}
Solubility:
Octanol/water PC:
LD_{50}: rat orl 800 mg/kg; rat orl 1920 mg/kg; mus orl 1730 mg/kg; mus ipr 200 mg/kg; mus ivn 178 mg/kg.
40CFR180.178

CAS RN 38260-54-7
$C_{10}H_{17}N_2O_4PS$

Etrimfos (ACN)
Phosphorothioic acid, O-(6-ethoxy-2-ethyl-4-pyrimidinyl) O,O-dimethyl ester
Ekamet
Ekamet G
Ekamet ULV
Etrimphos
O-(6-Ethoxy-2-ethyl-4-pyrimidinly) O,O-dimethyl phosphorothioate
O,O-Dimethyl O-(2-ethyl-4-ethoxy-pyrimidinyl-6) thionophosphate
O,O-Dimethyl O-(2-ethyl-6-ethoxy-4-pyrimidyl)phosphorothioate
Phosphorothioic acid, O,O-dimethyl O-(6-ethoxy-2-ethyl-4-pyrimidinyl) ester
San 197 I
SAN 197

C_2H_5O N C_2H_5 N O P S CH_3O OCH_3

Merck Index No. 3847
MP: -3.35°
BP:
Density: 1.195^{20}
Solubility: 40ppm water 23°, miscible with ethanol, dimethylsulfoxide, EtOAc, diethyl ether, acetone, chloroform, xylene, hexane
Octanol/water PC: >2000
LD_{50}: rat orl 1800 mg/kg; rat skn >2000 mg/kg; mus orl 437 mg/kg; rbt skn >500 mg/kg;
40CFR180.(temp 1976)

CAS RN 109-94-4
$C_3H_6O_2$

Ethyl formate (ACN)
Formic acid, ethyl ester (8CI9CI)
Aethylformiat (German)
Anozol
Areginal
Ethyl formic ester
Ethyl methanoate
Ethyle(formiate d') (French)
Ethylformiaat (Dutch)
Etile(formiato di) (Italian)
ENT-407
Formic ether
Mrowczan etylu (Polish)
Neantine
Palatinol A
Placidol E
Solvanol

H O C_2H_5 O

Merck Index No. 3763
MP: -80°
BP: 53 - 54°
Density: 0.917^{20}
Solubility: sol. in 10 parts water, misc. alcohol, ether
Octanol/water PC:
LD_{50}: rat orl 1850 mg/kg; rat ihl LCLo 8000 ppm/4H; rbt orl 2075 mg/kg; rbt skn 20000 mg/kg; rbt scu LDLo 1000 mg/kg; gpg orl 1110 mg/kg.
40CFR185.2900

CAS RN 75-21-8
C_2H_4O

Ethylene oxide (ACN)(DOT)
Oxirane
α,β-Oxidoethane
Aethylenoxid (German)
Anprolene
Dihydrooxirene
Dimethylene oxide
Epoxyethane (French)
Ethane, 1,2-epoxy-
Ethene oxide
Ethyleenoxide (Dutch)
Ethylene (oxyde d') (French)
Etilene (ossido di) (Italian)
Etylenu tlenek (Polish)
ENT-26263
NCI-C50088
Oxacyclopropane
Oxane (VAN)
Oxidoethane
Oxiraan (Dutch)
Oxirene, dihydro-
Oxyfume
1,2-Epoxyaethan (German)
1,2-Epoxyethane

Merck Index No. 3758
MP: -111°
BP: 10.7°
Density: 0.891^{4}, 0.887^{7}, 0.882^{10}
Solubility: sol. water, alcohol, ether
Octanol/water PC:
LD_{50}: hmn ihl TCLo 12500 ppm/10S; wmn ihl TCLo 500 ppm/2M; rat orl 72 mg/kg; rat ihl LC_{50} 800 ppm/4H; rat scu 187 mg/kg; rat unr LDLo 200 mg/kg; mus ihl LC_{50} 836 ppm/4H; mus ipr 175 mg/kg; mus ivn 290 mg/kg; dog ihl LC_{50} 960 ppm/4H; dog ivn 330 mg/kg; cat scu LDLo 100 mg/kg; rbt ivn LDLo 175 mg/kg; gpg orl 270 mg/kg; gpg ihl LC_{50} 1500 mg/m^3/4H.
40CFR180.151

CAS RN 52-85-7
$C_{10}H_{16}NO_5PS_2$

Famphur (ACN)
Phosphorothioic acid, O-[4-[(dimethylamino)sulfonyl]phenyl] O,O-dimethyl ester
Phosphorothioic acid, O,O-dimethyl ester, O-ester with *p*-hydroxy-N,N-dimethyl benzene- sulfonamide
American Cyanamid 38,023
AC 38023
Bo-Ana
Cyflee
CL 38023
Dovip
ENT 25,644
Famfos
Famfur
Famophos
Famphos
Warbex

Merck Index No. 3882
MP: 52.5 - 53.5°
BP:
Density:
Solubility:
Octanol/water PC:
LD_{50}: rat orl 35mg/kg; rat orl 28 mg/kg; rat skn 400 mg/kg; mus orl 9.5 mg/kg; mus ipr 11.6 mg/kg; rbt skn 1460 mg/kg; dom orl 400 mg/kg; dom ims 59 mg/kg; mam ims 64 mg/kg; bwd orl 1.8 mg/kg.
40CFR180.233

CAS RN 85-34-7
$C_8H_5Cl_3O_2$

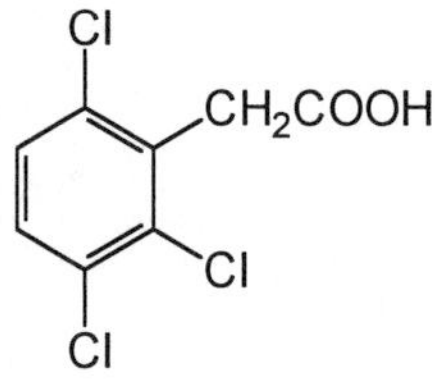

Chlorfenac
Benzeneacetic acid, 2,3,6-trichloro-
Acetic acid, (2,3,6-trichlorophenyl)-
(2,3,6-Trichlorophenyl)acetic acid
Fenac
Fenae
Fenatrol
Kanepar
Phenylacetic acid, 2,3,6-trichloro-
Tri-fene
Tri-Fen
Trifene
TCPA
2,3,6-Trichlorobenzeneacetic acid
2,3,6-Trichlorophenylacetic acid (ACN)
2,3,6-Trichlorphenylessigsaeure (German)

Merck Index No. 2085
MP: 161°
BP:
Density:
Solubility: insol. water, sol. alcohol, acetone, ether
Octanol/water PC:
LD_{50}: rat orl 1780 mg/kg; rat orl 3000mg/kg; rbt skn 1440 mg/kg; mam unr 1700 mg/kg.
40CFR180.283

CAS RN 60168-88-9
$C_{17}H_{12}Cl_2N_2O$

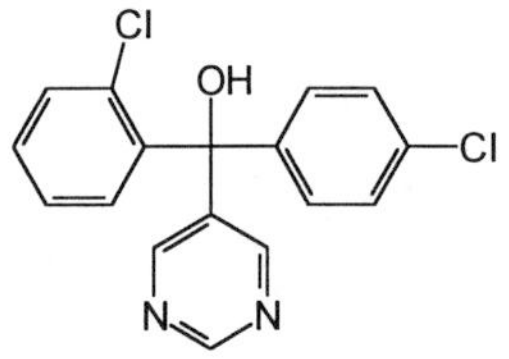

Fenarimol (ACN)
5-Pyrimidinemethanol, α-(2-chlorophenyl)-α-(4-chlorophenyl)-
α-(2-Chlorophenyl)-α-(4-chlorophenyl)-5-pyrimidinemethanol (ACN)
Bloc
Compound 56722
EL 222
Rimidin
Rubigan
2,4'-Dichloro-α-(5-pyrimidinyl)benzhydryl alcohol

Merck Index No. 3903
MP: 117 - 119°
BP:
Density:
Solubility: 13.7 ppm water ph7 25, >250 g/l acetone, 125 g/l methanol, 50 g/l xylene, sol. most org. solvents, sl. sol. hexane
Octanol/water PC: 4900 (octanol/water pH 7 25)
LD_{50}: rat orl 2500 mg/kg; mus orl 4500 mg/kg; dog orl >200 mg/kg; rbt skn >2000 mg/kg; bwd orl >200 mg/kg.
40CFR180.421

CAS RN 22224-92-6
$C_{13}H_{22}NO_3PS$

Fenamiphos
Phosphoramidic acid, (1-methylethyl)-, ethyl 3-methyl-4-(methylthio)phenyl ester
Phosphoramidic acid, isopropyl-, ethyl 4-(methylthio)-*m*-tolyl ester (8CI)
(Isopropylamino) O-ethyl-(4-methylmercapto-3-methylphenyl) phosphate
m-Cresol, 4-(methylthio)-, ethyl isopropylphosphoramidate
o-Aethyl-o-(3-methyl-4-methylthiophenyl)-isopropylamido-phosphorsaeureester (German)
B 68138
Bay 68138
Bayer 68138
BAY 68138
Ethyl 3-methyl-4-(methylthio)phenyl (1-methylethyl)phoshoramidate
Ethyl 3-methyl-4-(methylthio)phenyl (1-methylethyl)phosphoramidate (ACN)
Ethyl 3-methyl-4-(methylthio)phenyl isopropylphosphoramidate
Ethyl 3-methyl-4-(methylthio)phenyl 1-(methylethyl) phosphoramidate
Ethyl 3-methyl-4-(methylthio)phenyl 1-(methylethyl)phosphoramidate
Ethyl 3-methyl-4-(methylthio)phenyl 1-methylethylphosphoramidate
Ethyl 4-(methylthio)-*m*-tolyl isopropylphosphoramidate (ACN)
Ethyl 4-methylthio-*m*-tolyl isopropylphosphoramidate
Ethyl-4-(methylthio)-*m*-tolyl isopropylphosphoramidate
ENT 27572
Nemacur
Nemacur P
O-Ethyl O-(3-methyl-4-methylthiophenyl) isopropylphosphoramidate
O-Ethyl 4-(methylthio)-*m*-tolyl isopropylphosphoramidate
Phenamiphos
Phosphoramidic acid, (1-methylethyl)-, ethyl [3-methyl-4-(methylthio)phenyl] ester
Phosphoramidic acid, isopropyl-, 4-(methylthio)-m-tolyl ethyl ester
1-(Methylethyl) ethyl 3-methyl-4-(methylthio)phenyl phosphoramidate

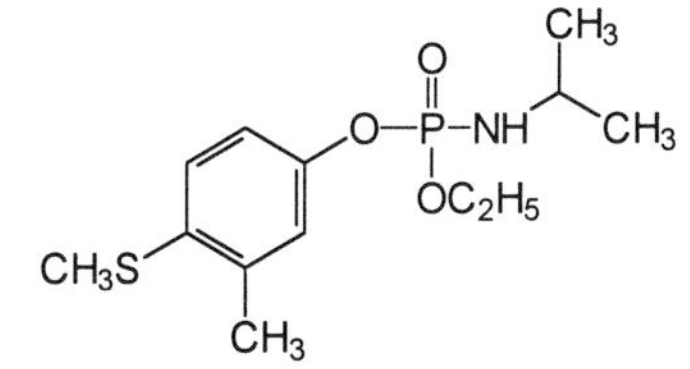

Merck Index No.3901
MP: 49°
BP:
Density: 1.15^{20}
Solubility: 329 mg/l water, 700 mg/l water 20, readily sol. dichloromethane, isopropanol, toluene, spar. sol. hexane
Octanol/water PC:
LD_{50}: rat orl 8 mg/kg; rat orl 10-25 mg/kg; rat orl 15.3 mg/kg; rat ihl LC_{50} 91 mg/m^3/4H; rat skn 80 mg/kg; mus orl 22.7 mg/kg; dog orl 10 mg/kg; cat orl 10 mg/kg; rbt orl 10 mg/kg; rbt skn 178 mg/kg; ckn orl 12 mg/kg; qal orl 1 mg/kg; dck orl 1.68 mg/kg; dck skn 24 mg/kg; brd orl 2.4 mg/kg.
40CFR180.349

CAS RN 122-14-5
$C_9H_{12}NO_5PS$

Fenitrothion
Phosphorothioic acid, O,O-dimethyl O-(3-methyl-4-nitrophenyl) ester
Phosphorothioic acid, O,O-dimethyl O-(4-nitro-m-tolyl) ester
ac-47300
m-Cresol, 4-nitro-, O-ester with O,O-dimethyl phosphorothioate
Accothion
Aceothion
Agria 1050
Agriya 1050
Agrothion
Akotion
Arbogal
AC 47300
AC-47300
Bay 41831
Bayer S 5660
Bayer 41831
BAY 41831
Cyfer
Cytel
CL 47300
CP 47114
Dimethyl 3-methyl-4-nitrophenyl phosphorothionate
Dybar
EI 47300
ENT 25,715
Falithion
Fenitox
Fentrothione
Folithion
Kotion
Metathio e-50
Metathion

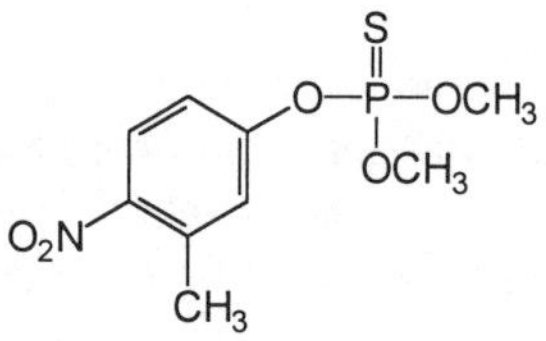

Merck Index No. 3922
MP:
BP: $118^{\circ 0.05}$, $164^{\circ 1}$
Density: 1.3227^{25}
Solubility: 30 mg/l water 21°, 24 g/l hexane, insol. water, low sol, hydrocarbons, sol. most organics
Octanol/water PC: 2700
LD_{50}: wmn orl TDLo 800 mg/kg; rat orl 250 mg/kg; rat orl 570-800 mg/kg; rat ihl LC_{50} 378 mg/m^3/4H; rat skn 2500 mg/kg; rat ipr 300 mg/kg; rat scu 1300 mg/kg; rat ivn 33 mg/kg; rat itr 950 mg/kg; rat unr 290 mg/kg; mus orl 870 mg//kg; mus orl 229 mg/kg; mus skn 2500 mg/kg; mus ipr 280 mg/kg; mus scu 1000 mg/kg; mus ice 1000 mg/kg; mus unr 1250 mg/kg; cat orl 142 mg/kg; rbt skn 1250 mg/kg; gpg orl 500 mg/kg; gpg ivn 112 mg/kg; ckn orl 280 mg/kg; qal orl 56.2 mg/kg; dck orl 1190 mg/kg; dck skn 504 mg/kg; ctl unr 217 mg/kg; mam unr 142 mg/kg; bwd orl 11 mg/kg.
40CFR180.(temp 1978)

CAS RN 115-90-2
$C_{11}H_{17}O_4PS_2$

Fensulfothion

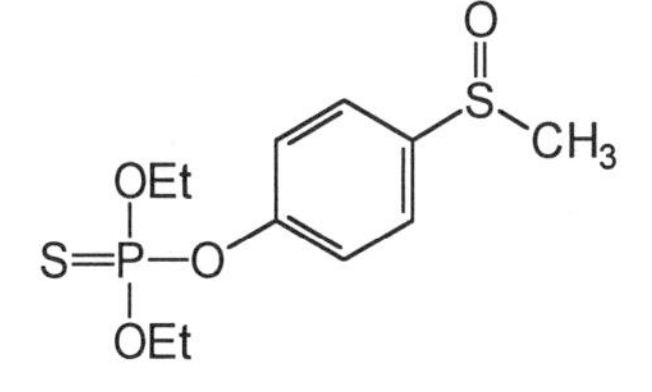

Phosphorothioic acid, O,O-diethyl O-[4-(methylsulfinyl)phenyl] ester (9CI)
Phosphorothioic acid, O,O-diethyl O-[*p*-(methylsulfinyl)phenyl] ester (8CI)
B 25141
Bayer s767
Bayer 25141
BAY 25141
Chemagro 25141
Daconit
Dasanit
Desanit
DMSP
ENT 24,945
O,O-Diaethyl-O-4-methylsulfinyl-phenyl-monothiophosphat (German)
O,O-Diethyl O-[*p*-(methylsulfinyl)phenyl] phosphorothioate (ACN)
S 767
Terracur P
VUAgt 108
VUAgt 96
VUAgT

Merck Index No. 3943
MP:
BP: 138 - 141$^{\circ 0.01}$
Density: 1.202^{20}
Solubility: 1.54 g/l water 25°, misc. most org. solvents except aliphatics
Octanol/water PC:
LD_{50}: rat orl 2.2 mg/kg, rat orl 10.5 mg/kg; rat skn 3 mg/kg; rat ipr 1.5 mg/kg; rat unr 11 mg/kg; mus ipr 7 mg/kg; mus ice 2 mg/kg; gpg orl 9 mg/kg; gpg ipr 5.4 mg/kg; pgn orl 0.56 mg/kg; qal orl 1.2 mg/kg; dck orl 0.747 mg/kg; dck skn 3 mg/kg; bwd orl 0.240 mg/kg; bwd skn 0.42 mg/kg.
40CFR180.234

CAS RN 55-38-9
$C_{10}H_{15}O_3PS_2$

Fenthion (ACN)
Phosphorothioic acid, O,O-dimethyl O-[3-methyl-4-(methylthio)phenyl] ester (9CI)
Phosphorothioic acid, O,O-dimethyl O-[4-(methylthio)-*m*-tolyl] ester (8CI)
m-Cresol, 4-(methylthio)-, O-ester with O,O-dimethyl phosphorothioate
B 29493
Batex
Baycid
Bayer S-1752
Bayer 29493
Bayer 9007
Baytex
BAY 29493
Dimethyl methylthiotolyl phosphorothioate
Dimethyl [3-methyl-4-(methylthio)phenyl] phosphorothionate
DMTP
Ekalux
Entex
ENT 25540
Fenthion 29493
Fenthion-R
Lebaycid
Leboycid
Mercaptofos
Mercaptophos (VAN)
MPP (VAN)
NCI-C08651
O,O-Dimethyl O-(3-methyl-4-methylmercaptophenyl) phosphorothioate
O,O-Dimethyl O-[(4-methylmercapto)-3-methylphenyl]thionophosphate
O,O-Dimethyl O-[(4-methylthio)-*m*-tolyl] phosphorothioate
O,O-Dimethyl O-[3-methyl-4-(methylthio)phenyl] phosphorothioate
O,O-Dimethyl O-[3-methyl-4-(methylthio)phenyl] thiophosphate
OMS 2

OCH3
S=P—O
OCH3
S
CH3
CH3

Merck Index No. 3945
MP: 7.5°
BP: $87^{o0.01}$
Density: 1.246^{20}
Solubility: 2 mg/l water 20°, readily sol. dichloromethane, isopropanol, toluene, sl. sol. hexane
Octanol/water PC:
LD_{50}: man orl TDLo 257 mg/kg; wmn orl TDLo 525 mg/kg; hmn unr LDLo 50 mg/kg; rat orl 180 mg/kg; rat orl 190-315 mg/kg; rat orl 245-615 mg/kg; rat ihl LCLo 1000 mg/m^3/2H; rat skn 330 mg/kg; rat ipr 260 mg/kg; mus orl 88.1 mg/kg; mus ihl LCLo 1000 mg/m^3/2H; mus skn 500 mg/kg; mus ipr 125 mg/kg; mus scu 144 mg/kg; mus ivn 320 mg/kg; mus ice 50 mg/kg; dog ims 40 mg/kg; rbt orl 150 mg/kg; rbt ihl LCLo 1000 mg/m^3/2H; gpg orl 260 mg/kg; gpg ihl LCLo 1000 mg/m^3/2H; gpg ipr 310 mg/kg; pgn orl 1.78 mg/kg; ckn orl 20 mg/kg; ckn ipr 20 mg/kg; qal orl 11 mg/kg; dck orl 5.9 mg/kg; dcl skn 44 mg/kg; dom ims 46 mg/kg; mam orl 105 mg/kg; mam ims 46.2 mg/kg; bwd orl 1.3 mg/kg.
40CFR180.214

CAS RN 76-87-9
$C_{18}H_{16}OSn$

Fentin hydroxide
Triphenyltin hydroxide (ACN)
Stannane, hydroxytriphenyl- (8CI9CI)
Dowco 186
Du-Ter
Du-Ter W-50
Dur-ter
Duter
Duter extra
Erithane
ENT 28009
Fenolovo
Fintin hydroxid (German)
Fintin hydroxyde (Dutch)
Fintin idrossido (Italian)
Fintine hydroxyde (French)
Gidrookis fenolovo (Russian)
Griffin super tin 4l
Hydroxotriphenyltin(IV)
Hydroxyde de triphenyl-etain (French)
Hydroxytriphenyl stannane
Hydroxytriphenylstannane
Hydroxytriphenyltin
Idrossido di stagno trifenile (Italian)
K 19 (VAN)
K19
NCI-C00260
Stannol, triphenyl-
Suzu H
Tin, hydroxytriphenyl-
Tptoh
Trifenyl-tinhydroxyde (Dutch)
Triphenyl-zinnhydroxid (German)
Triphenylstannanol
Triphenyltin oxide
Tubotin
TPTH
TPTOH
Vancide ks

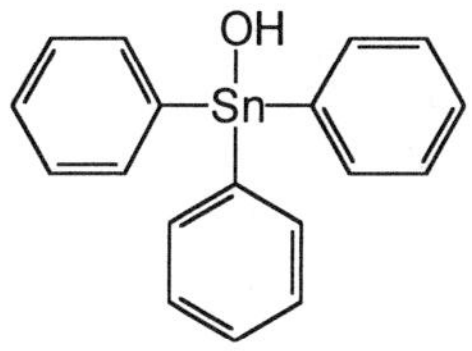

Merck Index No. 9659
MP: 118 - 120°
BP:
Density: 1.54^{20}
Solubility: 1 mg/l water pH 7 20°, 10 g/l ethanol, 171 g/l dichloromethane, 28 g/l diethyl ether, 50 g/l acetone
Octanol/water PC: 2700
LD_{50}: rat orl 171 mg/kg; rat orl 110 mg/kg; rat orl 46 mg/kg; rat ipr LDLo 100 mg/kg; mus orl 245 mg/kg; mus orl 209 mg/kg; mus ipr 8.5 mg/kg; rbt skn 1600 mg/kg; mam unr 500 mg/kg.
40CFR180.236

CAS RN 51630-58-1
$C_{25}H_{22}ClNO_3$

Fenvalerate
α-Cyano-*m*-phenoxybenzyl-2-(*p*-chlorophenyl)-3-methylbutyrate
α-Cyano-3-phenoxybenzyl 2-(4-chlorophenyl)-3-methylbutyrate
Belmark
Cyano(3-phenoxyphenyl)methyl 4-chloro-α-(methylethyl)benzeneacetate
Cyano(3-phenoxyphenyl)methyl 4-chloro-α-(1-methylethyl)benzeneacetate (ACN)
Cyano(3-phenoxyphenyl)methyl 4-chloro-a-(l-methylethyl)benzeneacetate
Cyano(3-phenoxyphenyl)methyl-4-chloro-α-(1-methylethyl)benzeneacetate
Ectrin
Phenvalerate
Pydrin
Pyrethroid
S 5602
S-5602
Sumicide
Sumicidin
SD 43775
SD-43775
WL-43775
4-Chloro-α-(1-methylethyl)benzeneacetic acid, cyano(3-phenoxyphenyl)methyl ester

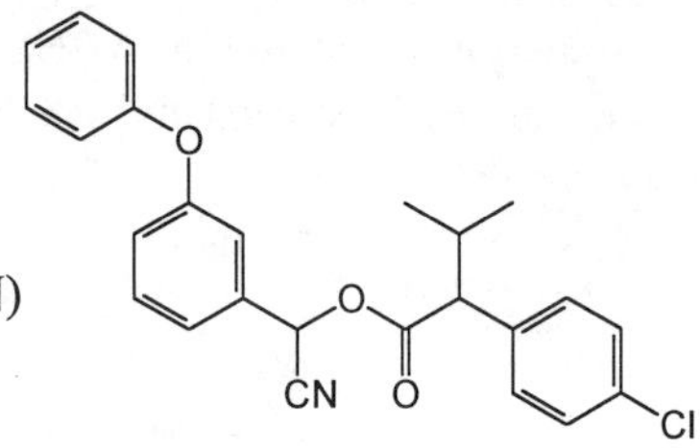

Merck Index No. 3952
MP:
BP: (dec)
Density: 1.15^{25}
Solubility: <1 mg/l water 20°, 155 g/kg hexane, >1 kg/kg acetone, ethanol, chloroform, cyclohexanone, xylene
Octanol/water PC: 12000 (octanol/water 23)
LD_{50}: rat orl 70.2 mg/kg; rat orl 451 mg/kg; rat skn >5000 mg/kg; rat ivn LDLo 50 mg/kg; rat unr 202 mg/kg; mus orl 185 mg/kg; mus ipr >500 mg/kg; mus ice LDLo 0.2 mg/kg; dog orl >1000 mg/kg; rbt skn 2500 mg/kg; qal orl >4000 mg/kg; mam unr 451 mg/kg.
40CFR180.379

CAS RN 14484-64-1
$C_9H_{18}FeN_3S_6$

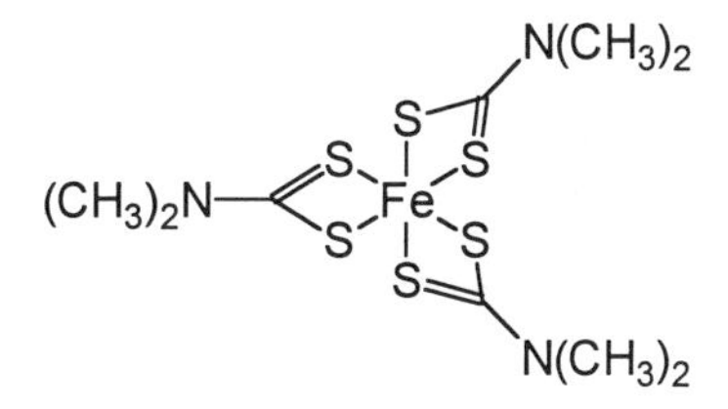

Ferbam (ACN)
Iron, tris(dimethylcarbamodithioato-S,S')-, (OC-6-11)- (9CI)
Iron, tris(dimethyldithiocarbamato)- (VAN8CI)
Aafertis
Aaferzimag
Bercema Fertam 50
Carbamate
Carbamic acid, dimethyldithio-, iron(3+) salt
Carbamodithioic acid, dimethyl-, iron(3+) salt
Cormate
Dimethyldithiocarbamic acid, ferric salt
Dimethyldithiocarbamic acid, iron salt
Dimethyldithiocarbamic acid, iron(3+) salt
Dimethyldithiocarbamic acid,ferric salt
ENT-14689
F 40
Ferbam 50
Ferbam, iron salt
Ferban
Ferbeck
Ferberk
Fermacide
Fermate
Fermate Ferbam fungicide
Ferradow (VAN)
Ferric dimethyl dithiocarbamate (ACN)
Ferric dimethyldithioate (ACN)
Ferric dimethyldithiocarbamate (VAN)
Fuklasin Ultra
Fuklazin
Hexaferb
Iron dimethyldithiocarbamate (VAN)
Karbam Black (VAN)
Liromate
Mixt. contg. ferbam
Namate-90%, iron salt
Stauffer ferbam
Tris(dimethyldithiocarbamato)iron
Tris(N,N-dimethyldithiocarbamato)iron(III)

Merck Index No. 3954
MP: dec. >180°
BP:
Density:
Solubility: 130 mg/l water 20°, sol. pyridine, chloroform, acetonitrile, acetone
Octanol/water PC:
LD_{50}: rat orl >4000 mg/kg; rat orl 1130 mg/kg; rat ipr 2700 mg/kg; mus orl 3400 mg/kg; mus ipr 3000 mg/kg; rbt orl LDLo 3000 mg/kg; rbt ipr LDLo 1500 mg/kg; gpg orl LDLo 2000 mg/kg; gpg ipr LDLo 2300 mg/kg.
40CFR180.114

CAS RN 15457-05-3
$C_{13}H_7F_3N_2O_5$

Fluorodifen (ACN)
Benzene, 2-nitro-1-(4-nitrophenoxy)-4-(trifluoromethyl)- (9CI)
Ether, *p*-nitrophenyl α,α,α-trifluoro-2-nitro-*p*-tolyl (8CI)
p-Nitrophenyl α,α,α-trifluoro-2-nitro-*p*-tolyl ether (ACN)
p-Nitrophenyl 2-nitro-4-(trifluoromethyl)phenyl ether
C 6929
C 6989
C6989
Ether, (*p*-nitrophenyl) (α,α,α-trifluoro-2-nitro-*p*-tolyl)
Ether, 2,4'-dinitro-4-trifluoromethyldiphenyl
Fluordifen
Fluorodiphen
Preforan
Soyex
2-Nitro-1-(4-nitrophenoxy)-4-(trifluoromethyl)benzene
2,4-Dinitro-*p*-trifluoromethyl phenyl ether
2,4'-Dinitro-4-(trifluoromethyl)diphenyl ether
4-Nitrophenyl α,α,α-trifluoro-2-nitro-*p*-tolyl ether
4-Nitrophenyl 4-(trifluoromethyl)-2-nitrophenyl ether
4-Trifluoromethyl-2,4'-dinitrodiphenyl ether
4-Trifluoromethyl-2,4'-dinitrophenyl ether

O_2N CF_3 O NO_2

Merck Index No.
MP: 94°
BP:
Density:
Density: Sparingly sol. water, sol. org. solvents
Octanol/water PC:
LD_{50}: rat orl 9000 mg/kg; rat ihl LC_{50} >990 mg/m^3/6H; rat skn >3000 mg/kg; mus orl >15000 mg/kg; rbt skn >10000 mg/kg.
40CFR180.290 (revoked 5/4/88)

CAS RN 59756-60-4
$C_{19}H_{14}F_3NO$

Fluridone (ACN)
4(1H)-Pyridinone, 1-methyl-3-phenyl-5-[3-(trifluoromethyl)phenyl]- (9CI)
1-methyl-3-phenyl-5-[3-(trifluoromethyl)phenyl]-4(1H)-pyridinone
1-Methyl-3-phenyl-5-(α,α,α-trifluoro-*m*-tolyl)-4-pyridone
1-Methyl-3-phenyl-5-(3-trifluoromethylphenyl)-4-(1H)-pyridinone
1-Methyl-3-phenyl-5-[3-(trifluoromethyl)phenyl]-4(1H)-pyridinone (ACN)
1-Methyl-3-phenyl-5-[3-(trifluoromethyl)phenyl]-4-(1H)-pyridinone
1-Methyl-3-phenyl-5-[3-(trifluromethyl)phenyl]-4(1H)-pyridinone
4(1H)-Pyrimidinone, 1-methyl-3-phenyl-5-[3-(trifluoromethyl)-phenyl]-
EL 171

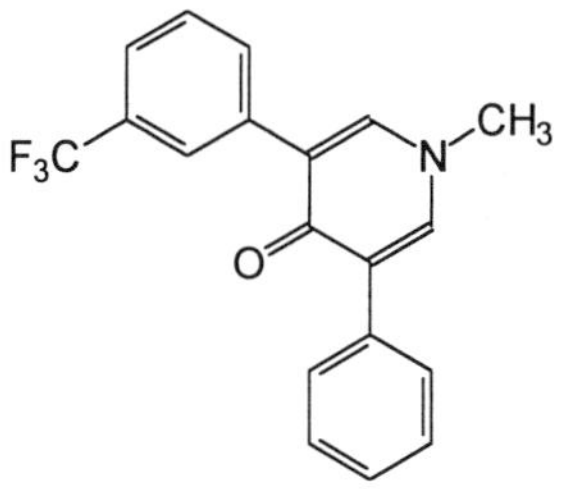

Merck Index No.
MP: 154 - 155°
BP:
Density:
Solubility: 12 mg/l water pH 7, >10 g/l MeOH, >10 g/l chloroform, >5 g/l EtOAc, >1 g/l diethyl ether, >0.5 g/l hexane
Octanol/water PC: 74 (octanol/water 25, pH 7)
LD_{50}: rat orl >10000 mg/kg; mus orl >10000 mg/kg; dog orl >500 mg/kg; cat orl >250 mg/kg; rbt skn >500 mg/kg; qal orl >2000 mg/kg.
40CFR180.420

CAS RN 133-07-3
$C_9H_4Cl_3NO_2S$

Folpet (ACN)
1H-Isoindole-1,3(2H)-dione, 2-[(trichloromethyl)thio]- (9CI)
Phthalimide, N-[(trichloromethyl)thio]- (8CI)
ENT-26539
Faltan
Folpan
Folpel
Ftalan
Fungitrol 11
N-(Trichloromethylthio)phthalimide
N-[(Trichlormethyl)thio]phthalimide
N-[(Trichloromethyl)thio]phthalamide
N-[(Trichloromethyl)thio]phthalimide (ACN)
Orthophaltan
Phaltan
Phaltane
Phthaltan
Spolacid
Thiophal
2-[(Trichloromethyl)thio]-1H-isoindole-1,3(2H)-dione
Merck Index No. 4142
MP: 177°
BP: 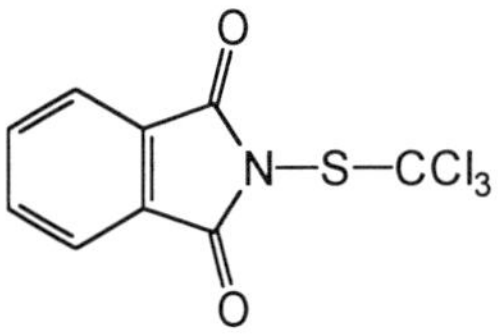
Density:
Solubility: 1 mg/l water 25°, 87 g/l chloroform, 22 g/l benzene, 12.5 g/l isopropanol
Octanol/water PC:
LD_{50}: rat orl 7540 mg/kg; rat orl >5000 mg/kg; rat ihl LC_{50} >5000 mg/m^3/2H; rat ipr 68.4 mg/kg; rat unr >10000 mg/kg; mus orl 1546 mg/kg; mus ihl LC_{50} >6000 mg/m^3/2H; mus skn 5000 mg/kg; mus ipr 80 mg/kg; mus unr 750 mg/kg; rbt orl 1115 mg/kg; rbt skn >22600 mg/kg.
40CFR180.191

CAS RN 50-00-0
CH_2O

Formaldehyde (ACN)(8CI9CI)
component of Odix
Aldehyde formique (French)
Aldeide formica (Italian)
BFV
Fannoform
Formaldehyd (CZECH, POLISH)
Formaldehyde solution
Formaldehyde, gas
Formaldehyde, 37%, methanol-free
Formalin
Formalin-loesungen (German)
Formalina (Italian)
Formaline (German)
Formalith
Formic aldehyde
Formol
Fyde
FA
FYDE
Karsan
Lysoform
Methaldehyde
Methan 21
Methanal
Methyl aldehyde
NCI-C02799
Oplossingen (Dutch)
Oxomethane
Oxymethylene
Paraform (VAN)
Superlysoform

$H_2C{=}O$

Merck Index No. 4148, 4150
MP: -92°
BP: -19.5°
Density: 1.081 - 1.085^{25}
Solubility: 55% water, misc. acetone, alcohol, ethers
Octanol/water PC:
LD_{50}: man orl TDLo 643 mg/kg; wmn orl LDLo 108 mg/kg; hmn ihl TCLo 17 mg/m^3/30M; man orl TDLo 646 mg/kg; man ihl TCLo 0.300 mg/m^3; man unr LDLo 477 mg/kg; rat orl 100 mg/kg; rat orl 550-800 mg/kg; rat ihl LC_{50} 203 mg/m^3; rat scu 420 mg/kg; rat ivn 87 mg/kg; mus orl 42 mg/kg; mus ihl LC_{50} 400 mg/m^3/2H; mus ipr LDLo 16 mg/kg; mus scu 300 mg/kg; dog scu LDLo 350 mg/kg; dog ivn LDLo 70 mg/kg; cat ihl LCLo 400 mg/m^3/2H; cat ivn LDLo 30 mg/kg; rbt skn 270 mg/kg; rbt scu LDLo 240 mg/kg; rbt ivn LDLo 48 mg/kg; gpg orl 260 mg/kg; frg par LDLo 800 mg/kg; mam ihl LC_{50} 92 mg/m^3.
40CFR180. (Exempt)

CAS RN 23422-53-9
$C_{11}H_{15}N_3O_2.ClH$

Formetanate hydrochloride (ACN)
Methanimidamide,
N,N-dimethyl-N'-[3-[[(methylamino)carbonyl]oxy]phenyl]-,
monohydrochloride (9CI)
Carbamic acid, methyl-, ester with N'-(*m*-hydroxyphenyl)-
N,N-dimethylformamidine,
monohydrochloride (8CI)
m-[[(Dimethylamino)methylene]amino]phenyl methylcarbamate hydrochloride
m-[[(Dimethylamino)methylene]amino]phenyl methylcarbamate
monohydrochloride
m-[[(Dimethylamino)methylene]amino]phenyl methylcarbamate, hydrochloride
Carbamic acid, methyl- ester with N'-(*m*-hydroxyphenyl)-N,N-dimethylformamidine,
hydrochloride
Carzol
Carzol SP
Dicarzol
ENT-27566
Formetanate monohydrochloride

Merck Index No.
MP: 200 - 202°
BP:
Density:
Solubility: (hydrochloride) >500 g/l water, 250 g/l methanol, <1 g/l acetone, chloroform, dichloromethane, ethyl acetate, hexane.
Octanol/water PC: <0.002 (hydrochloride)
LD_{50}: rat orl 20 mg/kg; rat skn >5600 mg/kg; rat ipr 4.7 mg/kg; mus orl 18 mg/kg; dog orl 19 mg/kg; rbt skn 10200 mg/kg; ckn orl 21.5 mg/kg; qal orl 42 mg/kg; dcl orl 12 mg/kg; mam ihl LC_{50} 2800 mg/m^3/4H.
40CFR180.276

CAS RN 961-11-5
$C_{10}H_9Cl_4O_4P$

Gardona
Phosphoric acid, 2-chloro-1-(2,4,5-trichlorophenyl)ethenyl dimethyl ester (9CI)
Phosphoric acid, 2-chloro-1-(2,4,5-trichlorophenyl)vinyl dimethyl ester (8CI)
trans-2-Chloro-1-(2,4,5-trichlorophenyl)vinyl dimethyl phosphate
Benzyl alcohol, 2,4,5-trichloro-α-(chloromethylene)-, dimethyl phosphate
Dimethyl 2,4,5-trichloro-α-(chloromethylene)benzyl phosphate

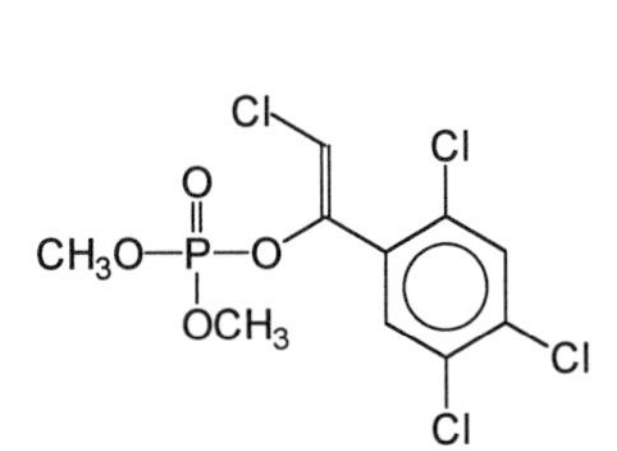

Appex
CVMP
Dietreen
ENT 25,841
Gardcide
NCI-C00168
Rabon
Stirofos
SD 8447
Tetrachlorvinphos
2-Chloro-1-(2,4,5-trichlorophenyl)vinyl dimethyl phosphate (ACN)

Merck Index No.
MP:
BP:
Density:
Solubility:
Octanol/water PC:
LD_{50}: rat orl 4000 mg/kg; rat skn >10000 mg/kg; rat ipr 1160 mg/kg; rat scu >15000 mg/kg; mus orl 4200 mg/kg; mus ihl LC50 >1500 mg/m3/4H; mus skn >7500 mg/kg; mus ipr 1170 mg/kg; mus scu 15000

mg/kg; bwd orl 100 mg/kg.
40CFR180.252

CAS RN 556-22-9
$C_{20}H_{40}N_2.C_2H_4O_2$

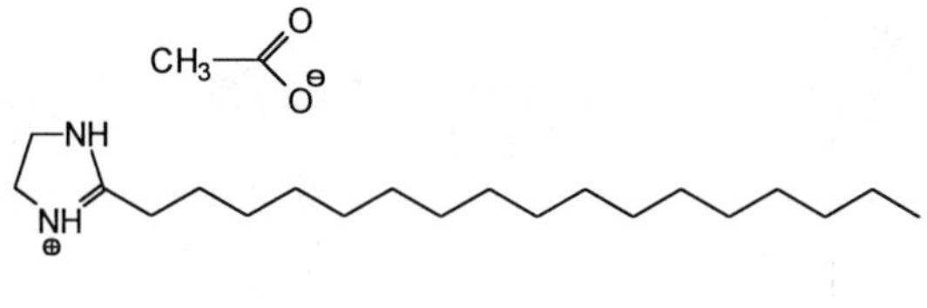

Glyodin (ACN)
1H-Imidazole, 2-heptadecyl-4,5-dihydro-, monoacetate (9CI)
2-Imidazoline, 2-heptadecyl-, monoacetate (8CI)
Acetic acid, 2-heptadecylglyoxalidine
Crag Fruit Fungicide 341
Crag 341
Experimental Fungicide 341
Glyodin acetate
Glyoxide
Glyoxide Dry
2-Heptadecyl glyoxalidine acetate
2-Heptadecyl imidazoline acetate
2-Heptadecyl-2-imidazoline acetate (ACN)
2-Heptadecyl-4,5-dihydro-1H-imidazolyl monoacetate
2-Imidazoline, 2-heptadecyl-, acetate

Merck Index No. 4404
MP: 62 - 68° (free base 94°)
BP:
Density: 1.035^{20}
Solubility: Soluble iPrOH. Insol. water, acetone, toluene.
Octanol/water PC:
LD_{50}: rat orl 4600 mg/kg; mam unr 100 mg/kg.
40CFR180.124

CAS RN 557-30-2
$C_2H_4N_2O_2$

Glyoxal, dioxime (8CI)
Ethanedial, dioxime (9CI)
Cga-22911
Ethanedial dioxime (ACN)
Ethanedione dioxime
Glyoxime
Pik-off

Merck Index No.
MP: 176 - 178°;178°(dec)
BP:
Density:
Solubility: Very sol. water, alchol, diethyl ether
Octanol/water PC:
LD_{50}: rat orl 119 mg/kg; rbt skn 1580 mg/kg.
40CFR180. (temp. 1976)

CAS RN 1071-83-6
$C_3H_8NO_5P$

Glyphosate (ACN)
Glycine, N-(phosphonomethyl)- (8CI9CI)
CP 67573
CP 70139
CP-67573
CP-70139
Glyphosate, isopropylamine salt
Glyphosphate
Isopropylamine glyphosate
Isopropylamine salt N-(Phosphonomethyl)glycine (ACN)
MON 0573
MON 2139
MON 39
MON 8000
N-(Phosphonomethyl)glycine (ACN)
N-Phosphonomethylglycine
Polado
Roundup
Sodium salt N-(Phosphonomethyl)glycine (ACN)

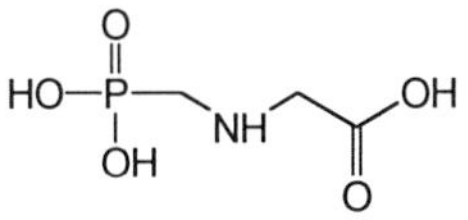

Merck Index No. 4408
MP: 200°, 230° (dec)
BP:
Density:
Solubility: 12 g/l water 25°, insoluble in common organic solvents.
Octanol/water PC:
LD_{50}: man orl LDLo 2143 mg/kg; man orl TDLo 1214 mg/kg; wmn orl LDLo 4000 mg/kg; rat orl 4873 mg/kg; rat ipr 235 mg/kg; mus orl 1568 mg/kg; mus ipr 130 mg/kg; rbt orl 3800 mg/kg; rbt skn 7940 mg/kg; qal orl >4640 mg/kg.
40CFR180.364

CAS RN 1024-57-3
$C_{10}H_5Cl_7O$

Heptachlor epoxide
2,5-Methano-2H-indeno[1,2-b]oxirene, 2,3,4,5,6,7,7-heptachloro-1a,1b,5,5a,6,6a-hexahydro-, (1aα,1bβ,2α,5α,5aβ,6β,6aα)-
4,7-Methanoindan, 1,4,5,6,7,8,8-heptachloro-2,3-epoxy-3a,4,7,7a-tetrahydro- (8CI)
β-Heptachlorepoxide
Epoxyheptachlor
ENT 25,584
Heptachlorepoxide
HCE
Velsicol 53-CS-17
1,4,5,6,7,8,8-Heptachloro-2,3-epoxy-2,3,3a,4,7,7a-hexahydro-4,7-methanoindene
1,4,5,6,7,8,8-Heptachloro-2,3-epoxy-2,3,3a,4,7,7a-tetrahydro-4,7-methanoindene

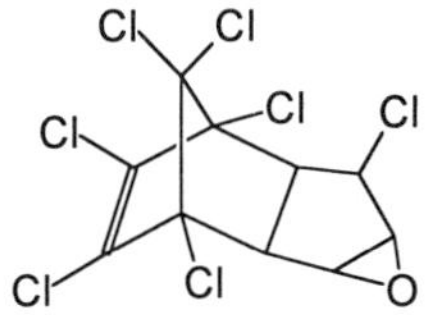

Merck Index No.
MP: 157 - 159°
BP:
Density:
Solubility:
Octanol/water PC:
LD_{50}: rat orl 15 mg/kg; mus orl 39 mg/kg; mus ivn LDLo 10 mg/kg; rbt orl 144 mg/kg.
40CFR180.104

CAS RN 76-44-8
$C_{10}H_5Cl_7$

Heptachlor (ACN)
4,7-Methano-1H-indene, 1,4,5,6,7,8,8-heptachloro-3a,4,7,7a-tetrahydro- (9CI)
4,7-Methanoindene, 1,4,5,6,7,8,8-heptachloro-3a,4,7,7a-tetrahydro- (8CI)
racemic-1,4,5,6,7,8,8-Heptachloro-3a,4,7,7a-tetrahydro-4,7-methanoindene
Aahepta
Agroceres
Dicyclopentadiene, 3,4,5,6,7,8,8a-heptachloro-
Drinox
E 3314
Eptacloro (Italian)
ENT 15,152
GPKh
H
H-34
Hepta
Heptachloor (Dutch)
Heptachlorane
Heptachlore (French)
Heptachloro-tetrahydro-endo-methanoindene
Heptachlorodicyclopentadiene
Heptachlorotetrahydro-endo-methanoindene
Heptachlorotetrahydro-4,7-endo-methanoindene
Heptachlorotetrahydro-4,7-methanoindene (ACN)
Heptagran
Heptamul
NCI-C00180
OMS 193
Rhodiachlor
Velsicol heptachlor
Velsicol 104
1(3a),4,5,6,7,8,8-Heptachloro-3a(1),4,7,7a-tetrahydro-4,7-methanoindene
1-Chlorochlordene
1,4,5,6,7,8,8-Eptacloro-3a,4,7,7a-tetraidro-4,7-endo-metano-indene (Italian)
1,4,5,6,7,8,8-Heptachloor-3a,4,7,7a-tetrahydro-4,7-endo-methano-indeen (Dutch)
1,4,5,6,7,8,8-Heptachlor-3a,4,7,7a-tetrahydro-4,7-endo-methano-inden (German)

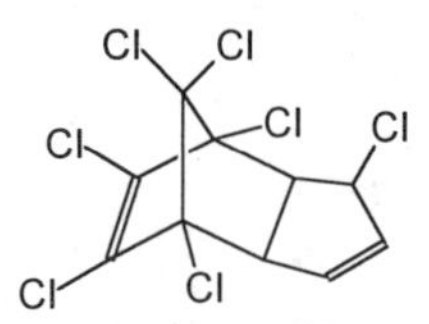

Merck Index No. 4576
MP: 95 - 96°
BP:
Density:
Solubility: 75 g/100ml Me2CO, 106 g/100ml C6H6, 112g/100ml CCl4, 119g/100ml cyclohexanone, 4.5g/100ml EtOH, 102 g/100ml xylene
Octanol/water PC:
LD_{50}: rat orl 100, 162mg/kg; rat orl 40 mg/kg; rat skn 119 mg/kg; rat ipr 27 mg/kg; mus orl 68 mg/kg; mus ipr 130 mg/kg; mus ivn LDLo 20 mg/kg; cat orl LDLo 50 mg/kg; cat ihl LCLo 150 mg/m3/4H; rbt skn >2000 mg/kg; gpg orl 116 mg/kg; gpg skn LDLo 1000 mg/kg; ham orl 100 mg/kg; mam ihl LCLo 200 mg/m3/4H; mam unr 60 mg/kg .
40CFR180.104

CAS RN 70-30-4
$C_{13}H_6Cl_6O_2$

Hexachlorophene
Phenol, 2,2'-methylenebis[3,4,6-trichloro- (8CI9CI)
component of Soy-Dome
pHisohex
pHisoHex
Acigena
Almederm
Almerdem
AT-17
AT-7
AT7
B 32 (VAN)
Bilevon
Bis(2-hydroxy-3,5,6-trichlorophenyl)methane
Bis(3,5,6-trichloro-2-hydroxyphenyl)methane
Bis-2,3,5-trichlor-6-hydroxyfenylmethan (Czech)
B32
Compound G-11
Cotofilm
Derl
Dermadex
Distodin
Exofene
Fesia-sin
Fomac
Fostril
G-Eleven
G-11
Gamophen
Gamophene
Gamphen
Germa-Medica
Gramophen
Hexabalm
Hexachlorofen
Hexachlorophane
Hexachlorophen
Hexachlorophenum
Hexaclorofeno
Hexafen

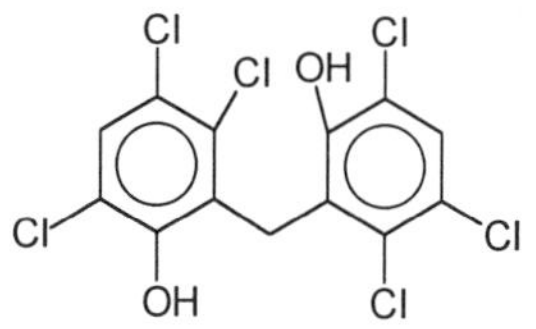

Merck Index No. 4602
MP: 164 - 165°
BP:
Density:
Solubility: Insol. water. Sol. acetone, ethanol, diethyl ether, chloroform, glycols.
Octanol/water PC:
LD_{50}: inf orl TDLo 257 mg/kg/7D-I; chd orl LDLo 250 mg/kg; wmn orl TDLo 600 mg/kg; rat orl 56 mg/kg; rat ihl LC50 340 mg/m^3; rat skn 1840 mg/kg; rat ipr 22 mg/kg; rat scu 7.65 mg/kg; rat ivn 7.5 mg/kg; mus orl 67 mg/kg; mus ihl LC_{50} 290 mg/m^3; mus skn 270 mg/kg; mus ipr 20 mg/kg; dog orl LDLo 40 mg/kg; dog ivn LDLo 5 mg/kg; rbt orl 40.69 mg/kg; rbt ivn 8.5 mg/kg; gpg orl 60 mg/kg; gpg skn 1100 mg/kg; gpg ipr LDLo 25 mg/kg; dom orl 30 mg/kg.
40CFR180.302

CAS RN 54460-46-7
$C_{20}H_{38}O_2$

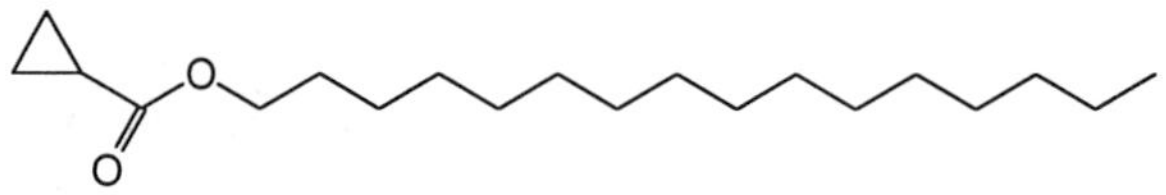

Cycloprate
Cyclopropanecarboxylic acid, hexadecyl ester (9CI)
Hexadecyl cyclopropanecarboxylate (ACN)
Zardex
ZR-856

Merck Index No.
MP:
BP: $154^{\circ 0.05}$
Density:
Solubility:
Octanol/water PC:
LD_{50}: rat orl 12200 mg/kg; dog orl 2500 mg/kg; rbt skn 2670 mg/kg.
40CFR180.(temp 1976)

CAS RN 51235-04-2
$C_{12}H_{20}N_4O_2$

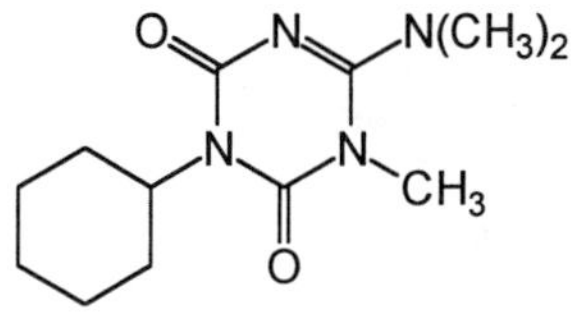

Hexazinone (ACN)
1,3,5-Triazine-2,4(1H,3H)-dione, 3-cyclohexyl-6-(dimethylamino)-1-methyl- (9CI)
s-Triazine-2,4(1H,3H)-dione, 3-cyclohexyl-6-(dimethylamino)-1-methyl-
DPX 367
DPX 3674
Velpar
Velpar weed killer
3-Cyclohexyl-1-methyl-6-(dimethylamino)-*s*-triazine-2,4(1H,3H)-dione
3-Cyclohexyl-6-(dimethylamino)-1-methyl-*s*-triazine-2,4(1H,3H)-dione
3-Cyclohexyl-6-(dimethylamino)-1-methyl-1,3,5-triazine-(1H,3H)-dione
3-Cyclohexyl-6-(dimethylamino)-1-methyl-1,3,5-triazine-2,4(1H,3H)-dione (ACN)
3-Cyclohexyl-6-(dimethylamino)-1-methyl-1,3,5-triazine-2,4(1H,3H)dione
3-Cyclohexyl-6-(dimethylamino)-1-methyl-1,3,5-triazine-2,4-(1H,3H)-dione
3-Cyclohexyl-6-(dimethylamino)-1-methyl-1,3,5-triazine-2,4-(1H,3H)dione
3-Cyclohexyl-6-dimethylamino-1-methyl-1,3,5-triazine-2,4-dione

Merck Index No. 4617
MP: 115 - 117°, 97 - 100.5°
BP: (dec)
Density: 1.25
Solubility: 33 g/kg water 25°, 330 g/l water 25, 3880 g/kg chloroform, 2650 g/kg methanol, 940 g/kg benzene, 836 g/kg dimethylformamide, 792 g/kg acetone, 386 g/kg toluene, 3 g/kg hexane.
Octanol/water PC:
LD_{50}: rat orl 1690 mg/kg; rat skn 5278 mg/kg; rat ipr 530 mg/kg; rbt skn >5278 mg/kg; gpg orl 860 mg/kg; qal orl 2258 mg/kg; qal ipr 2258 mg/kg.
40CFR180.396

CAS RN 74-90-8
CHN

Hydrogen cyanide
Hydrocyanic acid (ACN)(8CI9CI)
Acide cyanhydrique (French)
Acido cianidrico (Italian)
Aero liquid HCN
Aero@ Liquid HCN
Blausaeure (German)
Blauwzuur (Dutch)
Carbon hydride nitride (CHN)
Cyaanwaterstof (Dutch)
Cyanwasserstoff (German)
Cyclon
Cyclone B
Cyjanowodor (Polish)
Evercyn
ENT-31100
Formic anammonide
Formonitrile
Hydrocyanic acid (prussic) solution
Hydrocyanic acid (prussic), unstabilized
Hydrocyanic acid (solution)
Hydrocyanic acid (unstabilized)
Hydrocyanic acid gas
Hydrocyanic acid solution
Hydrocyanic acid, anhydrous
Hydrocyanic acid, liquefied (DOT)
Hydrocyanic acid, liquid
Hydrocyanic acid, solution (DOT)
Hydrocyanic acid, unstabilized (DOT)
HCN
Prussic acid
Prussic acid solution
Prussic acid, liquid
Prussic acid, unstabilized
Zaclondiscoids

H—C≡N

Merck Index No. 4722
MP: -15-13.4°
BP: 26.5°, 25.6°
Density: 0.69920
Solubility: Soluble in water, ethanol, diethyl ether.
Octanol/water PC:
LD_{50}: hmn orl LDLo 5.70 mg/kg; man ihl TCLo 500 mg/m3/3M-C; hmn ihl LCLo 120 mg/m3/1H; hmn ihl LCLo 200 mg/m3/10H; man ihl LCLo 400 mg/m3/2H; hmn scu LDLo 1 mg/kg; man ivn TDLo 0.055 mg/kg; man unr LDLo 1.471 mg/kg; rat orl 10-15 mg/kg (sodium salt); rat ihl LC_{50} 554 ppm/5M; rat ihl LC_{50} 160 ppm/30M; rat ivn 0.810 mg/kg; mus orl 3.7 mg/kg; mus ihl LC50 323 ppm/5M; mus ipr 2.99 mg/kg; mus scu LDLo 3 mg/kg; mus ivn 0.990 mg/kg; mus ims 2.7 mg/kg; dog orl LDLo 4 mg/kg; dog scu LDLo 1.70 mg/kg; dog ivn 1.34 mg/kg; mky ivn 1.30 mg/kg; cat ihl LC_{50} 850 mg/m^3/1H; cat scu 1.10 mg/kg; cat ivn 0.810 mg/kg; rbt orl LDLo 4 mg/kg; rbt ihl LC_{50} 850 mg/m^3/1M; rbt ipr 1.57 mg/kg; rbt scu 2.50 mg/kg; rbt ivn 0.66 mg/kg; rbt ims 0.486 mg/kg; rbt ocu 1.04 mg/kg; pig orl LDLo 2 mg/kg; gpg scu 0.10 mg/kg; gpg ivn 1.430 mg/kg; pgn orl LDLo 14 mg/kg; pgn scu LDLo 2.15 mg/kg; pgn ims LDLo 1.50 mg/kg; dck orl LDLo 3.28 mg/kg; frg scu LDLo 60 mg/kg; dom ihl LC_{50} 1300 mg/m^3/30S; dom ivn 0.66 mg/kg; mam ihl LCLo 200 ppm/5M; mam ihl LCLo 36 ppm/2H; brd orl LDLo 0.60 mg/kg; bwd orl LDLo 7.50 mg/kg; bwd scu LDLo 0.10 mg/kg; brd scu LDLo 0.10 mg/kg.
40CFR180.130

CAS RN 420-04-2
CH_2N_2

Cyanamide (ACN)(8CI9CI)
Amidocyanogen
Carbamonitrile
Carbimide
Carbodiimide
Cyanoamine
Cyanogen nitride
Cyanogenamide
Hydrogen cyanamide
N-Cyanoamine
USAF EK-1995

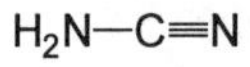

$H_2N{-}C{\equiv}N$

Merck Index No. 2691
MP: 45 - 46°
BP: $83^{\circ 0.5}$
Density: 1.28220
Solubility: 4.59 kg/l water 20°, 7.75 kg/l water 15, 10.0 kg/l water 43, 505 g/kg methyl ethyl ketone, 424 g/kg ethyl acetate, 288 g/kg n-butanol, 2.4 g/kg chloroform.
Octanol/water PC:
LD_{50}: rat orl 125 mg/kg; rat ihl LCLo 86 mg/m3/4H; rat skn 84 mg/kg; rat ipr LDLo 200 mg/kg; rat ivn 56 mg/kg; rat unr 280 mg/kg; mus orl 388 mg/kg; mus ipr 200 mg/kg; cat orl 100 mg/kg; rbt orl 150 mg/kg; rbt skn 590 mg/kg.
40CFR180. (temp 1988)

CAS RN 41096-46-2
$C_{17}H_{30}O_2$

Hydroprene (ACN)
2,4-Dodecadienoic acid, 3,7,11-trimethyl-, ethyl ester, (E,E)- (9CI)
Altozar
Altozar IGR
Dodeca-2,4-dienoic acid, 3,7,11-trimethyl-, ethyl ester, (2E,4E)
Entocone
Ethyl (E,E)-3,7,11-trimethyl-2,4-dodecadienoate
Ethyl (E,E)-3,7,11-Trimethyl-2,4-dodecadienoate (ACN)
Ethyl (2E,4E)-3,7,11-trimethyl-2,4-dodecadienoate
Ethyl (2E,4E)-3,7,11-trimethyldodeca-2-4-dienoate
Ethyl 3,7,11-trimethyldodeca-2,4-dienoate
Ethyl 3,7,11-trimethyldodecadienoate
ENT 70459
ZR 512

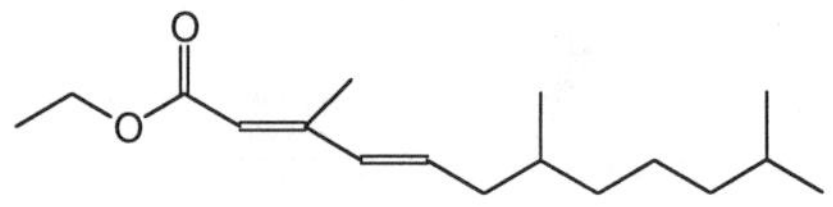

Merck Index No.
MP:
BP: $174^{\circ 19}$
Density: 0.8955^{20}
Solubility: 0.54 mg/l water 20°, soluble most organic solvents.
Octanol/water PC: 5700
LD_{50}: rat orl 34600 mg/kg; rat ihl LC_{50} >5400 mg/m^3; rat skn >5000 mg/kg; dog orl >10000 mg/kg; rbt skn 4550 mg/kg.
40CFR185.---

CAS RN 35554-44-0
$C_{14}H_{14}Cl_2N_2O$

Imazalil
1H-Imidazole, 1-[2-(2,4-dichlorophenyl)-2-(2-propenyloxy)ethyl]- (9CI)
(±)-1-[β-(Allyloxy)-2,4-dichlorophenethyl]-imidazole
Enilconazole
Fungaflor
R 23,979
R 23979
1-(β-Allyloxy-2,4-dichlorophenethyl)imidazole
1-(2-(2,4-Dichlorphenyl)-2-propenyloxy)aethyl)-1H-imidazol (German)
1-[2-(2,4-Dichlorophenyl)-2-(2-propenyloxy)ethyl]-1H-imidazole
1H-Imidazole, 1-[2-(2,4-dichlorophenyl)-2-(2-propenyloxy)ethyl]-, (.+-.)-

Merck Index No. 3537
MP: 50°
BP: (dec)
Density: 1.243^{23}
Solubility: 1.4 g/l water 20°, >500 g/l ethanol, methanol, benzene, xylene. Soluble in isopropanol, toluene, heptane, hexane, petroleum ether.
Octanol/water PC: 6600
LD_{50}: rat orl 320 mg/kg; rat orl 227 mg/kg; rat ihl LC_{50} 16000 mg/m^3/4H; rat skn 4200 mg/kg; rat ipr 155 mg/kg; dog orl >640 mg/kg; rbt skn 4200 mg/kg.
40CFR180.413

CAS RN 33820-53-0
$C_{15}H_{23}N_3O_4$

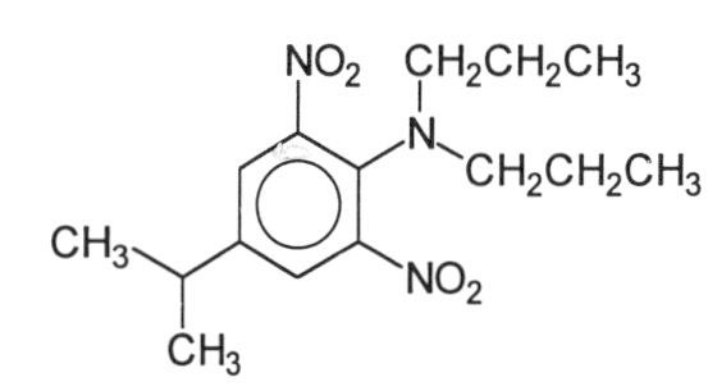

Isopropalin (ACN)
Benzenamine, 4-(1-methylethyl)-2,6-dinitro-N,N-dipropyl- (9CI)
Cumidine, 2,6-dinitro-N,N-dipropyl- (8CI)
Aniline, 2,6-dinitro-N,N-dipropyl-*p*-isopropyl-
EL 179
Isopropalin solution
Paarlan
2,6-Dinitro-N,N-dipropylcumidine (ACN)
4-(1-Methylethyl)-2,6-dinitro-N,N-dipropylbenzenamine
4-Isopropyl-2,6-dinitro-N,N-di-(*n*-propyl)aniline
4-Isopropyl-2,6-dinitro-N,N-dipropylaniline

Merck Index No. 5089
MP:
BP:
Density:
Solubility: 0.1 mg/l water 25°, >1 kg/l acetone, hexane, benzene, chloroform, diethyl ether, acetonitrile, methanol.
Octanol/water PC:
LD_{50}: rat orl 5000 mg/kg; mus orl >5000 mg/kg; dog orl LDLo >2000 mg/kg; rbt orl LDLo >2000 mg/kg; rbt skn >2000 mg/kg; ckn orl LDLo >2000 mg/kg; qal orl LDLo >1000 mg/kg; dck orl LDLo >2000 mg/kg;
40CFR180.313

CAS RN 50-35-1
$C_{13}H_{10}N_2O_4$

Thalidomide
1H-Isoindole-1,3(2H)-dione, 2-(2,6-dioxo-3-piperidinyl)- (9CI)
Phthalimide, N-(2,6-dioxo-3-piperidyl)- (8CI)
α-(N-Phthalimido)glutarimide
α-N-Phthalylglutaramide
α-Phthalimidoglutarimide
α-Thalidomide
Algosediv
Asidon 3
Asmadion
Asmaval
Bonbrain
Calmore
Calmorex
Contergan
Corronarobetin
Countergan-R
Distaval
Distaval-R
Distaxal
Distoval
Ectiluran
Enterosediv
Gastrinide
Glupan
Glutanon
Glutarimide, 2-phthalimido-
Grippex
Hippuzon
Imida-Lab
Imidan
Imidan (peyta)
Imidene
Isomin
K 17 (VAN)
K-17
Kedavon
Kevadon
Kevadon-R
Lulamin
N-(α-Glutarimido)phthalimide

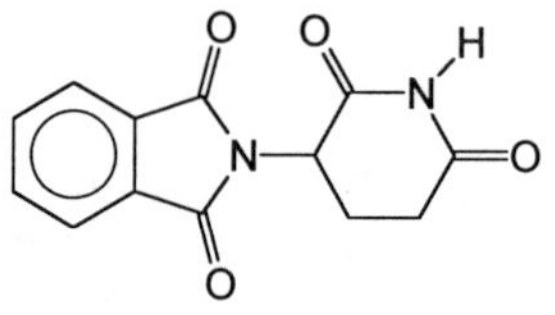

Merck Index No. 9182
MP: 269 - 271°
BP:
Density:
Solubility: Sparingly soluble in water, methanol, ethanol, acetone, ethyl acetate, butyl acetate, acetic acid. Very soluble in dioxane, dimethylformamide, pyridine. Insoluble in ether, chloroform, benzene.
Octanol/water PC:
LD_{50}: rat orl 113 mg/kg; rat skn 1550 mg/kg; rat ipr >8000 mg/kg; mus orl 2000 mg/kg; mus ipr LDLo 800 mg/kg; mus scu >5000 mg/kg; dog orl LDLo >1538 mg/kg.
40CFR180.261

CAS RN 732-11-6
$C_{11}H_{12}NO_4PS_2$

Imidan
Phosmet
Phosphorodithioic acid, S-[(1,3-dihydro-1,3-dioxo-2H-isoindol-2-yl)methyl] O,O-dimethyl ester (9CI)
Phosphorodithioic acid, O,O-dimethyl ester, S-ester with N-(mercaptomethyl)phthalimide (8CI)
APPA
Decemthion
Decemthion P-6
Decemtion P-6
ENT 25,705
Ftalophos
N-(Mercaptomethyl)phthalamide S-(O,O-dimethyl) phosphorodithioate
N-(Mercaptomethyl)phthalamide S-(O,O-dimethyl phosphorodithioate)
N-(Mercaptomethyl)phthalimide S-(O,O-dimethyl phosphorodithioate) (ACN)
N-(Mercaptomethyl)phthalimide S-(O,O-dimethyl) phosphorodithioate
O,O-Dimethyl phosphorodithioate S-ester with N-(mercaptomethyl)phthalimide
O,O-Dimethyl phthalimidiomethyl dithiophosphate
O,O-Dimethyl S-(phthalimidomethyl)dithiophosphate
O,O-Dimethyl S-(phthalimidomethyl)phosphorodithioate
O,O-Dimethyl S-(N-phthalimidomethyl)dithiophosphate
O,O-Dimethyl S-phthalidomethyl phosphorodithioate
Percolate
Phosphorodithioic acid, S-[(1,3-dihydro-1,3-dioxo-isoindol-2-yl)methyl] O,O-dimethyl ester
Phthalimide, N-(mercaptomethyl)-, S-ester with O,O-dimethyl phosphorodithioate
Phthalimidomethyl O,O-dimethyl phosphorodithioate
Phthalophos
Prolate
PMP (VAN)
PMP (pesticide)
R 1504
Safidon
Smidan
Stauffer R-1504

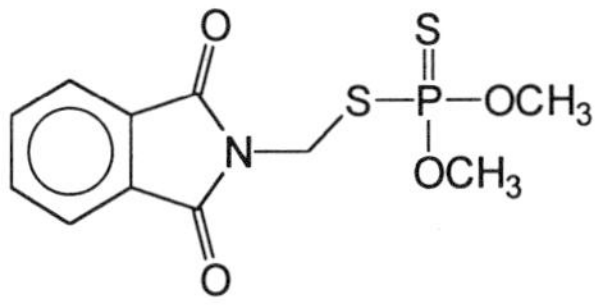

Merck Index No. 7311
MP: 71.9°, 66.5 - 69.5° (technical, 95-98%), 72 - 72.7°, 66 - 69° (technical).
BP: (dec)
Density:
Solubility: 25 mg/ml water 25°, 650 g/l acetone, 50 g/l methanol, 600 g/l benzene, 300 g/l toluene, 300 g/l methyl isobutyl ketone, 250 g/l xylene, <10 g/l kerosene.
Octanol/water PC: 1100
LD_{50}: hum orl LDLo 50 mg/kg; hmn ihl TCLo 2 mg/m^3/8H; rat orl 113 mg/kg; rat orl 160 mg/kg; rat orl 92.5 mg/kg; rat ihl LC_{50} 54 mg/m^3/4H; rat skn 1326 mg/kg; mus orl 26 mg/kg; cat ihl LC_{50} 65 mg/m^3/4H; rbt skn >5000 mg/kg; gpg orl 200 mg/kg; ckn orl 707 mg/kg; dcl orl 1830 mg/kg; mamunr 40 mg/kg; bwd orl 18 mg/kg.
40CFR180.261

CAS RN 36734-19-7
$C_{13}H_{13}Cl_2N_3O_3$

Iprodione
1-Imidazolidinecarboxamide, 3-(3,5-dichlorophenyl)-N-(1-methylethyl)-2,4-dioxo-
1-Isopropyl carbamoyl-3-(3,5-dichlorophenyl)-hydantoin
3-(3,5-Dichlorophenyl)-N-(1-methylethyl)-2,4-dioxo-1-imidazolidinecarboxamide (ACN)
Anfor
Chipco 26019
Glycophen
Glycophene
LFA 2043
MRC 910
Prodione
Promidione
Rop 500 F
Rovral
Rovrol
RP 26019

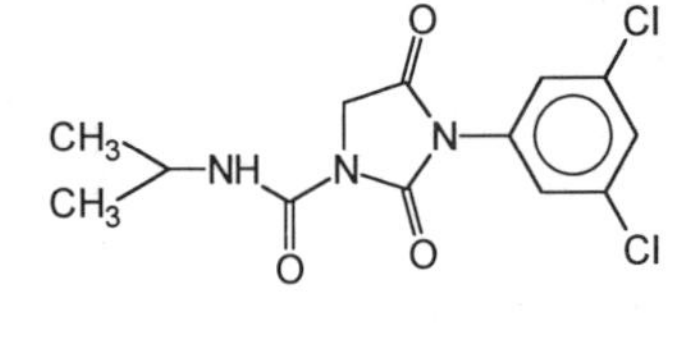

Merck Index No.
4964
MP: approx. 136°
BP:
Density:
Solubility: 13 mg/l water 20°, 25 g/l ethanol, 25 g/l methanol, 150 g/l acetonitrile, 150 g/l toluene, 200 g/l benzene, 300 g/l acetone, 500 g/l dichloromethane, 500 g/l dimethylformamide.
Octanol/water PC: 1260 (octanol/water 22)
LD_{50}: rat orl 3500 mg/kg; rat ihl LC_{50} >3300 mg/m^3/4H; rat skn >2500 mg/kg; rat ipr 1700 mg/kg; mus orl 4000 mg/kg; rbt skn >5000 mg/kg; qal orl 930 mg/kg; dck orl 10400 mg/kg.
40CFR180.399

CAS RN 25311-71-1
$C_{15}H_{24}NO_4PS$

Isofenphos
Benzoic acid, 2-[[ethoxy[(1-methylethyl)amino]phosphinothioyl]oxy]-, 1-methylethyl ester
Salicylic acid, isopropyl ester, O-ester with O-ethyl isopropylphosphoramidothioate
Amaze
BAY 92114
Isophenphos
Isopropyl salicylate O-ester of O-ethyl isopropylphosphoramidothioate
O-Ethyl O-[2-(isopropoxycarbonyl)phenyl]isopropylphosphoramidothioate
Oftanol
SRA 12869
1-Methylethyl 2-[[ethoxy(1-methylethyl)amino]phosphinothioylloxy]benzoate
1-Methylethyl 2-[[ethoxy[(1-methylethyl)amino]phosphinothioyl]oxy]benzoate (ACN)

Merck Index No. 5055
MP: < -12°
BP: $120^{\circ 0.01}$
Density: 1.134^{20}
Solubility: 23.8 mg/kg water 25°, >600 g/kg cyclohexanone, isopropanol, dichloromethane, toluene. Soluble in acetone, ether, benzene, xylene, kerosene.

Octanol/water PC:
LD_{50}: rat orl 28 mg/kg; rat ihl LC_{50} 144 mg/m^3/4H; rat skn 188 mg/kg; mus orl 91.3 mg/kg; rbt skn 162 mg/kg; ckn orl 3 mg/kg; qal orl 13 mg/kg.
40CFR180.387

CAS RN 600-25-9
$C_3H_6ClNO_2$

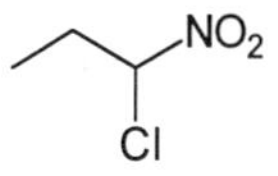

Chloronitropropane
Korax
Propane, 1-chloro-1-nitro- (8CI9CI)
Lanstan
Chloronitropropan (Polish)
Lastan
Niagara 5,961
NIA 5961
1-Chloro-2-nitropropane (ACN)

Merck Index No.
MP: 123.5°
BP: 141 - 145°, $67^{\circ 56}$
Density: 1.209^{20}

Solubility: Very soluble in glycols, diethyl ether, ethanol and oils. Soluble in water, chloroform.
Octanol/water PC:
LD_{50}: rat orl LDLo 50 mg/kg; mus orl 510 mg/kg; mus ihl LC_{50} 66 mg/m^3/3H; mus scu 165 mg/kg; rbt orl LDLo 50 mg/kg; rbt ihl LCLo 2000 mg/m^3/6H; gpg ihl LCLo 18000 mg/m^3/2H.
40CFR180.286

CAS RN 7784-40-9
$AsH_3O_4.Pb$

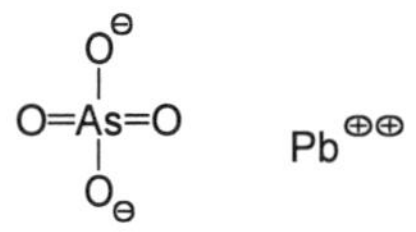

Lead arsenate (ACN)
Arsenic acid (H_3AsO_4), lead(2+) salt (1:1) (9CI)
Acid lead arsenate
Acid lead orthoarsenate
Arsenate of lead
Arsenic acid, lead salt
Arsenic acid, lead(2+) salt(1:1)
Arsinette
Basic lead arsenate
Dibasic lead arsenate
Gypsine
Lead acid arsenate
Lead arsenate (basic)
Lead arsenate (AsH_3O_4 . Pb)
Lead arsenate, basic (ACN)
Lead arsenate, solid (DOT)
Lead hydrogenarsenate
Ortho 110 dust
Ortho 140 dust
Plumbous arsenate
Schultenite
Security
Soprabel
Standard lead arsenate (ACN)
Talbot

Merck Index No. 5270
MP: 280° (dec)
BP:
Density: 5.79
Solubility: Insoluble in water. Soluble in HNO_3, caustic alkalies.
Octanol/water PC:
LD_{50}: rat orl 100 mg/kg; mus orl 1526 mg/kg; mus ipr 128 mg/kg; rbt orl 100 mg/kg; ckn orl 450 mg/kg.
40CFR180.194 (revoked 4/3/91)

CAS RN 58-89-9
$C_6H_6Cl_6$

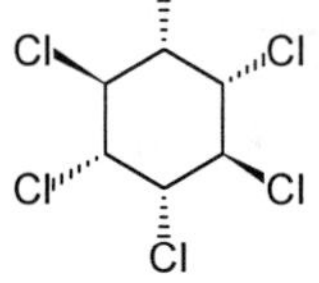

Lindane
Cyclohexane, 1,2,3,4,5,6-hexachloro-, (1α,2α,3β,4α,5α,6β)-
Cyclohexane, 1,2,3,4,5,6-hexachloro-, γ-
γ-(a,a,a,e,e,e)-1,2,3,4,5,6-Hexachlorocyclohexane
γ-isomer of Benzene hexachloride from lindane (ACN)
γ-isomer of 1,2,3,4,5,6-Hexachlorocyclohexane
γ-Benzene hexachloride
γ-BHC
γ-Hexa
γ-Hexachloran
γ-Hexachlorane
γ-Hexachlorobenzene
γ-Hexachlorocyclohexane
γ-HCH
γ-Lindane
γ-1,2,3,4,5,6-Hexachlorocyclohexane
γ Isomer of 1,2,3,4,5,6-hexachlorocyclohexane
γ1,2,3,4,5,6-Hexachlorocyclohexane
(1α,2α,3β,4α,5α,6β)-1,2,3,4,5,6-Hexachlorocyclohexane
component of Kwell
neo-Scabicidol
Aalindan
Aficide
Agrisol G-20
Agrocide
Agrocide III
Agrocide WP
Agrocide 2
Agrocide 6G
Agrocide 7
Agronexit
Ambocide
Ameisenatod
Ameisenmittel merck
Ameisentod
Antiscabbia, γ-
Aparasin
Aphtiria
Aplidal
Arbitex
Ben-hex
Benzex
BHC
Compound-666
Cyclohexane, 1,2,3,4,5,6-hexachloro-, (mixed isomers)
Cyclohexane, 1,2,3,4,5,6-hexachloro-, dl-isomers
Dolmix
DBH
DOL
ENT 8,601
FBHC
FHCH
Gammexane
Gyben
Hexablanc
Hexachlor
Hexachloran
Hexachlorocyclohexane
Hexafor
Hexaklor
Hexamul
Hexapurdre
Hexdow
Hexyclan
Hexylan
Hilbeech
HCCH
HCH
HEXA
HGI
Isaton
Kotol
Soprocide
Technical bhc

Merck Index No. 5379
MP: 112.5°
BP:
Density:
Solubility: 7.3 mg/l water 25°, 12 mg/l water 35°, 43.5 g/l acetone, 74 g/l methanol, 6.4 g/l ethanol, 28.9 g/l benzene, 27.6 g/l toluene, 24.7 g/l xylene, 20.8 g/l diethyl ether, 2.9 g/l pet. ether, 35.7 g/l ethyl acetate, 24.0 g/l chloroform, 6.7 g/l carbon tetrachloride, 36.7 g/l cyclohexanone, 31.4 g/l dioxane, 12.8 g/l AcOH
Octanol/water PC:
LD_{50}:
40CFR180.133

CAS RN 330-55-2
$C_9H_{10}Cl_2N_2O_2$

Linuron (ACN)
Urea, N'-(3,4-dichlorophenyl)-N-methoxy-N-methyl- (9CI)
Urea, 3-(3,4-dichlorophenyl)-1-methoxy-1-methyl- (8CI)
Afalon
Afalon inuron
Aphalon
Cephalon
Du Pont 326
Dupont Herbicide 326
DuPont Herbicide 326
Garnitan
H-326
Herbicide 326
HOE 2810
Linurex
Linuron (herbicide)
Lorex
Lorox
Lorox linuron weed killer
Methoxydiuron
N-(3,4-Dichlorophenyl)-N'-methoxy-N'-methylurea
N-(3,4-Dichlorophenyl)-N'-methyl-N'-methoxyurea
N'-(3,4-Dichlorophenyl)-N-methoxy-N-methylurea
Premalin
Sarclex
Scarclex
Sinuron
Urea, 1-(3,4-dichlorophenyl)-3-methoxy-3-methyl-
1-(3,4-Dichlorophenyl)3-methoxy-3-methyluree (French)
1-Methoxy-1-methyl-3-(3,4-dichlorophenyl)urea
3-(3,4-Dichloor-fenyl)-1-methoxy-1-methylureum (Dutch)
3-(3,4-Dichlor-phenyl)-1-methoxy-1-methyl-harnstoff (German)
3-(3,4-Dichloro-fenil)-1-metossi-1-metil-urea (Italian)
3-(3,4-Dichlorophenyl)-1-methoxy-1-methylurea (ACN)

Cl NH CH3 N O CH3 Cl O

Merck Index No. 5387
MP: 93 - 94°
BP:
Density:
Solubility: 81 mg/l water 25°, 500 g/kg acetone, 150 g/kg benzene, 150 g/kg ethanol, 130 g/kg xylene, 15 g/kg *n*-heptane. Readily soluble in dimethylformamide, chloroform, diethyl ether. Moderately soluble in aromatic hydrocarbons. Sparingly soluble in aliphatic hydrocarbons.
Octanol/water PC: 1010
LD_{50}: rat orl 1146 mg/kg; rat ihl LC_{50} 48 mg/m^3/4H; rat skn >2500 mg/kg; mus orl 2400 mg/kg; dog orl 500 mg/kg; rbt orl 2250 mg/kg; rbt skn >5000 mg/kg; ckn orl 3765 mg/kg.
40CFR180.184

CAS RN 124-58-3
CH_5AsO_3

MAA
Arsonic acid, methyl- (9CI)
Methanearsonic acid (8CI)
Methane arsonic acid
Methylarsinic acid
Methylarsonic acid
Monomethylarsinic acid

OH
CH3—As=O
OH

Merck Index No. 5864
MP: 140°, 154 - 155°, 160 - 161°
BP: 179 - 181^{o720}, 89^{o41}
Density: 2.66^{23}
Solubility:
Octanol/water PC:
LD_{50}: rat orl 961 mg/kg; mus scu 794 mg/kg.
40CFR180.289

CAS RN 12057-74-8
Mg_3P_2

Mg=P–Mg–P=Mg

Magnesium phosphide (Mg_3P_2) (8CI9CI)
Fosfuri di magnesio (Italian)
Magnesiumfosfide (Dutch)
Phosphure de magnesium (French)

Merck Index No.
MP:
BP:
Density:
Solubility:
Octanol/water PC:
$LD_{50:}$
40CFR180.375

CAS RN 121-75-5
$C_{10}H_{19}O_6PS_2$

Malathion
Butanedioic acid, [(dimethoxyphosphinothioyl)thio]-, diethyl ester (9CI)
Succinic acid, mercapto-, diethyl ester, S-ester with O,O-dimethyl phosphorodithioate
[(Dimethoxyphosphinothioyl)thio]butanedioic acid diethyl ester
[(Dimethoxyphosphinothioyl)thio]butanedioic acid, diethyl ester
American Cyanamid 4,049
American Cyanamide 4049
AC 26691
Calmathion
Carbethoxy malathion
Carbetox
Carbofos
Carbophos
Chemathion
Compound 4049
Cythion
Detmol MA
Dicarbethoxyethyl-O,O-dimethyldithiophosphate
Dicarboethoxyethyl O,O-dimethyl phosphorodithioate
Diethyl mercaptosuccinate O,O-dimethyl dithiophosphate ester
Diethyl mercaptosuccinate, O,O-dimethyl dithiophosphate, S-ester
Diethyl mercaptosuccinate, O,O-dimethyl phosphorodithioate
Diethyl mercaptosuccinate, O,O-dimethyl thiophosphate
Diethyl mercaptosuccinate, S-ester with O,O-dimethyl phosphorodithioate
Diethyl mercaptosuccinic acid, S-ester of O,O-dimethyl phosphorodithioate
Diethyl mercaptosuccinic acid, S-ester with O,O-dimethyl phosphorodithioate
Diethyl [(dimethoxyphosphinothioyl)thio]butanedioate
Diethyl 2-(dimethoxyphosphinothioylthio)succinate
Dithiophosphate de O,o-dimethyle et de S-(1,2-dicarboethoxyethyle) (French)
Dithiophosphate de O,O-dimethyle et de S-(1,2-dicarboethoxyethyle) (French)
Dithiophosphoric acid, S-(1,2-dicarboxyethyl)-O,O-dimethyl ester
Emmatos
Emmatos extra
Ethiolacar
Etiol
Experimental insecticide 4049
ENT 17,034
EPN
Fog 3
Formal
Fosfothion

Merck Index No. 5582
MP:
BP:
Density:
Solubility:
Octanol/water PC:
LD_{50}: man orl LDLo 471 mg/kg; wmn orl LDLo 246 mg/kg; rat orl 290 mg/kg; rat orl 1000 mg/kg; rat orl 1375 mg/kg; rat orl 1375-2800 mg/kg; rat ihl LC_{50} 43.79 mg/m^3/4H; rat skn >4444 mg/kg; rat ipr 250 mg/kg; rat scu 1000 mg/kg; rat ivn 50 mg/kg; rat unr 450 mg/kg; mus orl 775-3320 mg/kg; mus orl 190 mg/kg; mus skn 2330 mg/kg; mus ipr 193 mg/kg; mus scu 221 mg/kg; mus ivn 184 mg/kg; mus unr 375 mg/kg; dog ipr 1857 mg/kg; cat ihl LCLo 10 mg/m^3/4H; cat iat LDLo 1.82 mg/kg; rbt orl LDLo 1200 mg/kg; rbt skn 4100 mg/kg; gpg orl 570 mg/kg; gpg skn 6700 mg/kg; gpg ipr 550 mg/kg; ham ipr 2400 mg/kg; ckn orl 600 mg/kg; dck orl 1.485 mg/kg; dom orl 500 mg/kg; ctl orl 53 mg/kg; ctl unr 53 mg/kg; mam unr 500 mg/kg; bwd orl 400mg/kg.
40CFR180.119; 40CFR180.111

CAS RN 123-33-1
$C_4H_4N_2O_2$

Maleic hydrazide
3,6-Pyridazinedione, 1,2-dihydro-
Antergon
Antyrost
Burtolin
Chemform
De-Cut
De-Sprout
Drexel-Super P
ENT 18,870
Fair ps
Fair PS
Fair 30
KMH
Maintain 3
Malazide
Maleic acid, cyclic hydrazide
Maleic acid, hydrazide
Maleic hydrazine
Malein 30
Malzid
MAH
MG-
MH
MH 30
MH 36 bayer
MH-40
N,N-Maleoylhydrazine
Regulox
Regulox W
Regulox 36
Regulox 50 W
Retard
Royal mh-30
Royal Slo-Gro
Slo-Gro
Sprout-Stop
Sprout/off
Stuntman
Sucker-Stuff

Merck Index No.5587
MP: 292 - 298°, 260°(dec), >300°
BP:
Density: 1.60^{25}
Solubility: 6 g/kg/water 25°, 24 g/kg/ dimethylformamide, <1 g/kg ethanol, acetone, xylene.
Octanol/water PC: 0.011 (pH 7)
LD_{50}: rat orl 3800 mg/kg; rat ihl LC_{50} >20000 mg/m^3; rbt skn >4000 mg/kg; qal orl >10000 mg/kg; dck orl >10000 mg/kg.
40CFR180.175

CAS RN 12427-38-2
$C_4H_6MnN_2S_4$

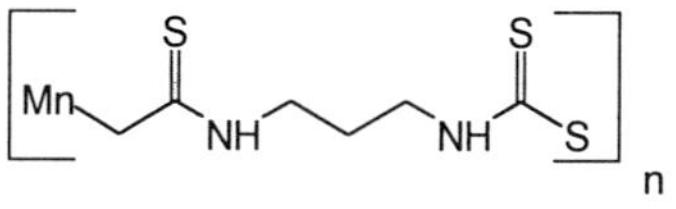

Maneb (ACN)
Manganese, [[1,2-ethanediylbis[carbamodithioato]](2-)]- (9CI)
Manganese, [ethylenebis[dithiocarbamato]]- (VAN8CI)
component of Dithane S-31
m-Diphar
[[1,2-Ethanediylbis[carbamodithioato]](2-)]manganese
[1,2-Ethanediylbis[carbamodithioic] acid], manganese complex
[1,2-Ethanediylbis[maneb]], manganese(2+) salt (1:1)
[1,2-Ethanediyl[biscarbamodithioic] acid], manganese(2+) salt (1:1)
Amangan
Carbamic acid, ethylenebis[dithio-, manganese salt
Carbamodithioic acid, 1,2-ethanediylbis-, manganese(2+) salt (1:1)
Chem neb
Chloroble M
CR 3029
Dithane M 22
Dithane M 22 special
Dithane M-45
Dithane S-31
Ethylene bis[dithiocarbamic acid], manganese salt
Ethylenebis(dithiocarbamato), manganese
Ethylenebis(dithiocarbamic acid), manganese salt
Ethylenebis(dithiocarbamic acid), manganous salt
Ethylenebisdithiocarbamate, manganese
Ethylenebis[dithiocarbamate], manganese salt
Ethylenebis[dithiocarbamic acid], manganese salt
Ethylenebis[dithiocarbamic acid], manganous salt
EBDC, manganese salt
ENT 14,875
F 10 (VAN)
F 10 (pesticide)
Griffin Manex
Kypman 80
Labilite
Lonocol M
M-Diphar
249
Manam
Maneb 80
Maneb-R
Maneba

Merck Index No. 5603
MP: 192 - 204°
BP:
Density:
Solubility: Insoluble in water and common solvents.
Octanol/water PC:
LD_{50}: rat orl 3000 mg/kg; rat unr 3000 mg/kg; mus orl 2600 mg/kg; gpg orl LDLo 6400 mg/kg; qal unr 467 mg/kg; mam unr 5000 mg/kg.
40CFR180.110

CAS RN 15339-36-3
$C_6H_{12}MnN_2S_4$

Manam
Manganous dimethyldithiocarbamate (ACN)
Manganese, bis(dimethylcarbamodithioato-S,S')-, (T-4)-
Manganese, bis(dimethyldithiocarbamato)- (8CI)
Carbamic acid, dimethyldithio-, manganese(2+) salt
Carbamodithioic acid, dimethyl-, manganese salt
Carbamodithioic acid, dimethyl-, manganese(2+) salt
Dimethyldithiocarbamic acid, manganese salt
Manganese dimethyldithiocarbamate
Manganese, bis(dimethylcarbamodithioato-S,S')-
Tennam

$N(CH_3)_2$ S S–Mn–S S $(CH_3)_2N$

Merck Index No.
MP:
BP:
Density:
Solubility:
Octanol/water PC:
LD_{50}: mus ivn 32 mg/kg.
40CFR180.

CAS RN 94-74-6
$C_9H_9ClO_3$

OCH_2COOH Cl CH_3

MCPA
Acetic acid, (4-chloro-2-methylphenoxy)- (9CI)
Acetic acid, [(4-chloro-*o*-tolyl)oxy]- (8CI)
(2-Methyl-4-chlorophenoxy)acetic acid (ACN)
(4-Chloro-*o*-cresoxy)acetic acid
(4-Chloro-*o*-toloxy)acetic acid
(4-Chloro-*o*-tolyloxy)acetic acid
(4-Chloro-2-methylphenoxy)acetic acid
[(4-Chloro-*o*-tolyl)oxy]acetic acid
Acetic acid, 2-methyl-4-chlorophenoxy-
Agroxohe
Agroxon
Agroxone
Anicon kombi
Anicon M
B-Selektonon M
Bh mcpa
Bordermaster
BH MCPA
Chiptox
Chlorotoloxyacetic acid (and salts)
Chwastox
Cornox-M
Ded-Weed
Dicopur-M
Dikotes
Dikotex
Emcepan
Empal
Hedapur M 52
Hedarex M
Hedonal
Hedonal M
Herbicide m
Herbicide M
Hormotuho
Hornotuho
Kilsem
Krezone
Leuna M
Linormone

Merck Index No. 5645
MP: 118 - 119°, 120°
BP:
Density:
Solubility: 825 mg/l water 25°, 1530 g/l ethanol, 770 g/l diethyl ether, 62 g/l toluene, 49 g/l xylene, 5 g/l heptane.
Octanol/water PC:
LD_{50}: man orl LDLo 814 mg/kg; rat orl 700 mg/kg; rat ihl LC_{50} 1370 mg/m^3/4H; rat skn >1000 mg/kg; mus orl 439 mg/kg; mus scu LDLo 28 mg/kg; mus ivn 28 mg/kg; rbt skn >2000 mg/kg; gpg orl 700 mg/kg; qal unr 377 mg/kg.
40CFR180.

CAS RN 94-81-5
$C_{11}H_{13}ClO_3$

MCPB
Butanoic acid, 4-(4-chloro-2-methylphenoxy)-
Butyric acid, 4-[(4-chloro-*o*-tolyl)oxy]-
γ-(2-Methyl-4-chlorophenoxy)butyric acid
γ-(4-Chloro-2-methylphenoxy)butyric acid
γ-MCPB
(2-Methyl-4-chlorophenoxy)butyric acid
(4-Chloro-2-methylphenoxy)butyric acid
[(4-Chloro-*o*-tolyl)oxy]butyric acid
Bexane
Bexone
Can-Trol
Legumex
MCP-Butyric
MCPB
PDQ
Thistrol
Thitrol
Trifolex
Tritrol
2,4-MCPB
2M 4KhM
2M-4XM
4-(MCB)
4-(2-Methyl-4-chlorophenoxy)butyric acid (ACN)
4-(4-Chlor-2-methyl-phenoxy)-buttersaeure (German)
4-(4-Chlor-2-methylphenoxy)-buttersaeure (German)
4MCPB

$OCH_2CH_2CH_2COOH$, Cl, CH_3

Merck Index No.
MP: 100°
BP:
Density:
Solubility: 44 mg/l water 25°, >200 g/l acetone, 160 g/l dichloromethane, 150 g/l ethanol, 65 g/l hexane, 8 g/l toluene.
Octanol/water PC:
LD_{50}: rat orl 680 mg/kg; mus orl 800 mg/kg.
40CFR180.318

CAS RN 53780-34-0
$C_{11}H_{13}F_3N_2O_3S$

Mefluidide (ACN)
Acetanilide, 2,4-dimethyl-5-[(trifluoromethyl)sulfonamido]-
Acetamide, N-[2,4-dimethyl-5-[[(trifluoromethyl)sulfonyl]amino]phenyl]- (9CI)
Embark
Embark plant growth regulator
Methafluoridamid
MBR 12325
Vistar
VEL 3793
VEL 3973
5'-(Trifluoromethanesulfonamido)acet-2',4'-xylidide

CH_3CONH, $NHSO_2CF_3$, CH_3, CH_3

MP: 183 - 185°
BP:
Density:
Solubility: 180 mg/l water 23°, 350 g/l acetone, 310 g/l methanol, 64 g/l acetonitrile, 50 g/l ethyl acetate, 17 g/l n-octanol, 3.9 g/l diethyl ether, 2.1 g/l dichloromethane, 0.31 g/l benzene, 0.12 g/l xylene.
Octanol/water PC:
LD_{50}: rat orl 4000 mg/kg; mus orl >1920 mg/kg; rbt skn >4000 mg/kg; dck orl >4640 mg/kg.
40CFR180.386

Merck Index No. 5684

CAS RN 24307-26-4
$C_7H_{16}N.Cl$

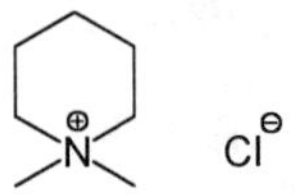

Mepiquat chloride
Piperidinium, 1,1-dimethyl-, chloride (8CI9CI)
Bas-08300W
BAS 083W
BAS 08301W
N,N-Dimethylpiperidinium chloride (ACN)
Pix
PIX
1,1-Dimethylpiperidinium chloride

Merck Index No. 5746
MP: 223°
BP:
Density:
Solubility: > 1 kg/kg/water 20°, 162 g/kg ethanol, <1 g/kg/acetone, 10.5 g/kg chloroform. Sparingly soluble in benzene, ethyl acetate and diethyl ether.
Octanol/water PC: 0.0015 (pH 7)
LD_{50}: rat orl 464 mg/kg; rat ihl LC_{50} >3900 mg/m^3; rat skn >7800 mg/kg; mus orl 780 mg/kg.
40CFR180.384

CAS RN 149-30-4
$C_7H_5NS_2$

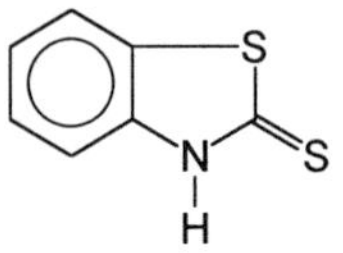

Mercaptobenzothiazole (VAN)
2(3H)-Benzothiazolethione
2-Benzothiazolethiol
Accel M
Accelerator M
Benzothiazole, mercapto-
Benzothiazole, 2-mercapto-
Benzothiazolethiol
Benzothiazoline-2-thione
Captax
Dermacid
Ekagom G
Kaptaks
Kaptax
Mebetizole
Mebithizol
Mercaptobenzenethiazole
Mercaptobenzothiazol
Mercaptobenzthiazole
Mertax
MBT
Nuodeb 84
Pneumax MBT
Rotax
Royal MBT
Soxinol M
Sulfadene
Thiotax
USAF gy-3
Usaf xr-29
Vulkacit mercapto C
Vulkacit M
Vulkacit Mercapto
Vulkacit Mercapto/C
2(3H)-Benzothiazolinethione
2-Benzothiazolinethione
2-Mercaptobenzothiazole (ACN)
2-Mercaptobenzthiazole
2-MBT

Merck Index No. 5759
MP: 180.2 - 181.7°; 170 - 175° (technical)
BP:
Density: 1.42
Solubility: Insoluble in water. 20 g/l ethanol, 10 g/l diethyl ether, 100 g/l acetone, 10 g/l benzene, < 2 g/l carbon tetrachloride, < 5 g/l naphtha.
Octanol/water PC:
LD_{50}: rat orl 100 mg/kg; rat ihl LC_{50} >1270 mg/m^3; rat ipr 300 mg/kg; mus orl 1158 mg/kg; mus ipr 100 mg/kg; rbt skn >7940 mg/kg.
40CFR180.160

CAS RN 150-50-5
$C_{12}H_{27}PS_3$

Merphos
Phosphorotrithious acid, tributyl ester
Butyl phosphorotrithioite ($(BuS)_3P$)
Chemagro B-1776
Deleaf defoliant
Easy off-D
Folex
Folex/Def
Phosphorotrithious acid, S,S,S-tributyl ester
S,S,S-Tributyl phosphorotrithioite
S,S,S-Tributyl trithiophosphite
Tributyl phosphorotrithioate
Tributyl phosphorotrithioite (ACN)
Tributyl trithiophosphite
Tributylphosphorotrithioite

Merck Index No.
MP: 100 - 101°
BP: $137^{\circ 0.7}$, 145 - $148^{\circ 1}$, 174 - $180^{\circ 15}$
Density: 1.0125^{20}, 1.042^{25}
Solubility: Sparingly soluble in water. Soluble in most organic solvents.
Octanol/water PC:
LD_{50}: rat orl 910 mg/kg; rat skn 615 mg/kg; rat ipr 70 mg/kg; rat unr 350 mg/kg; mus orl 635 mg/kg; mus ipr 1400 mg/kg; rbt orl 170 mg/kg; rbt skn >4600 mg/kg; gpg orl 850 mg/kg.
40CFR180.186

CAS RN 108-62-3
$C_8H_{16}O_4$

Metaldehyde (ACN)
1,3,5,7-Tetroxocane, 2,4,6,8-tetramethyl-
Acetaldehyde, tetramer
Antimilace
Ariotox
Corry's slug death
Metacetaldehyde
META
Namekil
Slug-Tox

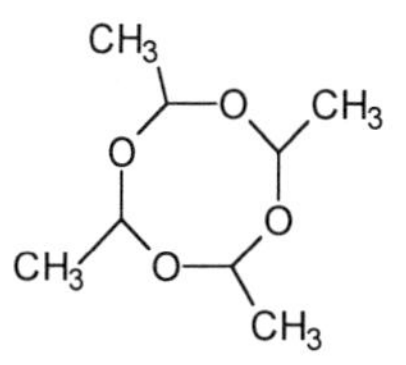

Merck Index No. 5827
MP: 246° (sublimes 112°)
BP:
Density:
Solubility: Insoluble in water. Soluble in benzene and chloroform. Sparingly soluble in ethanol and diethyl ether.
Octanol/water PC:
LD_{50}: chd orl LDLo 100 mg/kg; rat orl 227 mg/kg; rat ihl LC_{50} 203 mg/m^3/4H; rat skn 2275 mg/kg; mus orl 200 mg/kg; mus ihl LC_{50} 348 mg/m^3/2H; dog orl 600 mg/kg; cat orl 207 mg/kg; rbt orl 290 mg/kg; gpg orl 175 mg/kg; mam unr 600 mg/kg.
40CFR185.4025

CAS RN 57837-19-1
$C_{15}H_{21}NO_4$

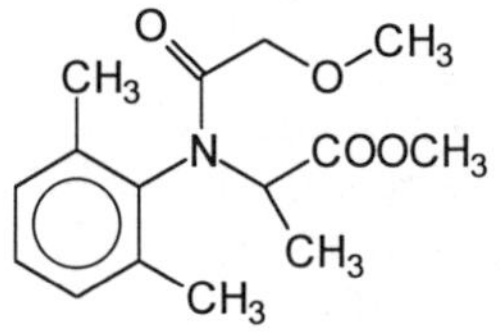

Metalaxyl
Methyl N-(2-methoxyacetyl)-N-(2,6-dimethylphenyl)alaninate
N-(2,6-Dimethylphenyl)-N-(methoxyacetyl)alanine, methyl ester
Methyl N-(2-methoxyacetyl)-N-(2,6-xylyl)alaninate
Ridomil

Merck Index No. 5826
MP: 71 - 72°
BP:
Density:
Solubility: 7.1 g/l water 20°. Readily soluble in most organic solvents.
Octanol/water PC:
LD_{50}: rat orl 566 mg/kg; rat skn >3100 mg/kg.
40CFR180.408

CAS RN 10265-92-6
$C_2H_8NO_2PS$

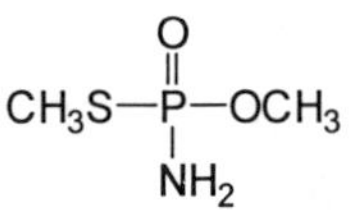

Methamidophos (ACN)
Phosphoramidothioic acid, O,S-dimethyl ester
Acephate-Met
Amidophos
Bayer 5546
Bayer 71628
BAY 71628
Chevron ortho 9006
Chevron 9006
ENT 27,396
Hamidop
Metamidophos
Monitor
O,S-Dimethyl ester amide of amidothioate
O,S-Dimethyl phosphoramidothioate (ACN)
O,S-Dimethyl phosphoramidothiolate
Ortho 9006
RE 9006
RE9006
SRA 5172
Tamaron
Thiophosphorsaeure-O,S-dimethylesteramid(German)

Merck Index No. 5858
MP: 46.1°, 54°
BP:
Density: 1.31^{20}
Solubility: >2 kg/l water 20°, 1400 g/l isopropanol, <100 g/l benzene, <100 g/l xylene, <25 g/l dichloromethane, <10 g/l kerosene, < 10 g/l hexane.
Octanol/water PC:
LD_{50}: man orl TDLo 257 mg/kg; wmn orl TDLo 360 mg/kg; rat orl 7.5 mg/kg; rat skn 50 mg/kg; rat ipr 15 mg/kg; rat ivn 10 mg/kg; mus orl 14 mg/kg; mus ipr 5.3 mg/kg; mus unr 27 mg/kg; rbt orl 10 mg/kg; rbt skn 118 mg/kg; gpg orl 30 mg/kg; gpg scu LDLo 10 mg/kg; ckn orl 25 mg/kg; qal orl 57.5 mg/kg; brd orl 8 mg/kg.
40CFR180.315

CAS RN 20354-26-1
$C_9H_6Cl_2N_2O_3$

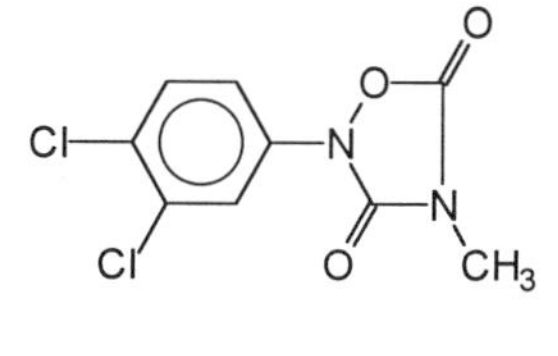

Methazole (ACN)
1,2,4-Oxadiazolidine-3,5-dione, 2-(3,4-dichlorophenyl)-4-methyl-
1,2,4-Oxadiazolidin-3,5-dione, 2-(3,4-dichlorophenyl)-4-methyl-
2-(3,4-Dichlorophenyl)-1,2,4-oxadiazoline-4-methyl-3,5-dione
2-(3,4-Dichlorophenyl)-4-methyl-1,2,4-oxadiazolidine-2,5-dione
2-(3,4-Dichlorophenyl)-4-methyl-1,2,4-oxadiazolidine-3,5-dione (ACN)
Bioxone
Chlormethazole
Metazol (VAN)
Metazole (VAN)
Oxydiazol
Paxilon
Probe
Probe 75
Tunic
VC 438
VCS 438

Merck Index No. 5876
MP: 123 - 124°
BP:
Density: 1.24^{25}
Solubility: 1.5 mg/l water 25°, 323 g/l dimethylformamide, 255 g/l dichloromethane, 171 g/l cyclohexanone, 40 g/l acetone, 55 g/l xylene, 6.5 g/l methanol.
Octanol/water PC: 386 ± 37 (25°)
LD_{50}: rat orl 777 mg/kg; rat ihl LC_{50} >200 mg/m^3/4H; rat skn 10200 mg/kg; mus ipr 600 mg/kg; rbt skn 12500 mg/kg.
40CFR180.357

CAS RN 40596-69-8
$C_{19}H_{34}O_3$

Methoprene (ACN)
Altosid
Altosid IGR
Altosid SR 10
Altoside
ENT 70460
2,4-Dodecadienoic acid, 11-methoxy-3,7,11-trimethyl-, 1-methylethyl ester, (E,E)- (9CI)
Isopropyl (E,E)-11-methoxy-3,7,11-trimethyl-2,4-dodecadienoate (ACN)
Isopropyl 11-methoxy-3,7,11-trimethyl-2,4-dodecadienoate
Isopropyl 11-methoxy-3,7,11-trimethyl-2,4-dodecadienoate, (E,E)
Isopropyl(E,E)-11-methoxy-3,7,11-trimethyl-2,4-dodecadienoate
Isopropyl(2E,4E)-11-methoxy-3,7,11-trimethyl-2,4-dodecadienoate
Kabat
Manta
ZR 515
2,4-Dodecadienoic acid, 11-methoxy-3,7,11-trimethyl-, (E,E), isopropyl ester

Merck Index No. 5906
MP:
BP: $100^{\circ 0.05}$
Density: 0.925^{20}
Solubility: 1.4 mg/l water 25°. Miscible with all organic solvents.
Octanol/water PC:
LD_{50}: rat orl 50 mg/kg; rat ihl LC_{50} >210000 mg/m^3; dog orl 5000 mg/kg; rbt orl 3038 mg/kg; rbt skn 3000 mg/kg; gpg ihl >210 000 mg/m^3.
40CFR180.359, 40CFR180.1033

CAS RN 950-37-8
$C_6H_{11}N_2O_4PS_3$

Methidathion (ACN)
Phosphorodithioic acid,
S-[(5-methoxy-2-oxo-1,3,4-thiadiazol-3(2H)-yl)methyl]
O,O-dimethyl ester
Phosphorodithioic acid, O,O-dimethyl ester, S-ester with 4-(mercaptomethyl)-2-methoxy- Δ^2-1,3,4-thiadiazolin-5-one
DMTP
ENT 27193
Geigy GS 13005
Geigy 13005
GS 13005
Medathion
O,O-dimethyl
S-[(5-methoxy-1,3,4-thiadiazol-2-(3H)-on-3-yl)methyl] dithiophosphate
O,O-Dimethyl phosphorodithioate S-ester with 4-(mercapto-methyl)-2-methoxy- Δ^2-1,3,4- thiadiazolin-5-one
O,O-Dimethyl phosphorodithioate, S-ester of 4-(mercapto-methyl)-2-methoxy-Δ^2-1,3,4- thiadiazolin-5-one
O,O-Dimethyl phosphorodithioate, S-ester of 4-(mercapto-methyl)-2-methoxy-Δ^2-1,3,4-thiadiazolin-5-one
O,O- Dimethyl S-(2-methoxy-1,3,4-thiadiazol-5(4H)-onyl-(4)-methyl) phosphorodithioate
O,O-Dimethyl S-(2-methoxy-1,3,4-thiadiazol-5-(4H)-onyl-(4)-methyl)-dithiophosphat (German)
O,O-DimethylS-(2,3-dihydro-5-methoxy-2-oxo-1,3,4-thia-diazol-3-ylmethyl)phosphorodithioate
O,O-Dimethyl S-(5-methoxy-1,3,4-thiadiazolinyl-3-methyl) dithiophosphate
O,O-Dimethyl S-[(2-methoxy-1,3,4 (4H)-thiodiazol-5-on-4-yl)-methyl]-dithiofosfaat (Dutch)
O,O-Dimethyl S-[(2-methoxy-1,3,4-thiadiazolin-5(4H)-on-4-yl)methyl] phosphorodithioate
O,O-Dimethyl S-[2-methoxy-1,3,4-thiadiazol-5(4H)-onyl-(4)-methyl] dithiophosphate
O,O-Dimethyl-S-(4-mercaptomethyl-2-methoxy-Δ^2-1,3,4-thiadiazolin-5- one)phosphorodithioate

Somonil
Supracid (VAN)
Supracide
Surpracide
Ultracid
Ultracid 40
Ultracide

Merck Index No. 5891
MP: 39 - 40°
BP:
Density:
Solubility: 240 mg/l water 20°, 850 g/kg cyclohexanone, 690 g/kg acetone, 600 g/kg xylene, 260 g/kg ethanol, 53 g/kg octanol.
Octanol/water PC:
LD_{50}: man orl TDLo 93 mg/kg; rat orl 20 mg/kg; rat ihl LC_{50} 50 mg/m^3/4H; rat skn 25 mg/kg; mus orl 25 mg/kg; rbt skn 196 mg/kg; mam orl 25 mg/kg; mam unr 25 mg/kg.
40CFR180.298

CAS RN 2032-65-7
$C_{11}H_{15}NO_2S$

Methiocarb
Phenol, 3,5-dimethyl-4-(methylthio)-, methylcarbamate
Carbamic acid, methyl-, 4-(methylthio)-3,5-xylyl ester
B 37344
Bay 9026
Bayer 37344
BAY 37344
BAY 5024
BAY 9026
Carbamic acid, N-methyl-, 4-(methylthio)-3,5-xylyl ester
Draza
DCR 736
Esurol
ENT 25,726
H 321
Mercaptodimethur
Mesurol
Mesurol phenol
Methyl carbamic acid, 4-(methylthio)-3,5-xylyl ester
Metmercapturan
Metmercapturon
OMS-93
SD 9228
3,5-Dimethyl-4-(methylthio)phenol methylcarbamate
3,5-Dimethyl-4-(methylthio)phenyl methylcarbamate
3,5-Dimethyl-4-(methylthio)phenyl N-methylcarbamate
3,5-Xylenol, 4-(methylthio)-, methylcarbamate
4-(Methyl thio)3,5-xylyl methyl carbamate
4-(Methylthio)-3,5-dimethylphenyl methylcarbamate
4-(Methylthio)-3,5-dimethylphenyl N-methylcarbamate
4-(Methylthio)-3,5-xylyl methylcarbamate (ACN)
4-(Methylthio)-3,5-xylyl N-methylcarbamate
4-(Methylthio)3,5-xylyl methylcarbamate
4-Methylmercapto-3,5-dimethylphenyl N-methylcarbamate
4-Methylmercapto-3,5-xylyl methylcarbamate
4-Methylthio-3,5-dimethylphenyl methylcarbamate

CH_3 CH_3S CH_3 O NH CH_3 O

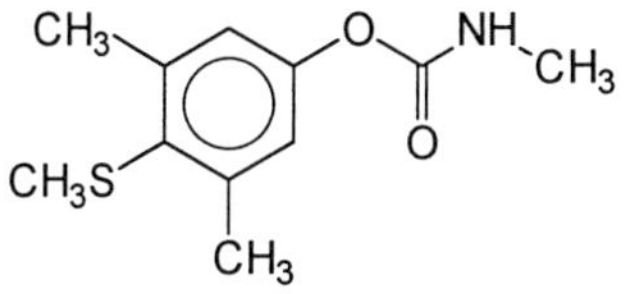

Merck Index No. 5893
MP: 119°, 121.5°
BP:
Density:
Solubility: 27 mg/l water 20°, >200 g/l dichloromethane, 53 g/l isopropanol, 50-100 g/l toluene, 2-5 g/l hexane.
Octanol/water PC:
LD_{50}: rat orl 20 mg/kg; rat skn 350 mg/kg; rat unr 100 mg/kg; mus orl 25.2 mg/kg; mus ipr 16 mg/kg; mus scu 940 mg/kg; gpg orl 40 mg/kg; pgn orl 13 mg/kg; ckn orl 179 mg/kg; qal orl 8.84 mg/kg; dck orl 13 mg/kg; bwd orl 2.4 mg/kg; bwd skn 100 mg/kg.
40CFR180.320

CAS RN 16752-77-5
$C_5H_{10}N_2O_2S$

Methomyl (ACN)
Ethanimidothioic acid, N-[[(methylamino)carbonyl]oxy]-, methyl ester
Acetimidic acid, N-[(methylcarbamoyl)oxy]thio-, methyl ester
Acetimidic acid, thio-N-[(methylcarbamoyl)oxy]-, methyl ester
Acetimidothioic acid, methyl-, N-(methylcarbamoyl) ester
Du Pont Insecticide 1179
Du Pont 1179
Insecticide 1,179
Insecticide 1179
IN 1179
Lannate
Mesomile
Methomyl N-[(methylcarbamoyl)oxy]thioacetimidate
Methomyl S-methyl N-[(methylcarbamoyl)oxy]thioacetimidate
Methomyl, N-[(methylcarbamoyl)oxy]thioacetimidate
Methyl N-[(methylamino)carbonyl]oxymethanimidothioate
Methyl N-[(methylcarbamoyl)oxy]thioacetamidate
Methyl N-[(methylcarbamoyl)oxy]thioacetimidate
Methyl N-[[[(methylamino)carbonyl]oxy]ethanimido]thioate
Methyl O-(methylcarbamoyl)thiolacetohydroxamate
Methyl O-(methylcarbamyl)thiolacetohydroxamate
N-[(Methylcarbamoyl)oxy]thioacetimidic acid methyl ester
Nudrin
S-Methyl N-(methylcarbamoyloxy)thioacetamidate
S-Methyl N-[(methylcarbamoyl)oxy]thioacetimidate (ACN)
S-Methyl N-[[(methylcarbamoyl)oxy]thio]acetimidate
S-Methyl-N-[(methylcarbamoyl)oxy]thioacetimidate
SD 14999
WL 18236
2-Methylthio-propionaldehyd-O-(methylcarbamoyl)-oxim (German)
3-Thiabutan-2-one, O-(methylcarbamoyl)oxime

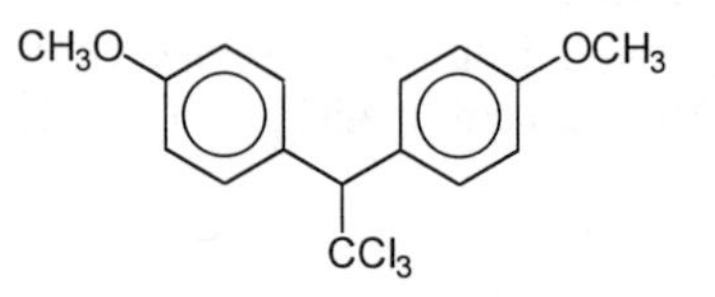

Merck Index No. 5905
MP: 78 - 79°
BP:
Density: 1.2946^{24}
Solubility: 57 g/l water 25°, 1000 g/kg methanol, 730 g/kg acetone, 420 g/kg ethanol, 220 g/kg isopropanol, 30 g/kg toluene. Sparingly soluble in hydrocarbons.
Octanol/water PC:
LD_{50}: rat orl 17 mg/kg; rat ihl LC_{50} 77 ppm; rat skn >1600 mg/kg; rat scu 9 mg/kg; mus orl 10 mg/kg; dog orl LDLo 30 mg/kg; mky orl LDLo 40 mg/kg; rbt skn 58.8 mg/kg; gpg orl LDLo 15 mg/kg; pgn orl 10 mg/kg; ckn orl 28 mg/kg; qal orl 23.7 mg/kg; qal ihl LC_{50} 3680 ppm; dcl orl 15 mg/kg; dck ihl LC_{50} 1890 ppm; bwd orl 10 mg/kg.
40CFR180.253

CAS RN 72-43-5
$C_{16}H_{15}Cl_3O_2$

Methoxychlor
Benzene, 1,1'-(2,2,2-trichloroethylidene)bis[4-methoxy- (9CI)
Ethane, 1,1,1-trichloro-2,2-bis(p-methoxyphenyl)- (8CI)
p,p'-(Dimethoxydiphenyl)trichloroethane
p,p'-DMDT
p,p'-Methoxychlor
Bis(para-methoxyphenyl)trichloroethane
Chemform
Di(*p*-methoxyphenyl)trichloromethyl methane
Dianisyl trichloroethane
Dimethoxy-DDT
Dimethoxy-DT
Dimethoxydiphenyltrichloroethane
DMDTG129
Ethane, 1,1-bis(*p*-methoxyphenyl)- 2,2,2-trichloro-
Ethane, 2,2-bis(*p*-anisyl)-1,1,1-trichloro-
ENT 1,716
ENT-1716
Maralate
Marlate
Methorcide
Methoxcide
Methoxo
Methoxy-DDT
Methoxychlor, technical (ACN)
Metoksychlor (Polish)
Metox
Moxie
NCI-C00497
OMS-466
1,1-Bis(*p*-methoxyphenyl)-2,2,2-trichloroethane
1,1,1-Trichloro-2,2-bis(*p*-anisyl)ethane
1,1,1-Trichloro-2,2-bis(*p*-methoxyphenyl)ethane
1,1,1-Trichloro-2,2-bis(*p*-methoxyphenyl)ethanol
1,1,1-Trichloro-2,2-bis(4-methoxyphenyl)ethane
1,1,1-Trichloro-2,2-di(4-methoxyphenyl)ethane
1,1'-(2,2,2-Trichloroethylidene)bis(4-methoxybenzene)
2,2-Bis(*p*-anisyl)-1,1,1-trichloroethane

CH_3O OCH_3 CCl_3

Merck Index No. 5913
MP: 78 - 78.2°, 89°, 77° (technical 88-90%), 86 - 88°
BP:
Density: 1.41^{25}
Solubility: 0.1 mg/l water 25°, 440 g/kg chloroform, 440 g/kg xylene, 50 g/kg methanol. Soluble in organic solvents.
Octanol/water PC:
LD_{50}: hmn orl LDLo 6430 mg/kg; hmn skn TDLo 2414 mg/kg; rat orl 5000 mg/kg; rat skn >6000 mg/kg; mus orl 1000 mg/kg; rbt orl >6000 mg/kg; ham ipr 500 mg/kg; dck orl >2000 mg/kg.
40CFR180.120

CAS RN 74-83-9
CH_3Br

Methyl bromide (ACN)
Methane, bromo-
Brom-methan (German)
Brom-o-gas
Brom-O-Gas
Bromometano (Italian)
Bromomethane
Bromosol
Bromure de methyle (French)
Bromuro di metile (Italian)
Broommethaan (Dutch)
Brozone
Celfume
Curafume
Dowfume
Dowfume MC-2
Dowfume MC-2 soil fumigant
Dowfume MC-33
Embafume
EDCO
Fumigant-1
Halon 1001
Haltox
Iscobrome
Kayafume
Me Br
Metafume
Meth-O-Gas
Methogas
Methyl bromide, liquid (DOT)
Methylbromid (German)
Metylu bromek (Polish)
Monobromomethane
MB
MBX
MEBR
Pestmaster
Profume
Rotox
Terabol

CH_3—Br

Merck Index No. 5951
MP: -93°
BP: 4.5°
Density: 1.732^0
Solubility: 13.4 g/l water 25°. Readily soluble in most organic solvents, *e.g.*, lower alcohols, ethers, esters, ketones, aromatic hydrocarbons, halogenated hydrocarbons and CS_2.
Octanol/water PC:
LD_{50}: man ihl LCLo 60000 ppm/2H; chd ihl LCLo 1000 mg/m^3/2H; hmn ihl TCLo 35 ppm; hmn skn TDLo 35000 mg/m^3/40M-I; rat orl 214 mg/kg; rat ihl LC_{50} 302 ppm/8H; rat scu 135 mg/kg; mus ihl LC_{50} 1540 mg/m^3/2H; rbt ihl LC_{50} 28900 mg/m^3/30M; gpg ihl LCLo 300 ppm/9H.
40CFR180.123

CAS RN 2631-37-0
$C_{12}H_{17}NO_2$

Promecarb
Phenol, 3-methyl-5-(1-methylethyl)-, methylcarbamate
Carbamic acid, methyl-, *m*-cym-5-yl ester
m-Cym-5-yl methylcarbamate (ACN)
Carbamic acid, N-methyl-, 3-methyl-5-isopropylphenyl ester
Carbamult
ENT 27,300-A
ENT 27300
EP 316
ITC
Minacide
Morton EP-316
OMS 716
Promecarbe
Sch 34615
Schering 34615
SN 34615
SN316
UC 9880
3-Isopropyl-5-methylphenyl methylcarbamate
3-Isopropyl-5-methylphenyl N-methylcarbamate
3-Methyl-5-(1-methylethyl)phenyl methylcarbamate

CH3 O NHCH3 O CH3 CH3

Merck Index No. 7794
MP: 87 - 87.5°
BP: $117^{\circ 0.01}$
Density:
Solubility: 91 mg/l water 25°, 400-600 g/l acetone, dimethylformamide, ethylene chloride, 200-400 g/l cyclohexanol, cyclohexanone, isopropanol, isobutanol, methanol, 100-200 g/l carbon tetrachloride, xylene.
Octanol/water PC: 1545 (pH 4)
LD_{50}: rat orl 35 mg/kg; rat skn 450 mg/kg; rat ipr 27.2 mg/kg; rat ivn 5 mg/kg; rat ims 44 mg/kg; mus orl 16 mg/kg; rbt skn >1000 mg/kg; gpg orl LDLo 25 mg/kg; gpg scu LDLo 25 mg/kg; qal orl 78 mg/kg; dcl orl 3.5 mg/kg; bwd orl 5 mg/kg.
40CFR180. temp (1973).

CAS RN 6597-78-0
$C_9H_8Cl_2O_3$

Methyl 3,6-dichloro-*o*-anisate (ACN)
Benzoic acid, 3,6-dichloro-2-methoxy-, methyl ester
o-Anisic acid, 3,6-dichloro-, methyl ester
Banvel 60CS16
Dicamba methyl ester
Dicamba-methyl
Disugran (ACN)
Methyl 2-methoxy-3,6-dichlorobenzoate
Methyl 3,6-dichloro-2-methoxybenzoate
Racusa
Racuza
2-Methoxy-3,6-dichlorobenzoic acid methyl ester
60CS16

Cl COOCH3 OCH3 Cl

Merck Index No.
MP:
BP: $115^{\circ 2.4}$
Density:
Solubility:
Octanol/water PC:
LD_{50}:
40CFR180. (pending)

CAS RN 7044-96-4
$C_7H_{12}O_4$

Methylene bispropionate (ACN)
Methanediol, dipropanoate
Methylene bispropanoate
Methylenebispropionate
Propanoic acid, methylene ester

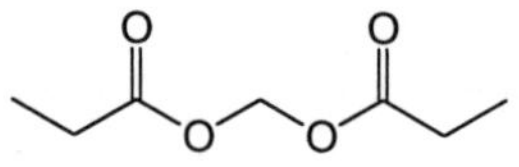

Merck Index No.
MP:
BP: $98^{\circ 1}$, 190 - $192^{\circ 745}$
Density: 1.053^{20}
Solubility:
Octanol/water PC:
LD_{50}:
40CFR180. (pending)

CAS RN 107-31-3
$C_2H_4O_2$

Methyl formate (ACN)(DOT)
Formic acid, methyl ester
Formiate de methyle (French)
Methyl methanoate
Methyle(formiate de) (French)
Methylformiaat (Dutch)
Methylformiat (German)
Metil(formiato di) (Italian)

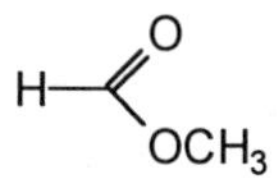

Merck Index No. 5994
MP: approx. -100°
BP: 31.5°
Density: 0.987^{15}
Solubility: Soluble in 3.3 parts water, miscible with alcohol.
Octanol/water PC:
LD_{50}: rat orl 2000 mg/kg; rat ihl LC_{50} >1100 mg/m^3/4H; rat skn >3000 mg/kg; rat ipr 430 mg/kg; mus orl 2098 mg/kg; mus ipr 847 mg/kg; dog orl >10200 mg/kg; rbt skn >10200 mg/kg; qal unr 565 mg/kg.
40CFR185.4300

CAS RN 3060-89-7
$C_9H_{11}BrN_2O_2$

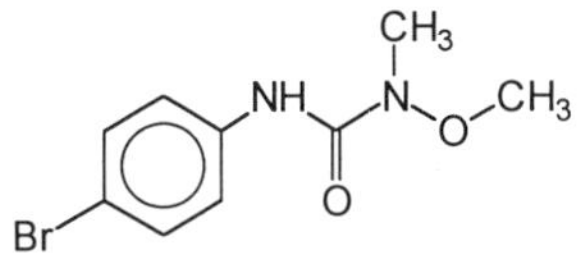

Metobromuron (ACN)
Urea, N'-(4-bromophenyl)-N-methoxy-N-methyl-
N-(4-Bromophenyl)-N'-methyl-N'-methoxy-harnstoff (German)
Urea, 3-(*p*-bromophenyl)-1-methoxy-1-methyl-
C 3126
Metbromuron
N'-(4-Bromophenyl)-N-methoxy-N-methylurea
Patoran
Pattonex
3-(*p*-Bromophenyl)-1-methoxy-1-methylurea (ACN)
3-(*p*-Bromophenyl)-1-methyl-1-methoxyurea

Merck Index No. 6061
MP: 95 - 96°
BP:
Density: 1.60^{20}
Solubility: 330 mg/l water 20°, 500 g/l acetone, 550 g/l dichloromethane, 240 g/l methanol, 100 g/l toluene, 70 g/l octanol, 62.5 g/l chloroform, 2.6 g/l hexane.
Octanol/water PC: 257
LD_{50}: rat orl 2603 mg/kg; rat orl 3875 mg/kg.
40CFR180.250

CAS RN 51218-45-2
$C_{15}H_{22}ClNO_2$

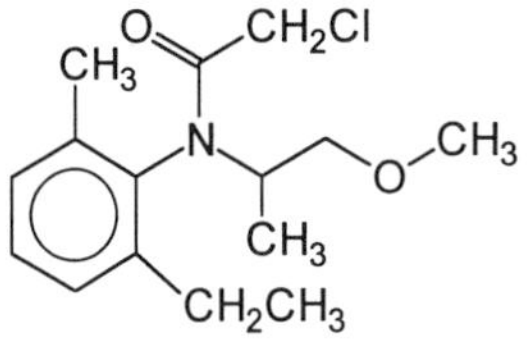

Metolachlor (ACN)
Acetamide,
2-chloro-N-(2-ethyl-6-methylphenyl)-N-(2-methoxy-1-methylethyl)- (9CI)
α-Chlor-6'-aethyl-N-(2-methoxy-1-methylaethyl)-acet-*o*-toluidin (German)
component of Milocep
o-Acetotoluidide, 2-chloro-6'-ethyl-N-(2-methoxy-1-methylethyl)-
Acetamide, 2-chloro-N-(6-ethyl-*o*-tolyl)-N-(2-methoxy-1-methylethyl)-
Bicep
CGA 24705
Dual
Metetilachlor
N-(2'-Methoxy-1'-methylethyl)-2'-ethyl-6'-methyl-2-chloroacetanilide
Primagram
Primextra
2-Aethyl-6-methyl-N-(1-methyl-2-methoxyaethyl)-chloracetanilid (German)
2-Chloro-N-(2-ethyl-6-methylphenyl)-N-(2-methoxy-1-methylethyl)acetamide (ACN)
2-Chloro-6'-ethyl-N-(2-methoxy-1-methylethyl)-*o*-acetotoluidide
2-Chloro-6'-ethyl-N-(2-methoxy-1-methylethyl)-*o*-acetotoluidine
2-Chloro-6'-ethyl-N-(2-methoxy-1-methylethyl)acet-*o*-toluidide

Merck Index No. 6067
MP:
BP: $100^{o0.001}$
Density: 1.12^{20}
Solubility: 530 mg/l water 20°, miscible with benzene, toluene, xylene, hexane, dimethylformamide, ethylene dichloride, cyclohexanone, methanol, octanol, and dichloromethane. Insoluble in ethylene glycol, propylene glycol and petroleum ether.
Octanol/water PC: 2820
LD_{50}: rat orl 2200 mg/kg; ihl rat LC_{50} >1750 mg/m^3/4H; rat skn 3170 mg/kg; rat ipr 620 mg/kg; rat scu >9000 mg/kg; rat unr 3170 mg/kg; mus orl 1150 mg/kg; mus ipr 410 mg/kg; mus scu 2400 mg/kg; rbt skn >10000 mg/kg; dck orl >2510 mg/kg.
40CFR180.368

CAS RN 21087-64-9
$C_8H_{14}N_4OS$

Metribuzin 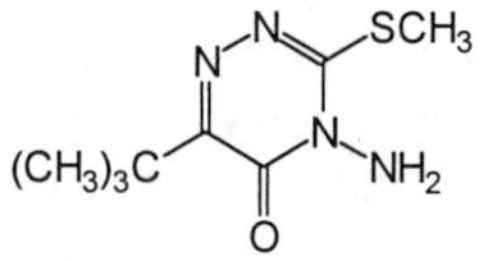
1,2,4-Triazin-5(4H)-one, 4-amino-6-(1,1-dimethylethyl)-3-(methylthio)-
as-Triazin-5(4H)-one, 4-amino-6-*tert*-butyl-3-(methylthio)-
As-Triazin-5(4H)-one, 4-amino-6-*tert*-butyl-3-(methylthio)-
Bayer 6159
Bayer 6159H
Bayer 6443H
Bayer 94337
BAY 6159
BAY 6159 H
BAY 6159H
BAY 61597
BAY 94337
DIC 1468
Lexone
Sencor
Sencoral
Sencorer
Sencorex
1,2,4-Triazin-5-one, 4-amino-6-*tert*-butyl-3-(methylthio)-
4-Amino-6-(1,1-dimethylethyl)-3-(methylthio)-1,2,4-triazin-5(4H)-one (ACN)
4-Amino-6-*tert*-butyl-3-(methylthio)-*as*-triazin-5(4H)-one (ACN)

SCH_3 / $(CH_3)_3C$ / N / N / N / NH_2 / O

Merck Index No. 6076
MP: 125.5 - 126.5°
BP:
Density: 1.31^{20}
Solubility: 1.05 g/l water 20°, 1780 g/kg/dimethylformamide, 1000 g/kg cyclohexanone, 850 g/kg chloroform, 820 g/kg acetone, 450 g/kg methanol, 333 g/kg dichloromethane, 220 g/kg benzene, 150 g/kg n-butanol, 190 g/kg ethanol, 90 g/kg xylene, 50-100 g/kg isopropanol, 2 g/kg hexane.
Octanol/water PC: 40
LD_{50}: rat orl 1100 mg/kg; rat ihl LC_{50} >860 mg/m^3/; rat skn 2000 mg/kg; rat ipr 239 mg/kg; rat scu 814 mg/kg; mus orl 564 mg/kg; mus skn >1000 mg/kg; mus ipr 210 mg/kg; mus scu 367 mg/kg; rbt skn >20000 mg/kg; gpg orl 250 mg/kg; qal unr 168 mg/kg.
CFR: 40CFR180.332

CAS RN 298-01-1 (7786-34-7)
$C_7H_{13}O_6P$

Mevinphos 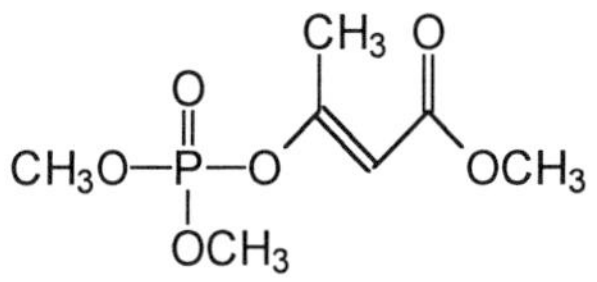
(α-2-Carbomethoxy-1-methylvinyl) dimethyl phosphate
(2-Methoxycarbonyl-1-methyl-vinyl)-dimethyl-fosfaat (Dutch)
(2-Metossicarbonil-1-metil-vinil)-dimetil-fosfato (Italian)
(2-Methoxycarbonyl-1-methyl-vinyl)-dimethyl-phosphat (German)
(*cis*-2-Methoxycarbonyl-1-methylvinyl) dimethyl phosphate
1-(Methoxycarbonyl)-1-propen-2-yl dimethyl phosphate
2-Butenoic acid, 3-[(dimethoxyphosphinyl)oxy]-, methyl ester, (E)
2-Butenoic acid, 3-[(dimethoxyphosphinyl)oxy]-, methyl ester
2-Carbomethoxy-1-methylvinyl dimethyl phosphate, α isomer and related *cis*-Phosdrin
cis-Phosdrin
CMDP
Compound 2046
compounds (ACN)
Crotonic acid, 3-hydroxy-, methyl ester, dimethyl phosphate, (E)
Crotonic acid, 3-hydroxy-, methyl ester, dimethyl phosphate, (E)
Crotonic acid, 3-hydroxy-, methyl ester, dimethyl phosphate
Dimethyl (1-carbomethoxy-1-propen-2-yl) phosphate
Dimethyl (1-methoxycarboxypropen-2-yl) phosphate
Dimethyl 1-methoxycarbonyl-1-propen-2-yl phosphate
Dimethyl phosphate of methyl 3-hydroxy-cis-crotonate
Dimethyl phosphate of 3-hydroxy-N-methyl-cis-crotonamide
Dimethyl (2-methoxycarbonyl-1-methylvinyl) phosphate
Dimethyl methoxycarbonylpropenyl phosphate
Dimethyl phosphate of methyl 3-hydroxycrotonate (α isomer)
ENT 22,374
Fosdrin
Gesfid
Gestid
Meniphos
Menite
Methyl 3-hydroxy-α-crotonate dimethyl phosphate
Mevinfos (Dutch)
Mevinphos
O,O-Dimethyl O-(2-carbomethoxy-1-methylvinyl) phosphate
O,O-Dimethyl O-(2-carbomethoxy)-1-methylvinyl phosphate
O,O-Dimethyl O-(1-methyl-2-carboxyvinyl) phosphate
OS-2046
Phosdrin
Phosfene
Phosphate de dimethyle et de 2-methoxycarbonyl-1 methylvinyle (French)
Phosphoric acid, (1-methoxycarboxypropen-2-yl) dimethyl ester
Phosphoric acid, dimethyl ester, ester with methyl 3-hydroxycrotonate

Merck Index No. 6089
MP: 21° (E isomer), 6.9° (Z isomer)
BP: 99 - 103$^{\circ 0.3}$
Density: 1.24, 1.235^{20} (E isomer), 1.245^{20} (Z isomer)
Solubility: Miscible with water, soluble in most organic solvents, slightly soluble in hydrocarbons.
Octanol/water PC:
LD_{50}: man orl TDL0 0.70 mg/kg/28D-I; rat orl 3000 mg/kg; rat ihl LC_{50} 14 ppm/1H; rat skn 4.2 mg/kg; rat ipr 0.80 mg/kg; rat scu 0.94 mg/kg; mus orl 4 mg/kg; mus skn 12 mg/kg; mus ipr 2 mg/kg; mus scu 1.18 mg/kg; mus ivn 0.68 mg/kg; rbt skn 4.7 mg/kg; pgn orl 4.21 mg/kg; qal orl 23.7 mg/kg; grb ipr 0.45 mg/kg; dck orl 4.6 mg/kg; dcl skn 11 mg/kg; bwd orl 1.4 mg/kg.
CFR: 40CFR180.157

CAS RN 113-48-4
$C_{17}H_{25}NO_2$

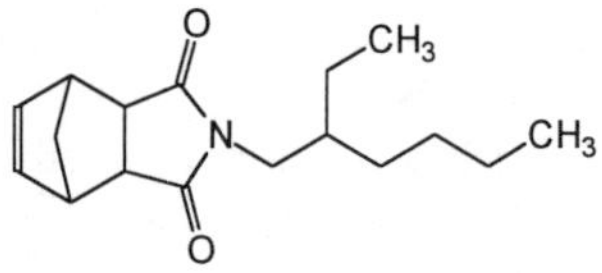

MGK 264
4,7-Methano-1H-isoindole-1,3(2H)-dione, 2-(2-ethylhexyl)-3a,4,7,7a-tetrahydro-
5-Norbornene-2,3-dicarboximide, N-(2-ethylhexyl)- (8CI)
endo Methylenetetrahydrophthalic acid, N-2-ethylhexyl imide
Bicyclo[2.2.1]heptene-2-dicarboxylic acid, 2-ethylhexylimide
ENT 8,184
MGK repellent 264
N-(2-Ethylhexyl)-5-norbornene-2,3-dicarboximide
N-(2-Ethylhexyl)-8,9,10-trinorborn-5-ene-2,3-dicarboximide
N-(2-Ethylhexyl)bicycloheptenedicarboximide (ACN)
N-(2-Ethylhexyl)bicyclo[2.2.1]-5-heptene-2,3-dicarboximide
N-(2-Ethylhexyl)bicyclo[2.2.1]hept-5-ene-2,3-dicarboximide
N-Octylbicycloheptenedicarboximide (ACN)
N-Octylbicyclo[2.2.1]-5-heptene-2,3-dicarboximide
N-2-Ethylhexylbicycloheptenedicarboximide
N-2-Ethylhexylimide endomethylenetetrahydrophthalic acid
Octacide 264
Pyrdone (obsolete)
Pyrodone
Sinepyrin 222
Synergist 264
Van Dyk 264
2-(2-Ethylhexyl)-3a,4,7,7a-tetrahydro-4,7-methano-1
H-isoindole-1,3(2H)-dione

Merck Index No.
MP: <-20°
BP: 156 - 158°
Density: 1.04
Solubility:
Octanol/water PC:
LD_{50}: rat orl 2800 mg/kg; rat skn 470 mg/kg; mus orl 1000 mg/kg; rbt skn 470 mg/kg.
CFR: 40CFR180.367

CAS RN 2212-67-1
$C_9H_{17}NOS$

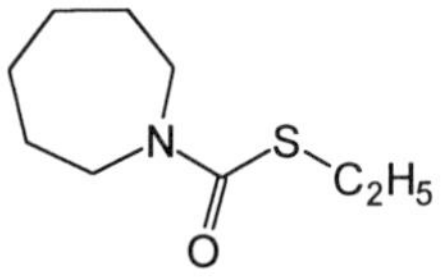

Molinate
1H-Azepine-1-carbothioic acid, hexahydro-, S-ethyl ester
Ethyl 1-hexamethyleneiminecarbothiolate
Felan
Hydram
Jalan
Molmate
Ordram
R-4572
S-Aethyl-N-hexahydro-1H-azepinthiolcarbamat (German)
S-Ethyl hexahydro-1H-azepine-1-carbothioate (ACN)
S-Ethyl hexahydroazepine-1-carbothioate
S-Ethyl N,N-hexamethylenethiocarbamate
S-Ethyl 1-hexamethyleneiminothiocarbamate
Stauffer R-4,572
Yalan

Merck Index No.
MP:
BP: 202^{o10}
Density: 1.063^{20}
Solubility: 880 mg/l water 20°, miscible with acetone, ethanol, methanol, kerosene, 4-methylpentan-2-one, benzene and xylene.
Octanol/water PC: 760
LD_{50}: rat orl 369 mg/kg; rat ihl LC_{50} >200 mg/m^3; rat skn 1167 mg/kg; rat scu 1167 mg/kg; rat unr 657 mg/kg; mus orl 530 mg/kg; mus unr 545 mg/kg; cat ihl LC_{50} >200 mg/m^3; rbt skn 3536 mg/kg; mam unr 720 mg/kg.
CFR: 40CFR180.228

CAS RN 2385-85-5
$C_{10}Cl_{12}$

Mirex
1,3,4-Metheno-1H-cyclobuta[cd]pentalene, 1,1a,2,2,3,3a,4,5,5,5a, 5b,6-dodecachloro-octahydro- (8CI9CI)
Bichlorendo
Cyclopentadiene, hexachloro-, dimer
CG-1283
Decane,perchloropentacyclo-
Dechloran Plus
Dechlorane
Dechlorane plus
Dechlorane plus 515
Dechlorane Plus
Dechlorane Plus 515
Dechlorane 4070
Dechlorane 515
Dechloranedechlorane 515
Dodecachlorooctahydro-1,3,4-metheno-1H-cyclobuta[cd]pentalene (ACN)
Dodecachlorooctahydro-1,3,4-metheno-2H-cyclobuta[c,d]pentalene
Dodecachlorooctahydro-1,3,4-metheno-2H-cyclobuta[cd]pentalene
Dodecachloropentacyclodecane
Dodecachloropentacyclo[3.2.2.$0^{2,6}$,$0^{3,9}$,$0^{5,10}$]decane
Dodecachloropentacyclo[3.2.2.$0^{2,6}$,$0^{3,9}$,$0^{5,10}$]]decane
Dodecachloropentacyclo[3.2.2.$0^{2,6}$,$0^{3,9}$,$0^{7,10}$]decane
Ent 25719
ENT 25,719
Ferriamicide
GC 1283
Hexachlorocyclopentadiene dimer
Hrs l276
HRS 1276
Mirex-R
NCI-C06428
Paramex
Pentacyclodecane, dodecachloro-
Perchlorodihomocubane
Perchloropentacyclodecane
Perchloropentacyclo[5.2.1.0(sup2,6).0(sup3,9).0(sup5,8)]decane
1,1a,-2,2,3,3a,4,5,5,5a,5b,6-Dodecachlorooctahydro-1,3,4-metheno-1H-cyclobuta[cd] pentalene
1,1a,2,2,3,3a,4,5,5,5a,5b,6-Dodecachlorooctahydro-1,3,4-metheno-H-cyclobuta[cd] pentalene

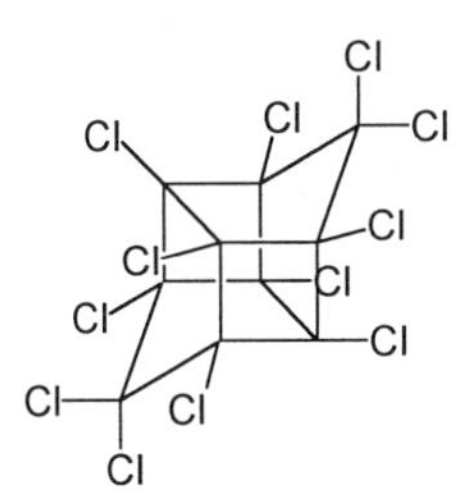

Merck Index No. 6126
MP: 485° (dec)
BP:
Density:
Solubility: Insoluble in water, 153 g/l dioxane, 143 g/l xylene, 122 g/l benzene, 72 g/l carbon tetrachloride, 56 g/l methyl ethyl ketone.
Octanol/water PC:
LD_{50}: rat orl 235 mg/kg; rat skn >2000 mg/kg; rbt skn 800 mg/kg; ham orl 125 mg/kg; dck orl 2400 mg/kg; brd ihl LC_{50} 1400 ppm.
CFR: 40CFR180.251 (removed)

CAS RN 2623-64-5
$C_{11}H_{18}N_2O_3P.I$

4-PPAM
Pyridinium, 1-methyl-4-[[[[methyl(1-methylethoxy)phosphinyl]oxy]imino]methyl]-, iodide Pyridinium, 4-formyl-1-methyl-, iodide, O-(hydroxymethylphosphinyl)oxime isopropyl ester
Phosphonic acid, methyl-, isopropyl ester, O-[(4-pyridylmethylene)amino] deriv., methiodide
Phosphonic acid, methyl-, isopropyl ester, 4-formyl-1-methylpyridinium iodide oxime derivative

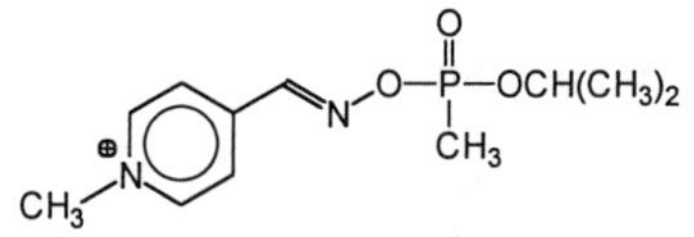

Merck Index No.
MP:
BP:
Density:
Solubility:
Octanol/water PC:
LD_{50}:
CFR: 40CFR180.

CAS RN 300-76-5
$C_4H_7Br_2Cl_2O_4P$

Naled (ACN)
Phosphoric acid, 1,2-dibromo-2,2-dichloroethyl dimethyl ester
o-Dibrom 8E
Alvora
Arthodibrom
Bromchlophos
Bromex (VAN)
Bromex 50
Dibrom
Dibromfos
Dimethyl 1,2-dibromo-2,2-dichloroethyl phosphate
Dimethyl 2,2-dichloro-1,2-dibromoethyl phosphate
ENT 24988
Hibro
Hibrom
O-(1,2-Dibrom-2,2-dichlor-aethyl)-O,O-dimethyl-phosphat (German)
O-(1,2-Dibromo-2,2-dicloro-etil)-O,O-dimetil-fostato (Italian)
O-(1,2-Dibroom-2,2-dichloor-ethyl)-O,O-dimethyl-fosfaat (Dutch)
Ortho 4355
Ortho-Dibrom
Orthodibromo
Phosphate de O,O-dimethle et de O-(1,2- dibromo-2,2-dichlorethyle) (French)
RE 4355
RE4355
1,2-Dibromo-2,2-dichloroethyl dimethyl phosphate (ACN)

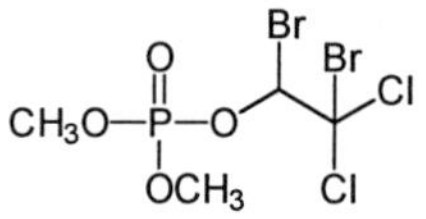

Merck Index No. 6272
MP: 26 - 27.5°, 26.5 - 27.5°
BP: $110^{\circ 0.5}$
Density: 1.96^{20}
Solubility: Insoluble in water. Readily soluble in aromatic and chlorinated solvents. Slightly soluble in aliphatic solvents and mineral oils.
Octanol/water PC:
LD_{50}: rat orl 92 mg/kg; rat skn 800 mg/kg; rat ipr 35 mg/kg; mus orl 222 mg/kg; mus skn 600 mg/kg; mus ipr 84 mg/kg; rbt skn 1100 mg/kg; dck orl 52 mg/kg; mam orl 430 mg/kg.
CFR: 40CFR180.215

CAS RN 86-86-2
$C_{12}H_{11}NO$

1-Naphthaleneacetamide (ACN)
α-Naphthalene acetamide (ACN)
α-Naphthaleneacetamide
α-Naphthaleneacetic acid amide
α-Naphthylacetamide
Acetamide, 2-(1-naphthyl)-,
Amid-Thin
Amid-Thin W
Dirigol N
Frufix
Fruitone
Naphthalene acetamide
Naphthylacetamide
NAAM
NAD
Rootone
Rosetone
Transplantone
1-Naphthylacetamide
1-Naphthylamine, N-acetyl-
2-(1-Naphthyl)acetamide

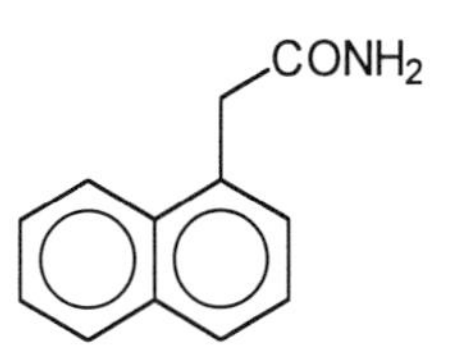

Merck Index No.
MP: 184°
BP:
Density:
Solubility: Sparingly soluble in water. Soluble in acetone, ethanol, isopropanol. Insoluble in kerosene.
Octanol/water PC:
LD_{50}: rat orl 1690 mg/kg; rat orl 6400 mg/kg; rbt skn >2000 mg/kg.
CFR: 40CFR180.309

CAS RN 86-87-3
$C_{12}H_{10}O_2$

1-Naphthaleneacetic acid (ACN)
α-Naphthaleneacetic acid
α-Naphthylacetic acid
α-Naphthyleneacetic acid
α-Naphthylessigsaeure (German)
α-NAA
Alphaspra
Appl-Set
ANA
ANU
Celmone
Fruitofix
Fruitone
Naphyl-1-essigsaeure (German)
Nu-Tone
NAA 800
Parmone
Phyomone
Pimacol-Sol
Plucker
Primacol

Merck Index No. 6290
MP: 132 - 133°
BP:
Density:
Solubility: 420 mg/l water 20°, 55 g/l xylene, 10.6 g/l carbon tetrachloride. Readily soluble in alcohols and acetone. Soluble in diethyl ether and chloroform.
Octanol/water PC:
LD_{50}: rat orl 1000 mg/kg; rat ihl LC_{50} >270 mg/m^3; rat ipr 100 mg/kg; mus orl 743 mg/kg; mis ipr 609 mg/kg; mus scu 733 mg/kg; rbt skn >5000 mg/kg.
CFR: 40CFR180.155

CAS RN 120-23-0
$C_{12}H_{10}O_3$

(2-Naphthyloxy)acetic acid (ACN)
Acetic acid, (2-naphthalenyloxy)- (9CI)
Acetic acid, (2-naphthyloxy)- (8CI)
β-Naphthoxyacetic acid
β-Naphthyloxyacetic acid
(β-Naphthalenyloxy)acetic acid
(β-Naphthoxy)acetic acid
(2-Naphthalenyloxy)acetic acid
(2-Naphthoxy)acetic acid
Betoxon
Bnoa
Gerlach 1396
Glycolic acid 2-naphthyl ether
Naphthoxyacetic acid
NOXA
O-(2-Naphthyl)glycolic acid
Phyomone
WLN: L66J CO1VQ
2-Naphthoxyacctic acid
2-Naphthyloxyacetic acid
2-NOXA

OCH_2COOH

Merck Index No. 6317
MP: 165°
BP:
Density:
Solubility: Sparingly soluble in water, soluble in ethanol, diethyl ether and acetic acid.
Octanol/water PC:
LD_{50}: rat orl 600 mg/kg.
CFR: 40CFR180.148

CAS RN 15299-99-7
$C_{17}H_{21}NO_2$

Napropamide
Propanamide, N,N-diethyl-2-(1-naphthalenyloxy)-
Propionamide, N,N-diethyl-2-(1-naphthyloxy)-
Devrinol
N,N-Diethyl-2-(1-naphthalenyloxy)propanamide (ACN)
N,N-Diethyl-2-(1-naphthalenyloxy)propionamide
N,N-Diethyl-2-(1-naphthyloxy)propionamide
Napromide
Napropamid
R-7475
Waylay
2-(α-Naphthoxy)-N,N-diaethyl-propionsaeureamid (German)
2-(α-Naphthoxy)-N,N-diethylpropionamide
2-(1-Naphthoxy)-N,N-diethylpropionamide
2-(1-Naphthyloxy)-N,N-diethylpropionamide

CH_3
$N(C_2H_5)_2$
O
O

Merck Index No. 6336
MP: 74.8 - 75.5°, 63 - 64°, 69.5° (technical)
BP:
Density:
Solubility: 73 mg/l water 20°, >1000 g/l acetone, >1000 g/l ethanol, 505 g/l xylene, 62 g/l kerosene, 15 g/l hexane.
Octanol/water PC: 2300^{25}
LD_{50}: rat orl 5000 mg/kg; rat par >5000 mg/kg; mus orl >5000 mg/kg; mus ipr >1000 mg/kg; mus scu >1000 mg/kg; rbt skn 4640 mg/kg.
CFR: 40CFR180.328

CAS RN 132-66-1
$C_{18}H_{13}NO_3$

Naptalam
Alanap (VAN)
Benzoic acid, 2-[(1-naphthalenylamino)carbonyl]- (9CI)
Phthalamic acid, N-1-naphthyl- (8CI)
α-Naphthylphthalamic acid
component of Dyanap
Alanap 1
Alanap 10G at
Alanap 3
Alanape
Analape
Aniline, 2-[(1-naphthalenylamino)carbonyl]-
ACP 322
Dyanap
Grelutin
Mor-cran
Mor-Cran
N-(α-Naphthyl)phthalamic acid
N-(1-Naphthyl)phthalamate
N-(1-Naphthyl)phthalamic acid
N-1-Naphthyl-phthalamidsaeure (German)
N-1-Naphthylphthalamic acid
Naphthalam
Naphthylphthalamic acid
Nip-a-thin
NPA (VAN)
Peach-Thin
Phthalamic acid, N-(1-naphthyl)-
PA
SOLO
1-(N-Naphthyl)phthalamic acid
1-Naphthylphthalamic acid
2-[(1-Naphthalenylamino)carbonylbenzoic acid
2-[(1-Naphthalenylamino)carbonyl]benzoic acid
6Q8

O COOH
NH

Merck Index No. 6338
MP: 185°, 203°, 175 - 180° (technical)
BP:
Density: 1.36^{20}, 1.40^{20}
Solubility: 200 mg/l water 20°, 5 g/kg acetone, 39 g/kg dimethylformamide, 43 g/kg dimethylsulfoxide, 4 g/kg methyl ethyl ketone, 2 g/kg isopropanol, 0.1 g/kg carbon tetrachloride, insoluble in benzene, xylene and hexane.
Octanol/water PC:
LD_{50}: rat orl 8200 mg/kg; rat ihl LC_{50} >2070 mg/m^3; rbt skn >2000 mg/kg.
CFR: 40CFR180.297

CAS RN 26896-20-8
$C_{10}H_{20}O_2$

Neodecanoic acid (8CI9CI)
Topper 5E
Wiltz 65
Wiltz-65

COOH

Merck Index No.
MP:
BP:
Density:
Solubility:
Octanol/water PC:
LD_{50}:
CFR: 40CFR180.248

CAS RN 54-11-5
$C_{10}H_{14}N_2$

Nicotine (ACN)(8CI)
Pyridine, 3-(1-methyl-2-pyrrolidinyl)-, (S)- (9CI)
β-Pyridyl-α-N-methylpyrrolidine
β-Pyridyl-N-methylpyrrolidine
(-)-Nicotine
(-)-3-(1-Methyl-2-pyrrolidyl)pyridine
(S)-3-(1-Methyl-2-pyrrolidinyl)pyridine
Black Leaf
Black Leaf 40
Destruxol
Destruxol orchid spray
Emo-Nik
Flux Maag
Fumetobac
L-Nicotine
L-3-(1-Methyl-2-pyrrolidyl)pyridine
Mach-Nic
Niagara p.a. dust
Niagara PA dust
Nic-Sal
Nico-Dust
Nico-Fume
Nicocide
Nicocide Ten dust
Nicotin
Nicotina (Italian)
Nicotine (and Salts)
Nicotine alkaloid
Nicotine, liquid (DOT)
Nikotin (German)
Nikotyna (Polish)
Ortho N-4 dust
Ortho N-5 dust
Pyridine, 3-(tetrahydro-1-methylpyrrol-2-yl)-
Pyridine, 3-(N-methyl-2-pyrrolidyl)-
Pyridine, 3-(1-methyl-2-pyrrolidinyl)-
Pyrrolidine, 1-methyl-2-(3-pyridal)-
Tendust

N
N
CH_3

Merck Index No. 6434
MP: -80°
BP: 246 - $247^{\circ 745}$, 123 - $125^{\circ 17}$
Density: 1.01^{20}
Solubility: Miscible with water below 60°. Readily soluble in most organic solvents.
Octanol/water PC:
LD_{50}: man unr LDLo 0.882 mg/kg; hmn rec TDLo 1.43 mg/kg; rat orl 50 mg/kg; rat skn 140 mg/kg; rat ipr 14.56 mg/kg; rat scu 25 mg/kg; rat ivn 7 mg/kg; rat ims LDLo 15 mg/kg; rat par 34 mg/kg; rat itr 19.30 mg/kg; rat idu LDLo 30 mg/kg; mus orl 3.34 mg/kg; mus ipr 5.9 mg/kg; mus scu 16 mg/kg; mus ivn 0.30 mg/kg; mus ims LDLo 8 mg/kg; dog orl 9.2 mg/kg; dog scu 20 mg/kg; dog ivn 5 mg/kg; dog ims LDLo 7.7 mg/kg; dog par LDLo 5.7 mg/kg; mky ims LDLo 6 mg/kg; cat scu LDLo 20 mg/kg; cat ivn LDLo 1.3 mg/kg; cat ims LDLo 9 mg/kg; rbt skn 50 mg/kg; rbt ipr 14 mg/kg; rbt scu LDLo 5 mg/kg; rbt ivn 6.25 mg/kg; rbt ims LDLo 30 mg/kg; pig ims LDLo >14 mg/kg; gpg ipr LDLo 15 mg/kg; gpg scu LDLo 15 mg/kg; gpg ivn 4.5 mg/kg; gpg ims LDLo 15 mg/kg; pgn scu LDLo 4.58 mg/kg; pgn ims LDLo 9 mg/kg; frg scu LDLo 6 mg/kg; mam ivn 8 mg/kg; hor ims LDLo 8.8 mg/kg; ctl ims LDLo 9 mg/kg; bwd orl 17.8 mg/kg; brd ims LDLo 8 mg/kg.
CFR: 40CFR180.167a

CAS RN 4726-14-1
$C_{13}H_{19}N_3O_6S$

Nitralin
Benzenamine, 4-(methylsulfonyl)-2,6-dinitro-N,N-dipropyl-
Aniline, 4-(methylsulfonyl)-2,6-dinitro-N,N-dipropyl-
Aniline, 2,6-dinitro-N,N-dipropyl-4-(methylsulfonyl)-
N,N-Dipropyl-4-methylsulfonyl-2,6-dinitroaniline
Niralin
Nitraline
Planavin
Planavin 75
Planuin
SD 11831
2,6-Dinitro-4-(methylsulfonyl)-N,N-dipropylaniline
4-(Methylsulfonyl)-2,6-dinitro-N,N-dipropylaniline (ACN)
4-Methylsulfonyl-2,6-dinitro-N,N-dipropylbenzenamine
4-Methylsulphonyl-2,6-dinitro-N,N-dipropylaniline

CH_3 NO_2 N CH_3 CH_3SO_2 NO_2

Merck Index No. 6486
MP: 150 - 151°
BP:
Density:
Solubility: 0.6 ppm in water at 22°, 36 g/100 ml acetone, 33 g/100 ml dimethyl sulfoxide. Poorly soluble in hydrocarbon and aromatic solvents and alcohol.
Octanol/water PC:
LD_{50}: rat orl >2000 mg/kg; rat skn >2000 mg/kg; mus orl >2000 mg/kg; rbt skn >2000 mg/kg.
CFR: 40CFR180.237

CAS RN 1929-82-4
$C_6H_3Cl_4N$

Nitrapyrin (ACN)
Pyridine, 2-chloro-6-(trichloromethyl)-
Dowco 163
Dowco-163
N-Serve
N-Serve nitrogen stabilizer
N-Serve TG
2-Chloro-6-(trichloromethyl)pyridine (ACN)
2-Chloro-6-trichloromethylpyridine
2-Picoline, α,α,α,6-tetrachloro-
6-Chloro-2-(trichloromethyl)pyridine

Cl N CCl_3

Merck Index No. 6490
MP: 62.5 - 62.9°
BP: 136 - 137.5°[11]
Density:
Solubility: 40 mg/kg water 22°, 540 g/kg anhydrous ammonia 20°, 300 g/kg ethanol 20°, 1.98 g/kg acetone 26°, 1.85 g/kg dichloromethane 26°, 1.04 g/kg xylene.
Octanol/water PC:
LD_{50}: rat orl 940 mg/kg; rat unr 3675 mg/kg; mus orl 710 mg/kg; mus unr 3113 mg/kg; rbt orl 713 mg/kg; rbt skn 850 mg/kg; ckn orl 235 mg/kg.
CFR: 40CFR180.350

CAS RN 1836-75-5
$C_{12}H_7Cl_2NO_3$

Nitrofen
Benzene, 2,4-dichloro-1-(4-nitrophenoxy)-
Ether, 2,4-dichlorophenyl *p*-nitrophenyl
FW 925
Niclofen
Nitofen
Nitraphen
Nitrochlor
Nitrofene (French)
Nitrophen (VAN)
Nitrophene (VAN)
NCI-C00420
NIP
Phenyl ether, 2,4-dichloro-4'-nitro-
Tok E-25
Tokkorn
Trizlin
TOK
TOK E 25
2,4-Dichloro-1-(4-nitrophenoxy)benzene
2,4-Dichloro-4'-nitrodiphenyl ether
2,4-Dichlorophenyl *p*-nitrophenyl ether (ACN)
2,4-Dichlorophenyl 4-nitrophenyl ether
4-(2,4-Dichlorophenoxy)nitrobenzene

Merck Index No. 6519
MP: 70 - 71°
BP:
Density:
Solubility: 0.7 - 1.2 ppm water 22°
Octanol/water PC:
LD_{50}: rat orl 740 mg/kg; rat skn 5000 mg/kg; rat unr 3000 mg/kg; mus orl 450 mg/kg; cat orl LDLo 300 mg/kg; cat ihl LCLo 620 mg/m^3/4H; rbt orl 1620 mg/kg.
CFR: 40CFR180.223 (revoked 9/8/85)

CAS RN 2163-79-3 (18530-56-8)
$C_{13}H_{22}N_2O$

Norea (ACN)
Urea, N,N-dimethyl-N'-(octahydro-4,7-methano-1H-inden-5-yl)-
Urea, 3-(hexahydro-4,7-methanoindan-5-yl)-1,1-dimethyl-
ENT-28205
Herban
Hercules 7531
Narea
Nores
Noruron
1-(Tetrahydrodicyclopentadienyl)-3,3-dimethylurea
1,1-Dimethyl-3-tetrahydrodicyclopentadienylurea
3-(Hexahydro-4,7-methanoindan-5-yl)-1,1-dimethylu
rea (ACN)
3-[5-(3a,4,5,6,7,7a-Hexahydro-4,6-methanoindanyl)]
-1,1-dimethylurea

Merck Index No. 6611
MP: 176 - 178°
BP:
Density:
Solubility: Soluble in acetone, cyclohexanone. Slightly soluble in benzene, toluene. Insoluble in water, hexane.
Octanol/water PC:
LD_{50}: rat orl 2000 mg/kg; rat skn 23000 mg/kg; mus orl 4600 mg/kg; dog orl 3700 mg/kg; rbt skn 723mg/kg.
CFR: 40CFR180.260

CAS RN 27314-13-2
$C_{12}H_9ClF_3N_3O$

Norflurazon (ACN)
3(2H)-Pyridazinone, 4-chloro-5-(methylamino)-2-[3-(trifluoromethyl)phenyl]-
3(2H)-Pyridazinone, 4-chloro-5-(methylamino)-2-(α,α,α-trifluoro-*m*- tolyl)-
Evital
H 9789
Monometflurazon
Monometflurazone
Norfluazon
Norflurazone
San 9789 H
Sandoz 9789
Solicam
SAN 9789
Zorial
4-Chloro-5-(methylamino)-2-(α,α,α-trifluoro-*m*-tolyl)-3(2H)-pyridazinone
4-Chloro-5-(methylamino)-2-(α,α,α-trifluoro-*m*-tolyl)-3-(2H)-pyridazinone
4-Chloro-5-(methylamino)-2-(α,α,α-triflurORO-*m*-tolyl)-3(2H)-pyridazinone
4-Chloro-5-(methylamino)-2-[3-(trifluoromethyl)phenyl]-3(2H)-pyridazinone
4-Chloro-5-methylamino-2-(α,α,α-trifluoro-*m*-tolyl)-3-pyridazinone

CH_3 NH N N CF_3 O Cl

Merck Index No. 6618
MP: 183 - 185°, 174 - 180°
BP:
Density:
Solubility: 28 mg/l water 25°, 142 g/l ethanol, 50 g/l acetone, 2.5 g/l xylene. Sparingly soluble in hydrocarbons.
Octanol/water PC: 280 (pH 6.5, 25°)
LD_{50}: rat orl 8000 mg/kg; rbt skn >20000 mg/kg; qal unr >1250 mg/kg; dck unr >1250 mg/kg.
CFR: 40CFR180.356

CAS RN 63284-71-9
$C_{17}H_{12}ClFN_2O$

Nuarimol (ACN)
5-Pyrimidinemethanol, α-(2-chlorophenyl)-α-(4-fluorophenyl)-
α-(2-Chlorophenyl)-α-(4-fluorophenyl)-5-pyrimidinemethanol
EL 228
EL 2289
Trimidal
Triminol
2-Chloro-4'-fluoro-α-(5-pyrimidinyl)benzhydryl alcohol

Cl OH F N N

Merck Index No.
MP: 126 - 127°
BP:
Density:
Solubility: 26 mg/l water 25° pH 7, 170 g/l acetone, 55 g/l methanol, 20 g/l xylene. Readily soluble acetonitrile, bezene, chloroform. Slightly soluble hexane.
Octanol/water PC: 1500 (pH 7)
LD_{50}: rat orl 1250 mg/kg; mus orl 2500 mg/kg; dog orl 500 mg/kg; rbt skn >2000 mg/kg; qal orl 200 mg/kg; brd orl 200 mg/kg.
CFR: 40CFR180. (pending 1983)

CAS RN 26530-20-1
$C_{11}H_{19}NOS$

Octhilinone (ACN)
Pancil-T
3(2H)-Isothiazolone, 2-octyl-
4-Isothiazolin-3-one, 2-octyl-
Kathon
Kathon LP
Kathon LP Preservative
Microbicide M-8
Octyl-4-isothiazol-3-one
RH 893
RH-893
Skane M 8
2-*n*-Octyl-3-isothiazolone
2-*n*-Octyl-4-isothiazolin-3-one
2-Octyl-3(2H)-isothiazolone (ACN)
2-Octyl-3-isothiazolinone
2-Octyl-3-isothiazolone
2-Octyl-3-isothioazolone
2-Octyl-4-isothiazolin-3-one

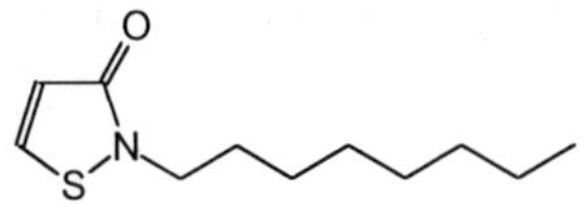

Merck Index No. 6677
MP:
BP: $120^{o0.01}$
Density:
Solubility:
Octanol/water PC:
LD_{50}: rat orl 550 mg/kg; rbt skn 690 mg/kg;
CFR: 40CFR180.366

CAS RN 7778-39-4
AsH_3O_4

Arsenic acid (ACN)(VAN)
Arsenic acid (H_3AsO_4)
o-Arsenic acid
Acide arsenique liquide (French)
Arsenate
Arsenic acid solution (DOT)
Arsenic acid, (solution)
Arsenic oxide
Arsenic pentoxide
Desiccant l-10
Desiccant L-10
Dessicant L-10
Hi-yield desiccant h-10
Orthoarsenic acid (ACN)(VAN)
True arsenic acid
Zotox
Zotox Crabgrass Killer

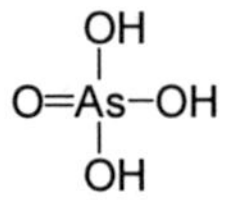

Merck Index No. 821
MP:
BP:
Density:
Solubility: Freely soluble in water, alcohol or glycerol.
Octanol/water PC:
LD_{50}: rat orl 48 mg/kg; dog orl LDLo 10 mg/kg; rbt orl LDLo 5 mg/kg; pgn orl LDLo 100 mg/kg; ckn orl LDLo 125 mg/kg.
CFR: 40CFR180.180

CAS RN 19044-88-3
$C_{12}H_{18}N_4O_6S$

Oryzalin (ACN)
Benzenesulfonamide, 4-(dipropylamino)-3,5-dinitro-
Sulfanilamide, 3,5-dinitro-N4,N4-dipropyl-
Compound 67019
Dirimal
EL 119
EL-119
N^4,N^4-Dipropyl-3,5-dinitrosulfanilamide
Rycelan
Rycelon
Ryzelan
Sulfanilamide, 3,5-dinitro-N^4,N^4-dipropyl-
Surflan
3,5-Dinitro-N^4,N^4-dipropylsulfanilamide (ACN)
3,5-Dinitro-N',N'-dipropylsulfanilamide
4-(Dipropylamino)-3,5-dinitrobenzenesulfonamide

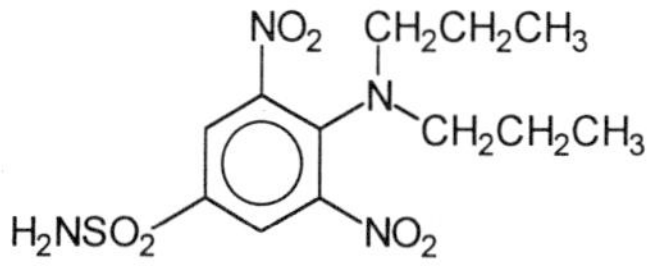

Merck Index No. 6840
MP: 141 - 142°, 137 - 138°, 141 - 142° (technical)
BP:
Density:
Solubility: 2.5 mg/l water 25°, >500 g/l acetone, 500 g/l methyl cellosolve, >150 g/l acetonitrile, 50 g/l methanol, >30 g/l dichloromethane, 4 g/l benzene, 2 g/l xylene.
Octanol/water PC: 5420 (pH 7)
LD_{50}: rat orl 10000 mg/kg; dog orl >1000 mg/kg; cat orl 1000 mg/kg; rbt skn >2000 mg/kg; ckn orl 1000 mg/kg; grb orl >10000 mg/kg.
CFR: 40CFR180.304

CAS RN 50-50-0
$C_{25}H_{28}O_3$

***β*-Estradiol 3-benzoate**
Estra-1,3,5(10)-triene-3,17-diol (17β)-, 3-benzoate
Estradiol, 3-benzoate
β-Estradiol benzoate
β-Oestradiol benzoate
β-Oestradiol 3-benzoate
(17β)-Estra-1,3,5(10)-triene-3,17-diol-3-benzoate
Benovocylin
Benzhormovarine
Benzo-Gynoestryl
Benzoate d'oestradiol (French)
Benzoato de estradiol
Benzoestrofol
Benzofoline
Benzoic acid estradiol
De graafina
Diffollisterol
Difolliculine
Dihydroestrin benzoate
Dimenformon benzoate
Dimenformone
Diogyn B
Eston-B
Estra-1,3,5(10)-triene-3,17β-diol, 3-benzoate
Estra-1,3,5(10)-triene-3,17-diol (17β), 3-benzoate 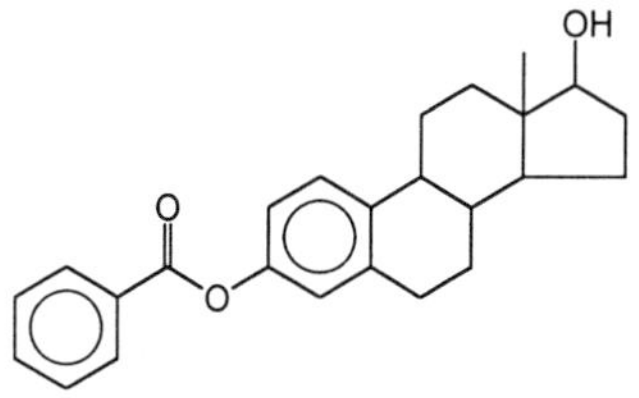
Estra-1,3,5(10)-triene-3,17-diol, (17β)-, 3-benzoate
Estradiol benzoate (VAN)
Estradiol monobenzoate
Estradiol 3-benzoate
Estradiol-17β-3-benzoate
Estradioli benzoas
EBZ
Femestrone

Merck Index No. 3655
MP: 191 - 196°
BP:
Density:
Solubility: Soluble in alcohol, acetone, dioxane. Slightly soluble in ether and vegetable oils.
Octanol/water PC:
LD_{50}:
CFR: 40CFR180. (revoked 5/4/88)

CAS RN 19666-30-9
$C_{15}H_{18}Cl_2N_2O_3$

Oxadiazon (ACN)
1,3,4-Oxadiazol-2(3H)-one, 3-[2,4-dichloro-5-(1-methylethoxy)phenyl]-5-(1,1-dimethylethyl)- (9CI)
Δ^2-1,3,4-Oxadiazolin-5-one, 2-*tert*-Butyl-4-(2,4-dichloro-5-isopropoxyphenyl)
Δ^2-1,3,4-Oxadiazolin-5-one, 2-*tert*-butyl-4-(2,4-dichloro-5-isopropoxyphenyl)-
tert-Butyl-2-(dichloro-2,4-isopropyloxy-5-phenyl)-4-oxo-5 oxadiazoline-1,3,4
Δ^2-1,3,4-Oxadiazolin-5-one, 2-*tert*-butyl-4-(2,4-dichloro-5-isopropyloxyphenyl)-
1,2,4-Oxadiazolidin-3,5-dione, 2-*tert*-butyl-4-(2,4-dichloro-5-isopropoxyphenyl)-
G 315
Oxadiazone
Ronstar
RP 17,623
RP-17623
17623 RP

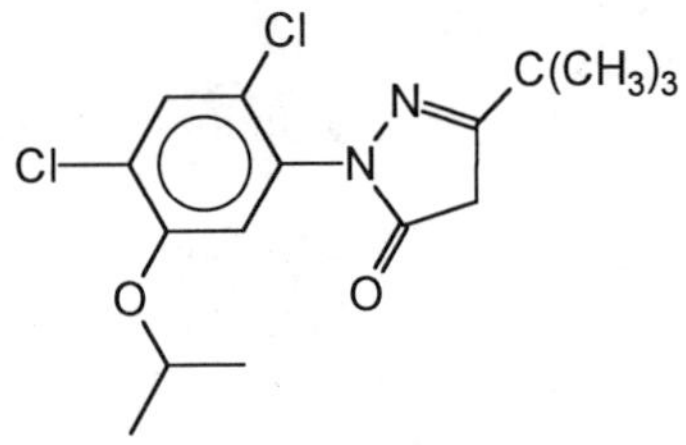

Cl
C(CH3)3
N
Cl
N
O
O

Merck Index No. 6860
MP: 88 - 90°
BP:
Density:
Solubility: 0.7 mg/l water 20°, 100 g/l ethanol 20°, 600 g/l acetone 20°,1000 g/l benzene 20°.
Octanol/water PC:
LD_{50}: rat orl 3500 mg/kg; rat ihl LC_{50} >200 mg/m^3; rat skn 5200 mg/kg; mus orl 12000 mg/kg; rbt skn >2000 mg/kg; qal orl 6000 mg/kg; dck orl 1000 mg/kg.
CFR: 40CFR180.346

CAS RN 23135-22-0
$C_7H_{13}N_3O_3S$

Oxamyl (ACN)(VAN)
Ethanimidothioic acid, 2-(dimethylamino)-N-[[(methylamino)carbonyl]oxy]-2-oxo-, methyl ester
Oxamimidic acid, N',N'-dimethyl-N-[(methylcarbamoyl)oxy]-1-thio-, methyl ester
D 1410
D-1410
Dioxamyl
Du Pont 1410
Dupont 1410
DPX 1410 (VAN)
DPX 1410 L
DPX 1410L
Insecticide-nematicide 1410
Methyl N',N'-dimethyl-N-[(methylcarbamoyl)oxy]-1-thiooxamidate
Methyl N',N'-dimethyl-N-[(methylcarbamoyl)oxy]-1-thiooxamimidate
Methyl 2-(dimethylamino)-N-[[(methylamino)carbonyl]oxy]-2-oxoethanimidothioate
Methyl-N',N'-dimethyl-N-[(methylcarbamoyl)oxy]-1-thiooxamidate
Methyl-N',N'-dimethyl-N-[(methylcarbamoyl)oxy]-1-thiooxamimidate
N,N-Dimethyl-2-methylcarbamoyloxyimino-2-(methylthio)acetamide
Nematicide 1410
Oxamimidic acid, N',N'-dimethyl-N-[(methylcarbamoyl)oxy]-1-(methylthio)-
Oxamyl (pesticide)
S-Methyl 1-(dimethylcarbamoyl)-N-[(methylcarbamoyl)oxy]thioformimidate
Thioxamyl
Vydate
Vydate L oxamyl insecticide/nematocide
2-(Dimethylamino)-1-(methylthio)glyoxal O-(methylcarbamoyl) monoxime

CH_3 CH_3 S N N O NH CH_3 CH_3 O O

Merck Index No. 6873
MP: 108 - 110° (phase change 100-102°)
BP: (dec)
Density: 0.97^{25}
Solubility: 280 g/l water 25°, 1440 g/l methanol, 330 g/l ethanol, 110 g/l isopropanol, 670 g/l acetone, 10 g/l toluene.
Octanol/water PC:
LD_{50}: rat orl 2.5 mg/kg; rat ihl LC_{50} >170 mg/m^3/1H; rat skn LDLo 300 mg/kg; rat ipr 4 mg/kg; mus orl 2.30 mg/kg; mus ipr 2.30 mg/kg; dog orl LDLo 30 mg/kg; rbt skn 740 mg/kg; gpg orl 7 mg/kg; gpg ipr LDLo 5.1 mg/kg; qal orl 4.18 mg/kg; dcl orl 2.6 mg/kg.
CFR: 40CFR180.303

CAS RN 301-12-2
$C_6H_{15}O_4PS_2$

Oxydemeton-methyl
Phosphorothioic acid, S-[2-(ethylsulfinyl)ethyl] O,O-dimethyl ester (8CI9CI)
Bayer 21097
BAY 21097
Demeton S Methylsulfoxide
Demeton-methyl sulfoxide
Demeton-methyl sulphoxide
Demeton-O-methyl sulfoxide
Demeton-S-methyl sulfoxide
Demeton-S-methyl-sulfoxid (German)
Demeton-S-methylsulfoxid
Demeton-S-methylsulfoxide
Dimethyl S-(2-eththionylethyl) thiophosphate
Ethanethiol, 2-(ethylsulfinyl)-, S-ester with O,O-dimethyl phosphorothioate
ENT 24,964
Isomethylsystox sulfoxide
Metaisosystox sulfoxide
Metaisosystoxsulfoxide
Metasystemox
Metasystox R
Metasystox-R
Methyl demeton-O-sulfoxide
Methyl demeton-R
Methyl oxydemeton
Metilmercaptofosoksid
O-O-Dimethyl S-2-(ethylsulfinylethyl)phosphorothioate
O,O-Dimethyl S-[(ethylsulfinyl)ethyl] phosphorothioate
O,O-Dimethyl S-[(ethylsulfinyl)ethyl] phosphorothiolate
O,O-Dimethyl S-[2-(eththionyl)ethyl] phosphorothioate
O,O-Dimethyl S-[2-(ethylsulfinyl)ethyl] monothiophosphate
O,O-Dimethyl-S-(2-aethylsulfinyl-aethyl)-monothiophosphat (German)
O,O-Dimethyl-S-(2-ethylsulfinyl-ethyl)-monothiofosfaat (Dutch)
O,O-Dimethyl-S-(3-oxo-3-thia-pentyl)-monothiophosphat (German)
O,O-Dimetil-S-(2-etil-solfinil-etil)-monotiofosfato (Italian)
Oxydemeton-metile (Italian)
Phosphorothioic acid, O,O-dimethyl S-[2-(ethylsulfinyl)ethyl] ester
R 2170 (VAN)
S-[2-(Ethylsulfinyl)ethyl] O,O-dimethyl phosphorothioate (ACN)

Merck Index No.
MP: $< -20°$
BP: $106^{o0.01}$
Density: 1.289^{20}
Solubility: misc. water, sol. org. solvents, except petroleum ether
Octanol/water PC:
LD_{50}: rat orl 30 mg/kg; rat orl 65-75 mg/kg; rat ihl LC_{50} >1500 mg/m^3/1H; rat skn 100 mg/kg; rat ipr 20 mg/kg; rat ivn 47 mg/kg; rat unr 65 mg/kg; mus orl 10 mg/kg; mus ipr 8 mg/kg; gpg orl 120 mg/kg; gpg ipr 30 mg/kg; pgn orl 15 mg/kg; ckn orl 80 mg/kg; qal orl 84 mg/kg; dcl orl 54 mg/kg; bwd orl 42 mg/kg.
40CFR180.330

CAS RN 79-57-2 (2058-46-0)
$C_{22}H_{24}N_2O_9$

Oxytetracycline (ACN)
2-Naphthacenecarboxamide, 4-(dimethylamino)-1,4,4a,5,5a,6,11,12a -octahydro-3,5,6,10,12,12a-hexahydroxy-6-methyl-1,11-dioxo-, [4S-(4α,4aα,5α,5aα,6β,12aα)]-
2-Naphthacenecarboxamide, 4-(dimethylamino)-1,4,4a,5,5a,6,11,12a-octahydro-3,5,6,10,12,12a-hexahydroxy-6-methyl-1,11-dioxo-
component of Terrastatin
Adamycin
Berkmycen
Biostat
Biostat PA
Dabicycline
Fanterrin
Geomycin
Geomycin (*Streptomyces rimosus*) (VAN)
Lenocycline
Liquamycin
Macocyn
Neoterramycin
NCI-C56473
NSC-9169
Oksisyklin
Oxitetracyclin
Oxy-Kesso-Tetra
Oxydon
Oxymycin
Oxymykoin
Oxypam
Oxyterracin
Oxyterracine
Oxyterracyne
Oxytetracyclin
Oxytetracycline hydrochloride
OTC (VAN)
Proteroxyna
Riomitsin
Ryomycin
Solkaciclina
Stevacin
Streptomyces rimosus antibiotic
Taomycin
Taomyxin

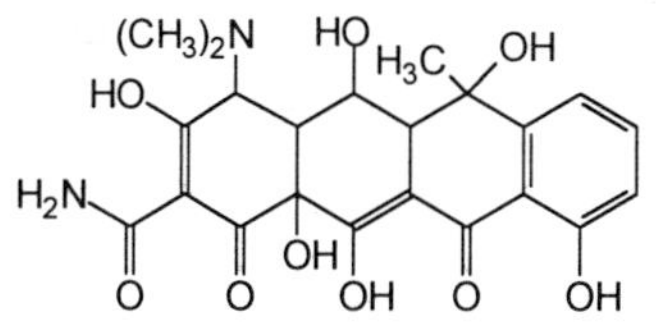

Merck Index No. 6931
MP: 181 - 182° (dec)
BP:
Density:
Solubility: 31.4 mg/ml water (pH 1.2), 4.6 mg/ml water (pH 2), 1.4 mg/ml water (pH 3), 0.5 mg/ml water (pH 5), 0.7 mg/ml water (pH 6), 1.1 mg/ml water (pH 7), 38.6 mg/ml water (pH 9), 12 mg/ml ethanol, 0.2 mg/ml 95% ethanol
Octanol/water PC:
LD_{50}: man orl TDLo 114 mg/kg/4D; inf par TDLo 136 mg/kg; rat orl 4800 mg/kg; rat ivn 260 mg/kg; mus orl 2240 mg/kg; mus ipr 5706 mg/kg; mus scu 700 mg/kg; mus ivn 140 mg/kg; dog ivn LDLo 220 mg/kg; gpg ipr LDLo 2250 mg/kg.
CFR: 40CFR180.337

CAS RN 2439-01-2
$C_{10}H_6N_2OS_2$

Chinomethionat
Oxythioquinox

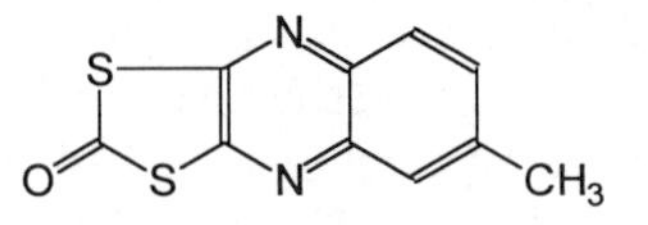

1,3-Dithiolo[4,5-b]quinoxalin-2-one, 6-methyl-
Carbonic acid, dithio-, cyclic S,S-(6-methyl-2,3-quinoxalinediyl) ester
Bayer ss2074
Bayer 36205
Bayer 4964
BAY 36205
Carbonic acid, dithio-, cyclic S,S-(6-methyl-2,3-quinoxalinediyl)ester
Chinomethionate
Cyclic S,S-(6-methyl-2,3-quinoxalinediyl) dithiocarbonate
Dithiolo(4,5-B)quinoxalin-2-one, 6-methyl-
Dithioquinox
Erade
Erazidon
ENT 25,606
Forestan
Forstan
Morestan
Morestane
MQD
Quinomethionate
SS 2074
2,3-Quinoxalinedithiol, 6-methyl-, cyclic carbonate
2,3-Quinoxalinedithiol, 6-methyl-, cyclic carbonate ester
2,3-Quinoxalinedithiol, 6-methyl-, cyclic dithiocarbonate (ester)
6-Methyl-quinoxaline-2,3-dithiolcyclocarbonate
6-Methyl-1,3-dithiolo[4,5-b]quinoxalin-2-one
6-Methyl-1,3-dithiolo[4,5-b]quinoxaline-2-one
6-Methyl-2-oxo-1,3-dithiolo[4,5-b]quinoxaline
6-Methyl-2-oxo-1,3-dithio[4,5-b]quinoxaline
6-Methyl-2,3-quinoxaline dithiocarbonate
6-Methyl-2,3-quinoxalinedithiol cyclic carbonate
6-Methyl-2,3-quinoxalinedithiol cyclic S,S-dithiocarbonate (ACN)

Merck Index No. 6933
MP: 169.8 - 170°
BP:
Density:
Solubility: 1 mg/l water 20°, 20-50 g/l toluene, 20-50 g/l dichloromethane, 1-2 g/l hexane, 0.1-1 g/l isopropanol, 18 g/l cyclohexanone, 10 g/l dimethylformamide, sol. hot benzene, dioxane
Octanol/water PC: 6020 20
LD_{50}: rat orl 2500-3000 mg/kg, rat orl 1100 mg/kg; rat orl 1800 mg/kg; rat ihl LC_{50} 3000 mg/m^3/4H; rat skn 500 mg/kg; rat ipr 95 mg/kg; mus orl LDLo 1070 mg/kg; mus ipr 473 mg/kg; cat orl >1000 mg/kg; rbt skn >2000 mg/kg; gpg orl 1500 mg/kg; ckn orl 980 mg/kg; mam unr 2500 mg/kg.
CFR: 40CFR180.338

CAS RN 110-88-3 (9002-81-7, 30525-89-4)
$C_3H_6O_3$

Paraformaldehyde (ACN)
1,3,5-Trioxane
s-Trioxane
sym-Trioxane
Formaldehyde, trimer
Metaformaldehyde
Metaformaldehyde trimer of formaldehyde
Polymerized formaldehyde
Triformol
Trioxan
Trioxane
Trioxin
Trioxymethylene
Trioxymethylene, *s*
1,3,5-Trioxacyclohexane
1,3,5-Trioxan

Merck Index No. 9646, 6974
MP: 64°
BP: $114.5°^{759}$
Density: 1.17^{65}
Solubility: 172 g/l water 18°, 211 g/l water 25°, sol. alcohols, ether, acetone, chlorinated hydrocarbons, aromatic hydrocarbons, sl. sol. pentane, petroleum ether.
Octanol/water PC:
LD_{50}: rat orl 800 mg/k; rbt skn LDLo 10000 mg/kg.
CFR: 40CFR185.4650

CAS RN 1910-42-5 (4685-14-7)
$C_{12}H_{14}N_2.2Cl$

Paraquat dichloride (ACN)
4,4'-Bipyridinium, 1,1'-dimethyl-, dichloride
o-Paraquat chloride
Bipyridinium, 1,1'-dimethyl-4,4'-, dichloride
Crisquat
Dextrone X
Dexuron
Esgram
Gramixel
Gramonol
Gramoxone dichloride
Gramuron
Methyl viologen
N,N'-Dimethyl-4,4'-dipyridylium dichloride
Para-Col
Paraquat
Paraquat chloride
Pathclear
Pillarquat
Pillarxone
Terraklene
Totacol
Toxer Total
Viologen, methyl-
Weedol
1,1'-Dimethyl-4,4'-bipyridinium chloride

Merck Index No. 6980
MP: dec. 300°
BP:
Density: $1.24 - 1.26^{20}$
Solubility: 700 g/l water 20°, sp. sol. lower alcohols, insol. most org. solvents
Octanol/water PC:
LD_{50}: hmn orl LDLo 214 mg/kg; wmn orl LDLo 3000 mg/kg; man orl TDLo 32 mg/kg; man orl LDLo 43 mg/kg; wmn orl LDLo 111 mg/kg; rat orl 57 mg/kg, rat ivn 21 mg/kg; mus orl 104 mg/kg; mus orl 120 mg/kg; mus ipr 20 mg/kg; mus scu 37 mg/kg; mus ivn 180 mg/kg; dog orl 25 mg/kg; dog ivn LDLo 34.5 mg/kg; mky orl 50 mg/kg; cat orl 35 mg/kg; rbt skn 325 mg/kg; rbt ipr 18 mg/kg; rbt ocu LDLo 100 mg/kg; pig orl 30 mg/kg; gpg orl 22 mg/kg; gpg ipr 3 mg/kg; ckn orl 362 mg/kg; dck orl 199 mg/kg; dck skn 600 mg/kg; frg par 260 mg/kg; dom orl 30 mg/kg; dom ivn 1 mg/kg.
CFR: 40CFR180.205

CAS RN 56-38-2
$C_{10}H_{14}NO_5PS$

Parathion 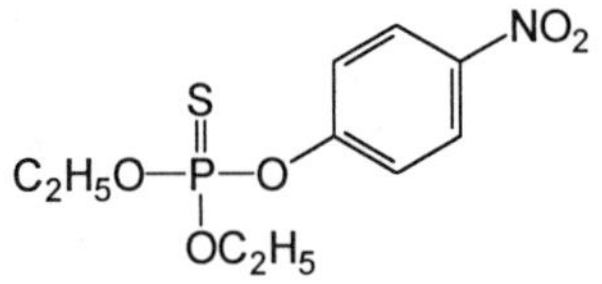
Phosphorothioic acid, O,O-diethyl O-(4-nitrophenyl) ester
Phosphorothioic acid, O,O-diethyl O-(*p*-nitrophenyl) ester
Alkron
Alleron
American Cyanamid 3422
Aphamite
Aralo
AAT
AATP
AC 3422
B 404
Bayer E-605
Bladan
Bladan F
BAY E-605
Compound 3422
Compound-3422
Corothion
Corthion
Corthione
Danthion
Diethyl *p*-nitrophenyl monothiophosphate
Diethyl *p*-nitrophenyl phosphorothioate
Diethyl *p*-nitrophenyl phosphorothionate
Diethyl *p*-nitrophenyl thionophosphate
Diethyl *p*-nitrophenyl thiophosphate
Diethyl parathion
Diethyl O-(4-nitrophenyl) phosphorothioate
Diethyl O-4-nitrophenyl phosphorothiote
Diethyl 4-nitrophenyl phosphorothionate
Diethyl 4-nitrophenyl thiophosphate
Diethyl-O-4-nitrophenyl phosphorothiote
Drexel parathion 8E
DNTP
DPP
E 605
E 605 f
Ecatox
Ekatox

Merck Index No. 6983
MP: 6.1°
BP: 150°$^{0.6}$, 375°^{760}, 157 - 162°$^{0.6}$
Density: 1.265^{25}

Solubility: 24 mg/l water 25°, misc. most org. solvents e.g. alcohols, acetone, benzene, chloroform, sl. sol. petroleum oils
Octanol/water PC:
LD_{50}: man orl TDLo 4.29 mg/kg/4D-I; hmn orl LDLo 0.171 mg/kg; wmn orl TDLo 5.67 mg/kg; hmn orl 3 mg/kg; hmn skn 7.143 mg/kg; hmn itr LDLo 0.714 mg/kg; man unr LDLo 1.471 mg/kg; rat orl 2 mg/kg; rat orl 3.6-13 mg/kg; rat ihl LC_{50} 84 mg/m^3/4H; rat skn 6.8 mg/kg; rat ipr 2 mg/kg; rat ivn 3.8 mg/kg; rat ims 6 mg/kg; mus orl 5 mg/kg; mus orl 12 mg/kg; mus ihl LCLo 15 mg/m^3; mus skn 19 mg/kg; mus ipr 3 mg/kg; mus scu 10 mg/kg; mus ivn 13 mg/kg; mus ims 7.2 mg/kg; dog orl 3 mg/kg; dog ipr 12 mg/kg; dog ivn 12 mg/kg; cat orl 0.93 mg/kg; cat ipr 3 mg/kg; cat ivn 3 mg/kg; rbt orl 10 mg/kg; rbt ihl LCLo 50 mg/m^3/2H; rbt skn 15 mg/kg; gpg orl 8 mg/kg; gpg ihl LCLo 14 mg/m^3/2H; gpg skn 45 mg/kg; gpg ipr 12 mg/kg; pgn orl 1.33 mg/kg; ckn orl 10 mg/kg; ckn ipr 2.5 mg/kg; qal orl 4.04 mg/kg; dck orl 2.1 mg/kg; dck skn 28 mg/kg; frg par 967 mg/kg; hor orl 5 mg/kg; dom orl LDLo 100 mg/kg; dom ims LDLo 20 mg/kg; mam orl 49 mg/kg; mam unr 6 mg/kg; bwd orl 1.33 mg/kg; bwd skn 1.8 mg/kg.
CFR: 40CFR180.121

CAS RN 82-68-8
$C_6Cl_5NO_2$

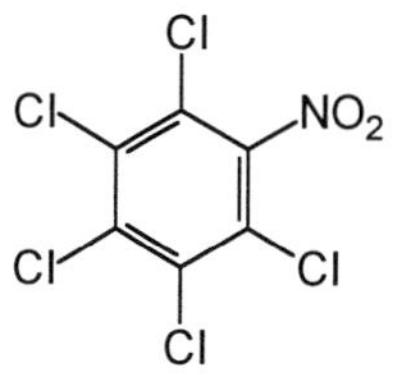

PCNB
Quintozene
Benzene, pentachloronitro-
Avicol
Avicol, pesticide (VAN)
Batrilex
Benzene, nitropentachloro-
Botrilex
Brassicol
Brassicol Super
Brassicol 75
Chinozan
Earthcide
Fartox
Folosan (VAN)
Fomac 2
Fungiclor
GC 3944-3-4
Kobutol
KOBU
KP 2
Liro-PCNB
Marisan forte
Nitropentachlorobenzene
NCI-C00419
Olpisan
Pentachlornitrobenzol (German)
Pentachloronitrobenzene (ACN)
Pentachloronitrobenzol
Pentagen
Phomasan
PKHNB
Quinosan
Quintocene
Quintozen
Saniclor 30
Terrachlor
Terraclor
Terraclor 30 G
Terrafun

Cl Cl NO2 Cl Cl Cl

Merck Index No. 8108
MP: 143 - 144°
BP: 328° (dec)
Density: 1.718^{25}
Solubility: 0.44 mg/l water 25°, 20g/kg ethanol, readily sol. aromatic solvents, ketones, chlorinated hydrocarbons, carbon disulfide
Octanol/water PC:
LD_{50}: rat orl >12000 mg/kg (aq), rat orl 1710mg/kg; rat orl 1650 mg/kg (maize oil); rat orl 1100 mg/kg; rat ihl LC_{50} 1400 mg/m^3; rat ipr 5000 mg/kg; mus orl 1400 mg/kg; mus ihl LC_{50} 2000 mg/m^3; mus ipr 4500 mg/kg; dog orl LD >2500 mg/kg; rbt orl 800 mg/kg.
CFR: 40CFR180.291

CAS RN 1114-71-2
$C_{10}H_{21}NOS$

Pebulate
Carbamothioic acid, butylethyl-, S-propyl ester
Carbamic acid, butylethylthio-, S-propyl ester
n-Propyl ethyl-*n*-butylthiolcarbamate
Butylethylthiocarbamic acid S-propyl ester
PEBC
R-2061 (VAN)
S-Propyl butylethylcarbamothioate
S-Propyl butylethylthiocarbamate (ACN)
Stauffer R-2061
Tillam

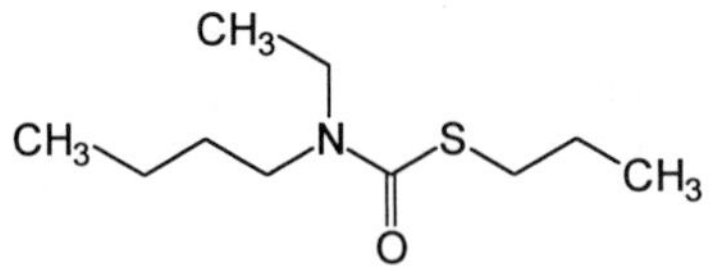

Merck Index No. 7007
MP:
BP: 142° ^{20}
Density: 0.956^{20}, 0.9458^{30}
Solubility: 60 mg/l water 20°, misc. acetone, benzene, toluene, xylene, methanol, isopropanol, kerosene
Octanol/water PC: 6918
LD_{50}: rat orl 921 mg/kg; rat orl 1120 mg/kg; rat skn 2000 mg/kg; rat unr 1125 mg/kg; mus orl 1652 mg/kg; mus unr 750 mg/kg; cat unr 750 mg/kg; rbt skn 4640 mg/kg; mam unr 713 mg/kg.
CFR: 40CFR180.238

CAS RN 40487-42-1
$C_{13}H_{19}N_3O_4$

Pendimethalin (ACN)
Benzenamine, N-(1-ethylpropyl)-3,4-dimethyl-2,6-dinitro-[N-(1-Ethylopropyl)-3,4-dimethyl-2,6-dinitrobenzenamine]
N-(1-Ethylpropyl)-3,4-dimethyl-2,6-dinitrobenzenamine (ACN)
Accotab
Aniline, 3,4-dimethyl-2,6-dinitro-N-(1-ethylpropyl)-
AC 92553
Benzenamine,
3,4-dimethyl-2,6-dinitro-N-(1-ethylpropyl)-
Cynoff
Go-Go-San
Herbadox
N-(1-Aethylpropyl)-2,6-dinitro-3,4-xylidin (German)
N-(1-Ethylpropyl)-2,6-dinitro-3,4-xylidine
Nicocyan
Pay-Off
Penoxaline
Penoxyalin
Penoxyn
Phenoxalin
Prowl
Stomp

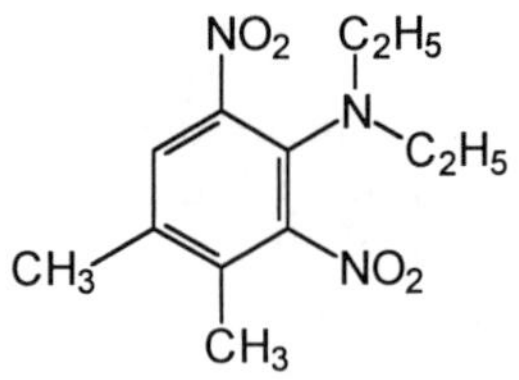

Merck Index No. 7026
MP: 54 - 58°
BP: (dec)
Density: 1.19^{25}
Solubility: 0.3 mg/l water 20°, 700 g/l acetone, 628 g/l xylene, 148 g/l corn oil, 138 g/l heptane, 77 g/l isopropanol, readily sol. benzene, toluene, chloroform, dichloromethane, sl. sol. petroleum ether, petrol
Octanol/water PC: 152000
LD_{50}: rat orl 1250mg/kg; rat orl 1050 mg/kg; rat ipr 500 mg/kg; mus orl 1620mg/kg; mus orl 1340 mg/kg; mus ipr 220 mg/kg; dog orl >5000 mg/kg; rbt skn 2260 mg/kg.
CFR: 40CFR180.361

CAS RN 37924-13-3
$C_{14}H_{12}F_3NO_4S_2$

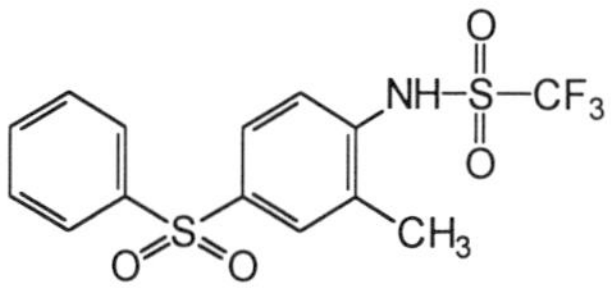

Perfluidone (ACN)
Methanesulfonamide, 1,1,1-trifluoro-N-[2-methyl-4-(phenyl sulfonyl)phenyl]-
o-Toluidine, 4-(phenylsulfonyl)-N-[(trifluoromethyl)sulfono]-
Destun
Diethanolamine perfluidone
Methanesulfonamide, N-(4-phenylsulfonyl-*o*-tolyl)-1,1,1-trifluoro-
Methanesulfonamide, N-[4-(phenylsulfonyl)-*o*-tolyl]-1,1,1-trifluoro-
Perfluidone, diethanolamine salt of
1,1,1-Trifluoro-N-(4-phenylsulfonyl-*o*-tolyl)methanesulfonamide
1,1,1-Trifluoro-N-[2-methyl-4-(phenylsulfonyl)phenyl]methanesulfonamide (ACN)
1,1,1-Trifluoro-N-[2-methyl-4-(phenylsulfonyl)phenyl]methanosulfonamide
2-Methyl-4-(phenylsulfonyl)trifluoromethanesulfonannilide
4-(Phenylsulfonyl)-N-[(trifluoromethyl)sulfono]-*o*-toluidine

Merck Index No. 7113
MP: 142 - 144°
BP:
Density:
Solubility: 60 mg/l water 22°, 750 g/l acetone, 11 g/l benzene, 162 g/l dichloromethane, 595 g/l methanol
Octanol/water PC:
LD_{50}: rat orl 633 mg/kg; mus orl 920 mg/kg; rbt skn >4000 mg/kg.
CFR: 40CFR180.165

CAS RN 116-06-3
$C_7H_{14}N_2O_2S$

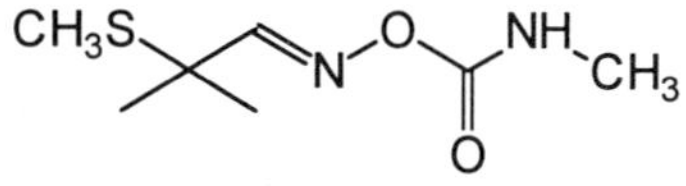

Aldicarb (ACN)
Propanal, 2-methyl-2-(methylthio)-, O-[(methylamino)carbonyl]oxime
Propionaldehyde, 2-methyl-2-(methylthio)-, O-(methylcarbamoyl)oxime
2-Methyl-2-(methylthio)propanal, O-[(methylamino)carbonyl]oxime
Aldecarb
Ambush
Carbanolate
ENT 27,093
Matadan
NCI-C08640
OMS 771
Permethrin
Pounce
Temik G10

Merck Index No. 216
MP: 99 - 100°
BP: Density: 1.195^{25}
Solubility: 6 g/l water 25°, 350 g/kg acetone, 300 g/kg dichloromethane, 150 g/kg benzene, 50 g/kg xylene, insol. heptane, mineral oils
Octanol/water PC:
LD_{50}: rat orl 0.93 mg/kg; rat orl 0.65 mg/kg; rat ihl LC_{50} 200 mg/m^3/5H; rat skn 2.5 mg/kg; rat ipr 0.28 mg/kg; rat scu 0.666 mg/kg; rat ivn 0.47 mg/kg; rat unr 0.93 mg/kg; mus orl 0.30 mg/kg; mus scu 0.25 mg/kg; rbt skn 1400 mg/kg; gpg skn 2400 mg/kg; pgn orl 3.16 mg/kg; ckn orl 8 mg/kg; qal orl 2 mg/kg; dck orl 3.4 mg/kg; dom orl LDLo 5 mg/kg; bwd orl 0.75 mg/kg.
CFR: 40CFR180.378

CAS RN 52645-53-1
$C_{21}H_{20}Cl_2O_3$

Permethrin (ACN)
Cyclopropanecarboxylic acid, 3-(2,2-dichloroethenyl)-2,2-dimethyl-, (3-phenoxyphenyl) methyl ester
(3-Phenoxyphenyl)methyl *cis,trans*-(±)-3-(2,2-dichloroethenyl)-2,2-dimethylcyclopropane carboxylate
(3-Phenoxyphenyl)methyl 3-(2,2-dichlorethenyl)-2,2-(dimethylcyclopropane)carboxylate
(3-Phenoxyphenyl)methyl 3-(2,2-dichloroethenyl)-2,2-dimethylcyclopropanecarboxylate
(3-Phenoxyphenyl)methyl 3-(2,2-dichloroethenyl)-2,2-dimethylcyclopropanecarboxylate(mixt. of 40% *cis* and 60% *trans* isomers)
A.I. 3-29158
Ambush
BW-21-Z
Cyclopropanecarboxylic acid, 3-(2,2-dichlorovinyl)-2,2-dimethyl-, 3-phenoxybenzyl ester (±)-, (*cis,trans*)-
Ectiban
Exmin
FMC 33297
FMC 41655
ICI-PP 557
Matadan
NDRC-143
NIA 33297
NRDC 143
Outflank-Stockade
Permethrin, mixed *cis,trans*
Permetrina (Portuguese)
Pounce
PP 557
S 3151
S-3151
SBP 1513
SBP-1513
WL 43479
3-Phenoxybenzyl (+)-*cis,trans*-3-(2,2-dichlorovinyl)-2,2-dimethylcyclopropane carboxylate
3-Phenoxybenzyl (±)-3-(2,2-dichlorovinyl)-2,2-(dimethylcyclopropane)carboxylate

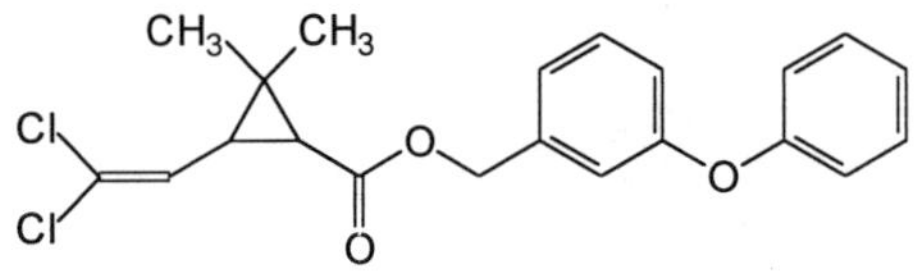

Merck Index No. 7132
MP: 34 - 35°
BP: 200° $^{0.01}$
Density: 1.19-1.27^{20}
Solubility: 0.2 mg/l water 20°, sol. most org. solvents, not ethylene glycol, >1000 g/kg Xylene hexane, 258 g/kg methanol
Octanol/water PC: 1260000
LD_{50}: man unr >4000 mg/kg; rat orl 383 mg/kg; rat orl 4000-6000 mg/kg; rat ihl LC_{50} 685 mg/m^3; rat skn 2500 mg/kg; rat scu 6600 mg/kg; mus orl 424 mg/kg; mus orl 4000 -6000 mg/kg; mus ihl LC_{50} 685 mg/m^3; mus skn >10000 mg/kg; mus ipr 429 mg/kg; mus scu 10000 mg/kg; mus ivn 31 mg/kg; mus ice LDLo 0.6 mg/kg; rbt orl 4000 mg/kg; rbt skn >2000 mg/kg; gpg orl 4000 mg/kg; ckn orl 7000 mg/kg; qal orl 13500 mg/kg; dck orl 11300 mg/kg; brd orl 32 mg/kg.
CFR: 40CFR180.378

CAS RN 72-56-0
$C_{18}H_{20}Cl_2$

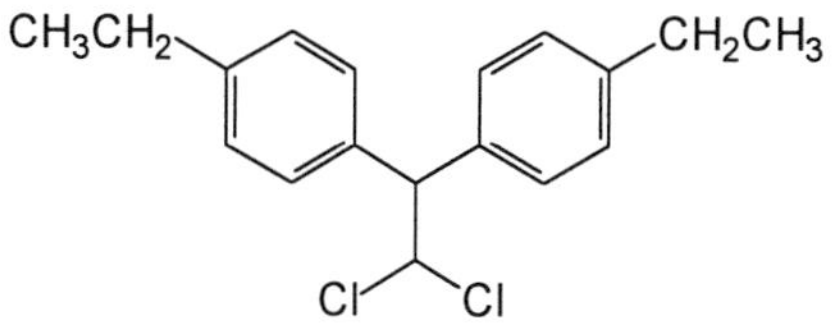

***p,p'*-Ethyl-DDD**
Perthane
Benzene, 1,1'-(2,2-dichloroethylidene)bis[4-ethyl-
Ethane, 1,1-dichloro-2,2-bis(*p*-ethylphenyl)-
(Diethyldiphenyl)dichloroethane
α,α-Dichloro-2,2-bis(*p*-ethylphenyl)ethane
p,p'-Ethyl DDD
Di(*p*-ethylphenyl)dichloroethane
Dichlorobis(*p*-ethylphenyl)ethane
Diethyl diphenyl dichloroethane, and related compounds (ACN)
Diethyldiphenyldichloroethane
Ethane, 2,2-bis(*p*-ethylphenyl)-1,1-dichloro-
Ethylan
ENT-17082
NCI-C02868
P,P-Ethyl DDD
P,P'-Ethyl-DDD
1,1-Bis(ethylphenyl)-2,2-dichloroethane (ACN)
1,1-Bis(*p*-ethylphenyl)-2,2-dichloroethane
1,1-Dichloro-2,2-bis(ethylphenyl)ethane
1,1-Dichloro-2,2-bis(*p*-ethylphenyl)ethane
1,1-Dichloro-2,2-bis(4-ethylphenyl)ethane
1,1'-(2,2-Dichloroethylidene)bis[4-ethylbenzene]
2,2-Bis(*p*-ethylphenyl)-1,1-dichloroethane
2,2-Dichloro-1,1-bis(*p*-ethylphenyl)ethane

CH_3CH_2 CH_2CH_3 Cl Cl

Merck Index No. 3050
MP: 56 - 57°
BP:
Density:
Solubility: insol. water, sol. acetone, kerosene, diesel fuel
Octanol/water PC:
LD_{50}: rat ro 6600 mg/kg; rat orl >4000 mg/kg; rat ivn 73 mg/kg; mus ivn 173 mg/kg; bwd orl 9 mg/kg.
CFR: 40CFR180.139

CAS RN 13684-63-4
$C_{16}H_{16}N_2O_4$

Phenmedipham (ACN)
Carbamic acid, (3-methylphenyl)-, 3-[(methoxycarbonyl)amino]phenyl ester
Carbanilic acid, *m*-hydroxy-, methyl ester, *m*-methylcarbanilate (ester)
Betanal
Carbamic acid, (3-methylphenyl)-3-((methoxycarbonyl)amino)phenyl ester
Carbanilic acid, *m*-hydroxy-, methyl ester, *m*-methylcarbanilate
EP-452
Fenmedifam
Methyl *m*-hydroxycarbanilate *m*-methylcarbanilate
Methyl *m*-Hydroxycarbanilate *m*-methylcarbanilate (ACN)
Methyl N-[3-[N-(3-methylphenyl)carbamoyloxy]phenyl]carbamate
Methyl 3-*m*-tolycarbamoyloxy phenylcarbamate
Methyl-*m*-hydroxycarbanilate-*m*-methylcarbanilate
Morton EP 452
S-4075
Schering 4072
Schering 4075
Schering-38584
Sn-38584
SN 4075
3-(Carbomethoxyamino)phenyl 3-methylcarbanilate
3-[(Methoxycarbonyl)amino]phenyl N-(3-methylphenyl)carbamate
3-[(Methoxycarbonyl)amino]phenyl N-(3'-methylphenyl)carbamate
3-[(Methoxycarbonyl)amino]phenyl 3-methylphenylcarbamate

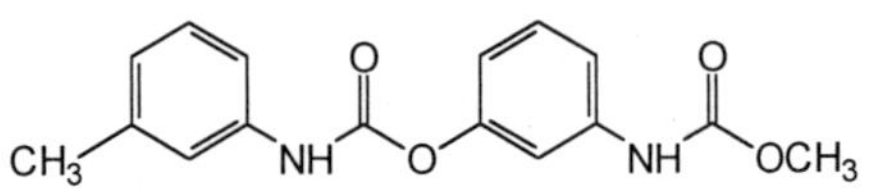

Merck Index No. 7199
MP: 143 - 144°, 139 - 142°, 143 - 144° (technical)
BP:
Density: $0.25\text{-}0.30^{20}$
Solubility: 4.7 mg/l water 25°, 200 g/l acetone, 200 g/l cyclohexanone, 50 g/l methanol, 20 g/l chloroform, 2.5 g/l benzene, 0.5 g/l hexane
Octanol/water PC: 3890 (octanol/water, pH 4)
LD_{50}: rat orl 4000 mg/kg; rat orl >8000 mg/kg; rat skn >500 mg/kg; rat unr 800 mg/kg; mus orl >8000 mg/kg; dog orl >4000mg/kg; gpg orl >4000 mg/kg; ckn orl >3000 mg/kg; mam unr 3000 mg/kg.
CFR: 40CFR180.278

CAS RN 92-84-2
$C_{12}H_9NS$

10H-Phenothiazine
Phenothiazine (ACN)
Afi-Tiazin
Agrazine
Antiverm
Contaverm
Dibenzo-p-thiazine
Dibenzo-1,4-thiazine
Dibenzothiazine
Early bird wormer
ENT 38
Fenothiazine (Dutch)
Fenotiazina (Italian)
Fenoverm
Fentiazin
Helmetina
Lethelmin
Nemazene
Nemazine
Nexarbol
Orimon
Padophene
Penthazine
Phenegic
Phenosan
Phenoverm
Phenovis
Phenoxur
Phenthiazine
Phenzeen
Reconox
Souframine
Thiodifenylamine (Dutch)
Thiodiphenylamin (German)
Thiodiphenylamine
Tiodifenilamina (Italian)
Vermitin
Wurm-Thional

Merck Index No. 7220
MP: 185.1°
BP: 371° ^{76}, 290° ^{40}, sublimes 130° ^{1}
Density:
Solubility: Insoluble in water, chloroform, petroleum ether. Readily soluble in benzene. Soluble in diethyl ether, AcOH. Slightly soluble in alcohol, mineral oils.
Octanol/water PC:
LD_{50}: chd orl LDLo 425 mg/kg/5D; mus orl 5000 mg/kg; mus ivn 178 mg/kg; rbt orl 4000 mg/kg; ctl orl 500 mg/kg; bwd unr LDLo 500 mg/kg.
CFR: 40CFR180.319

CAS RN 90-43-7
$C_{12}H_{10}O$

o-Phenylphenol (ACN)
[1,1'-Biphenyl]-2-ol
2-Biphenylol
o-Biphenylol
o-Diphenylol
o-Hydroxybiphenyl
o-Hydroxydiphenyl
o-Xenol
[1,1'-Biphenyl]-2-o1
Biphenyl-2-ol
Biphenyl, 2-hydroxy-
Dowcide 1
Dowicide
Dowicide 1
NCI-C50351
OPP
Phenol, *o*-phenyl-
Phenylphenol
Preventol O extra
Remol TRF
Torsite
Tumescal OPE
Usaf ek-2219
1-Hydroxy-2-phenylbenzene
2-Hydroxybifenyl (Czech)
2-Hydroxybiphenyl
2-Hydroxydiphenyl
2-Phenylphenol

HO

Merck Index No. 7276
MP: 57°, 55.5 - 57.5°
BP: 286°, 280 - 284°, 152 - 154° ^{15}
Density: 1.217^{25}
Solubility: 0.7 g/l water 25°, sol. most organic solvents e.g. ethanol, ethylene glycol, isopropanol, glycol ethers, polyglycols
Octanol/water PC:
LD_{50}: rat orl 2000 mg/kg; rat orl 2700 mg/kg, rat unr 2700 mg/kg; mus orl 1050 mg/kg; mus orl 2000 mg/kg; mus ipr 50 mg/kg; cat orl 500 mg/kg.
CFR: 40CFR180.129

CAS RN 298-02-2
$C_7H_{17}O_2PS_3$

Phorate (ACN)
Phosphorodithioic acid, O,O-diethyl S-[(ethylthio)methyl] ester
American Cyanamid 3,911
AC 3911
Dithiophosphate de O,o-diethyle et d'ethylthiomethyle (French)
Experimental Insecticide 3911
ENT 24,042
Foraat (Dutch)
Granutox
L 11/6
Methanethiol, (ethylthio)-, S-ester with O,O-diethyl phosphorodithioate
O,O-Diaethyl-S-(aethylthio-methyl)-dithiophosphat (German)
O,O-Diethyl ethylthiomethyl phosphorodithioate
O,O-Diethyl S-ethylmercaptomethyl dithiophosphate
O,O-Diethyl S-ethylthiomethyl dithiophosphate
O,O-Diethyl S-[(ethylmercapto)methyl] dithiophosphonate
O,O-Diethyl S-[(ethylthio)methyl] dithiophosphonate
O,O-Diethyl S-[(ethylthio)methyl] phosphorodithioate (ACN)
O,O-Diethyl S-[(ethylthio)methyl] thiothionophosphate
O,O-Diethyl-S-(ethylthio-methyl)-dithiofosfaat (Dutch)
O,O-Dietil-S-(etiltio-metil)-ditiofosfato (Italian)
Phorat (German)
Phorate 10G
Phorate-10G
Phosphorodithioic acid, O,O-diethyl S-(ethylthio)methyl ester
Rampart
Thimate
Thimet
Thimet G
Thimet 10G
Timet
Vegfru
Vergfru foratox

$C_2H_5O-P(=S)(OC_2H_5)-S-CH_2-S-CH_2-CH_3$

Merck Index No. 7305
MP: < -15° (technical >90%)
BP: 118 - 120° $^{0.8}$ (technical >90%), 75 - 78° $^{0.1}$, 118 - 120° $^{0.8}$, 125 - 127° ^{2}
Density: 1.167^{25} (technical >90%), 1.156^{25}
Solubility: 50 mg/l water 25°, misc. alcohols, ketones, ethers, esters, aromatic, aliphatic, and chlorinated hydrocarbons
Octanol/water PC: 8410
LD_{50}: rat orl 1.0 mg/kg; rat orl 3.7 mg/kg; rat orl 1.6 mg/kg; rat orl 1.1 mg/kg; rat orl 2.3 mg/kg; rat ihl LC_{50} 2.50 mg/m^3; rat ivn 1.2 mg/kg; mus orl 6.59 mg/kg; rbt skn 99 mg/kg; gpg skn 20 mg/kg; qal orl 7 mg/kg; grb ipr 1.866 mg/kg; dck orl 0.60 mg/kg; dck skn 203 mg/kg; bwd orl 1 mg/kg.
CFR: 40CFR180.206

CAS RN 2310-17-0
$C_{12}H_{15}ClNO_4PS_2$

Phosalone (ACN)
Phosphorodithioic acid, S-[(6-chloro-2-oxo-3(2H)-benzoxazolyl)methyl] O,O-diethyl ester
Phosphorodithioic acid, O,O-diethyl ester, S-ester with 6-chloro-3-(mercaptomethyl)-2-benzoxazolinone (8CI)
[3-(Diethyldithiophosphoryl)methyl]-6-chlorobenzoxazolone-2
Azofene
Benzophosphate
Benzphos
Chipman 11974
ENT 27,163
ENT 27163
Fosalone
Fozalon
Niagara 9241
NIA 9241
NIA-9241
NPH-1091
O,O-Diaethyl-S-(6-chlor-2-oxo-benz[b]-1,3-oxalin-3-yl)-methyl-dithiophosphat (German)
O,O-Diethyl dithiophosphorylmethyl-6-chlorbenzoxazolone
O,O-Diethyl phosphorodithioate, S-ester with 6-chloro-3-(mercaptomethyl)-2-benzoxazolinone
O,O-Diethyl S-(6-chlorobenzoxazolinyl-3-methyl) dithiophosphate
O,O-Diethyl S-(6-chlorobenzoxazolon-3-yl)methyl phosphorodithioate
O,O-Diethyl S-[(6-chloro-2-oxobenzoazolin-3-yl)methyl] phosphorodithioate
O,O-Diethyl S-[(6-chloro-2-oxobenzoxalin-3-yl)methyl] phosphorodithioate
O,O-Diethyl S-[(6-chloro-2-oxobenzoxazolin-3-yl)methyl] phosphorodithioate (ACN)
O,O-Diethyl S-[(6-chloro-2-oxobenzoxazolin-3-yl)methyl]phosphorodithioate
O,O-Diethyl S-[(6-chlorobenzoxazolone-3-yl)methyl] phosphorodithioate
O,O-Diethyl S-[-(6-chloro-2-oxo-benzoxazolin-3-ol)-methyl]phosphorodithioate
O,O-Diethyl S-[6-chloro-3-(mercaptomethyl)-2-benzoxazolinone] phosphorodithioate
O,O-Diethyl-S-[(6-chloor-2-oxo-benzoxazolin-3-yl)-methyl]-dithiofosfaat (Dutch)
O,O-Diethyl-S-[(6-chloro-2-oxo-benzoxazolin-3-yl)methyl] phosphorothiolothionate
O,O-Dietil-S-[(6-cloro-2-oxo-benzossazolin-3-il)-metil]-ditiofosfato (Italian)
P 974
P-974
Phasolon
Phosalon
Phozalon
Rhodia RP 11974

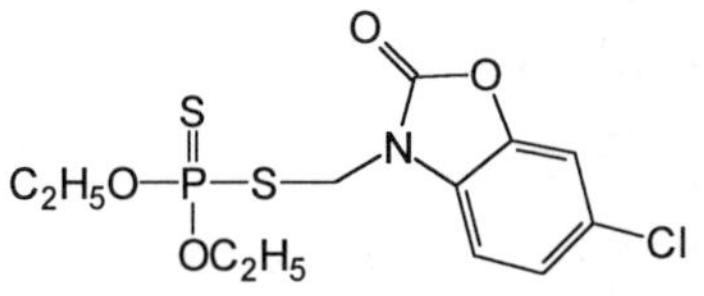

Merck Index No. 7308
MP: 45 - 48°
BP:
Density:
Solubility: 10 mg/l water 25°, ca. 1000g/l ethyl acetate, acetone, acetonitrile, benzene, chloroform, dichloromethane, dioxane, methyl ethyl ketone, toluene, xylene, 200 g/l methanol, ethanol
Octanol/water PC: 12600
LD_{50}: rat orl 85 mg/kg; rat orl 120-175 mg/kg; rat orl 135-170 mg/kg; rat skn 390mg/kg; rat unr 135 mg/kg; mus orl 73 mg/kg; mus orl 180 mg/kg; cat orl 112 mg/kg; rbt skn 1000 mg/kg; gpg orl 150 mg/kg; ckn orl 661 mg/kg.
CFR: 40CFR180.263

CAS RN 13171-21-6 (297-99-4)
$C_{10}H_{19}ClNO_5P$

Phosphamidon
Phosphoric acid, 2-chloro-3-(diethylamino)-1-methyl-3-oxo-1-propenyl dimethyl ester
Phosphoric acid, dimethyl ester, ester with 2-chloro-N,N-diethyl-3-hydroxycrotonamide
(2-Chloor-3-diethylamino-1-methyl-3-oxo-prop-1-en-yl)-dimethyl-fosfaat (Dutch)
(2-Chlor-3-diaethylamino-1-methyl-3-oxo-prop-1-en-yl)-dimethyl-phosphat (German)
(2-Cloro-3-dietilamino-1-metil-3-oxo-prop-1-en-il)-dimetil-fosfato (Italian)
C 570
Crotonamide, 2-chloro-N,N-diethyl-3-hydroxy-, dimethyl phosphate
Dimecron
Dimecron 100
Dimecron 50
Dimecron-20
Dimethyl phosphate of 2-chloro-N,N-diethyl-3-hydroxycrotonamide
Dimethyl 2-chloro-2-diethylcarbamoyl-1-methylvinyl phosphate
Dixon
ENT 25515
Famfos
Fosfamidon (Dutch)
Fosfamidone (Italian)
Merkon
ML 97
NCI-C00588
O,O-Dimethyl O-(1-methyl-2-chloro-2-diethylcarbamoylvinyl) phosphate
O,O-Dimethyl O-(2-chloro-2-(N,N-diethylcarbamoyl)-1-methylvinyl) phosphate
OR 1191
Phosphamidion
Phosphamidone
Phosphate de dimethyle et de (2-chloro-2-diethylcarbamoyl-1-methyl-vinyle) (French)
Sundaram 1975
1-Chloro-diethylcarbamoyl-1-propen-2-yl dimethyl phosphate
2-Chloro-N,N-diethyl-3-hydroxycrotonamide ester of dimethyl phosphate
2-Chloro-2-diethylcarbamoyl-1-methylvinyl dimethyl phosphate
2-Chloro-3-(diethylamino)-1-methyl-3-oxo-1-propenyl dimethyl phosphate

O Cl C_2H_5
CH_3O—P—O N C_2H_5
CH_3O CH_3 O

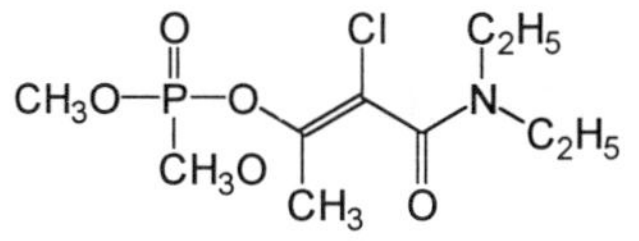

Merck Index No. 7312
MP: -45°
BP: $162°^{1.5}$, $120°^{0.001}$
Density: 1.21^{25}, 1.2132^{25}
Solubility: Miscible with water, acetone, dichloromethane, toluene, 32 g/l heptane.
Octanol/water PC:
LD_{50}: rat orl 8 mg/kg; rat orl 17.9-30 mg/kg; rat ihl LC_{50}135 mg/m^3/4H; rat skn 125 mg/kg; rat ipr 8.70 mg/kg; rat scu 15 mg/kg; mus orl 6 mg/kg; mus ihl LC_{50} 30 mg/m^3/1H; mus ipr 5.8 mg/kg; mus scu 13.2 mg/kg; mus ivn 6 mg/kg; rbt orl 70 mg/kg; rbt skn 80 mg/kg; gpg ihl LC_{50} 1300 mg/m^3/4H; dck orl 3.1 mg/kg; dck skn 26 mg/kg; bwd orl 1.8 mg/kg.
CFR: 40CFR180.239

CAS RN 1918-02-1
$C_6H_3Cl_3N_2O_2$

Picloram (ACN)
2-Pyridinecarboxylic acid, 4-amino-3,5,6-trichloro-
Picolinic acid, 4-amino-3,5,6-trichloro-
k-Pin
Amdon
ATCP
Borolin
Chloramp (Russian)
K-Pin
NCI-C00237
Picloram-R
Tordon
Tordon 10K
Tordon 101 mixture
Tordon 22K
Trichloropicloram
3,5,6-Trichloro-4-aminopicolinic acid
4-Amino-3,5,6-trichloro-2-picolinic acid
4-Amino-3,5,6-trichloro-2-pyridinecarboxylic acid
4-Amino-3,5,6-trichloropicolinic acid (ACN)
4-Amino-3,5,6-trichlorpicolinsaeure (German)
4-Aminotrichloropicolinic acid

Merck Index No. 7370
MP: dec 215°
BP:
Density:
Solubility: 430 mg/l water 25°, 19.8 g/l acetone, 10.5 g/l ethanol, 5.5 g/l isopropanol, 1.6 g/l acetonitrile, 1.2 g/l diethyl ether, 0.6 g/l dichloromethane, 0.2 g/l benzene, <0.0[illegible] g/l carbon disulfide
Octanol/water PC:
LD_{50}: rat orl 8200 mg/kg; rat orl 2898 mg/kg; mus orl 2000-4000 mg/kg; mus orl 1061 mg/kg; rbt orl 2000 mg/kg; rbt skn >4000 mg/kg; gpg orl 1922 mg/kg; ckn orl 4000 mg/kg; dom orl >100 mg/kg; mam orl 2000 mg/kg.
CFR: 40CFR180.292

CAS RN 23103-98-2
$C_{11}H_{18}N_4O_2$

Pirimicarb (ACN)
Carbamic acid, dimethyl-, 2-(dimethylamino)-5,6-dimethyl-4-pyrimidinyl ester
Aphox
Carbamic acid, dimethyl-, 2-(dimethylamino)-5,6-dimethyl-4-pyrimidyl ester
Fernos
Pirimor
Pyrimicarb
PP 062
Rapid
2-(Dimethylamino)-5,6-dimethyl-4-pyrimidinyl dimethylcarbamate (ACN)
5,6-Dimethyl-2-(dimethylamino)-4-pyrimidinyl dimethylcarbamate

Merck Index No. 7468
MP: 90.5°
BP:
Density:
Solubility: 2.7 g/l water 25°, 4.0 g/l acetone, 2.5 g/l ethanol, 2.9 g/l xylene, 3.3 g/l chloroform
Octanol/water PC: 50
LD_{50}: rat orl 100 mg/kg; rat orl 147 mg/kg; rat skn >500 mg/kg; rat unr 111 mg/kg; mus orl 107 mg/kg; mus unr 68 mg/kg; dog orl 100 mg/kg; pgn orl 20 mg/kg; ckn orl 25 mg/kg; qal orl 54 mg/kg; dck orl 17.2 mg/kg; bwd orl 30 mg/kg.
CFR: 40CFR180.365 (revoked 5/4/88)

CAS RN 51-03-6
$C_{19}H_{30}O_5$

Piperonyl butoxide
1,3-Benzodioxole, 5-[[2-(2-butoxyethoxy)ethoxy]methyl]-6-propyl-
Toluene, α-[2-(2-butoxyethoxy)ethoxy]-4,5-(methylenedioxy)-2-propyl-
α-[2-(2-Butoxyethoxy)ethoxy]-4,5-methylenedioxy-2-propyltoluene
α-[2-[2-(*n*-Butoxy)ethoxy]ethoxy]-4,5-methylenedioxy-2-propyltoluene
(Butylcarbityl) (6-propylpiperonyl) ether
(3,4-Methylenedioxy-6-propylbenzyl)butyldiethyleneglycol ether
Butacide
Butocide
Butoxide
Butyl carbitol 6-propylpiperonyl ether
Butylcarbityl (6-propylpiperonyl) ether
Butylcarbityl 6-propylpiperonyl ether
Ethanol butoxide
ENT 14,250
FMC 5273
NCI-C02813
NIA 5273
P.B. Dressing
Piperonyl butoxide, technical (ACN)
Pyrenone 606
PB
Toluene, α[2-(2-butoxyethoxy)ethoxy]-4,5-(methylenedioxy)-2-propyl-
1-(3,4-Methylenedioxy-6-propylbenzyloxy)-2-(2-butoxyethoxy)ethane
3,4-Methylendioxy-6-propylbenzyl-N-butyl-diaethylenglykolaether (German)
3,4-Methylenedioxy-6-propylbenzyl n-butyl diethyleneglycol ether
5-[[2-(2-Butoxyethoxy)ethoxy]methyl]-6-propyl-1,3-benzodioxole
5-Propyl-4-(2,5,8-trioxa-dodecyl)-1,3-benzodioxol (German)
6-(Propylpiperonyl)butylcarbityl ether
6-Propylpiperonyl butyl diethylene glycol ether

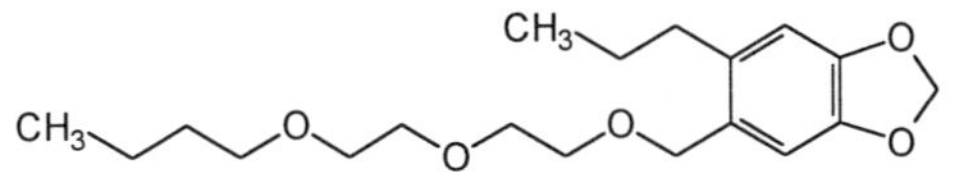

Merck Index No. 7446
MP:
BP: $180°^{1}$
Density: $1.04\text{-}1.07^{25}$
Solubility: Insoluble in water, soluble in all common organic solvents.
Octanol/water PC: 56200
LD_{50}: rat orl >7500 mg/kg; rat orl 6150 mg/kg; rat orl 7000 mg/kg; rat skn >7950 mg/kg; mus orl 2600 mg/kg; mus ipr LDLo 1000 mg/kg; rbt orl 2650 mg/kg; rbt skn 200 mg/kg.
CFR: 40CFR180.127

CAS RN 13121-70-5
$C_{18}H_{34}OSn$

Cyhexatin (ACN)
Stannane, tricyclohexylhydroxy-
Dowco 213
Ent 27,395-x
Hydroxytricyclohexylstannane
M 3180
Plictran
Plyctran
Tin, tricyclohexylhydroxy-
Tricyclohexylhydroxystannane (ACN)
Tricyclohexylhydroxytin
Tricyclohexyltin hydroxide
Tricyclohexylzinnhydroxid (German)
TCTH

Sn–OH

Merck Index No. 2767
MP: 195 - 198°
BP:
Density:
Solubility: <1 mg/l water 25°, 216 g/kg chloroform, 37 g/kg methanol, 34 g/kg dichloromethane, 28 g/kg carbon tetrachloride, 16 g/kg benzene, 10 g/kg toluene, 3.6 g/kg xylene, 1.3 g/kg acetone
Octanol/water PC:
LD_{50}: rat orl 180 mg/kg; rat orl 540 mg/kg; rat orl 779 mg/kg; rat orl 826 mg/kg; rat ihl LC_{50} 244 mg/m^3; rat skn 446 mg/kg; rat ipr 13 mg/kg; mos orl 275 mg/kg; mus unr 780 mg/kg; rbt orl 458 mg/kg; rbt skn 2422 mg/kg; pgp orl 780 mg/kg; ckn orl 654 mg/kg; dom orl LDLo 150 mg/kg.
CFR: 40CFR180.144

CAS RN 10124-50-2
$AsH_3O_3.xK$

Potassium arsenite (ACN)
Arsonic acid, potassium salt
Arsenious acid (H_3AsO_3), potassium salt
Arsenenous acid, potassium salt
Arsenious acid, potassium salt
Arsenite de potassium (French)
Fowler's solution
NSC 3060
Potassium arsenite, solid (DOT)
Potassium metaarsenite
Potassium Arsenite

HO, As, $O^{\ominus}$, OH, $K^{\oplus}$

Merck Index No. 7584
MP:
BP:
Density:
Solubility:
Octanol/water PC:
LD_{50}: hmn orl TDLo 74 mg/kg; rat orl 14 mg/kg; rat orl 15 mg/kg; rat skn 150 mg/kg; mus scu LDLo 16 mg/kg; dog orl LDLo 3 mg/kg; dog scu LDLo 0.70 mg/kg; dog ivn LDLo 2 mg/kg; cat scu LDLo 5 mg/kg; rbt scu LDLo 8 mg/kg; rbt ivn LDLo 6 mg/kg; gpg scu LDLo 9 mg/kg; pgn scu LDLo 12 mg/kg.
CFR: 40CFR180.334 (revoked 5-4-88)

CAS RN 20762-60-1
KN_3

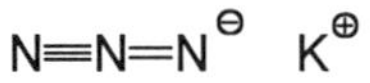

N≡N=N⁻ K⁺

Potassium azide (ACN)
Potassium azide ($K(N_3)$)
Kazoe
Potassium nitride (KN_3)
PPG 101

Merck Index No.
MP:
BP:
Density:
Solubility:
Octanol/water PC:
LD_{50}: rat orl 27 mg/kg; bwd orl 17.8 mg/kg;
CFR: 40CFR180. (temp 1976)

CAS RN 32889-48-8
$C_{10}H_{13}ClN_6$

Procyazine (ACN)
Propanenitrile, 2-[[4-chloro-6-(cyclopropylamino)-1,3,5-triazin-2-yl]amino]-2-methyl-
Propionitrile, 2-[[4-chloro-6-(cyclopropylamino)-*s*-triazin-2-yl]amino]-2-methyl-
Cycle
Cycle (herbicide)
CGA-18762
2-[[4-Chloro-6-(cyclopropylamino)-*s*-triazin-2-yl]amino]-2-methylpropionitrile
2-[[4-Chloro-6-(cyclopropylamino)-1,3,5-triazin-2-yl]amino]-2-methylpropanenitrile (ACN)
2-[[4-Chloro-6-(cyclopropylamino)-1,3,5-triazin-2-yl]amino]-2-methylpropionitrile

Merck Index No.
MP: 170 - 171°
BP:
Density:
Solubility:
Octanol/water PC:
LD_{50}: rat orl 290 mg/kg.
CFR: 40CFR180. temp

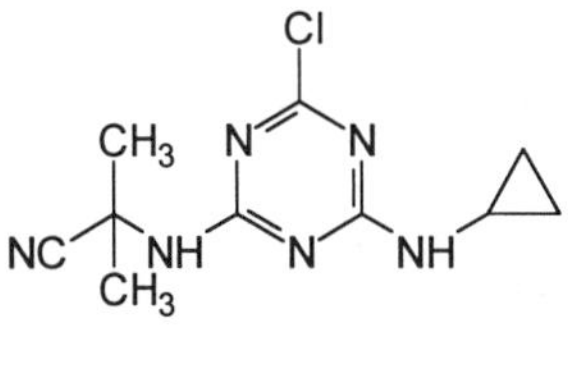

CAS RN 32809-16-8
$C_{13}H_{11}Cl_2NO_2$

Procymidone
N-(3',5'-Dichlorophenyl)-1,2-dimethylcyclopropane-1,2-dicarboximide
S 7131
Sumilex
Sumisclex
1,2-Cyclopropanedicarboximide, N-(3,5-dichlorophenyl)-1,2-dimethyl-
3-Azabicyclo[3.1.0]hexane-2,4-dione, 3-(3,5-dichlorophenyl)-1,5-dimethyl-

Merck Index No.
MP: 166 - 166.5°
BP:
Density: 1.452^{25} SG
Solubility: 4.5 mg/l water 25°, Sl.sol. alcohols, 180 g/l acetone, 180 g/l xylene, 210 g/l chloroform, 230 g/l dimethylformamide
Octanol/water PC: 1380 26°
LD_{50}: rat orl 6800 mg/kg; rat orl 7700 mg/kg; mus orl 7800 mg/kg;
CFR: 40CFR180.455

CAS RN 29091-21-2
$C_{13}H_{17}F_3N_4O_4$

Prodiamine (ACN)
1,3-Benzenediamine, 2,4-dinitro-N3,N3-dipropyl-6-(trifluoromethyl)-
Toluene-2,4-diamine, α,α,α-trifluoro-3,5-dinitro-N4,N4-dipropyl-
(2,4-Dinitro-N^3,N^3-dipropyl-6-(trifluoromethyl)-1,3-benzonediamine
m-Phenylenediamine, 2,4-dinitro-N^3,N^3-dipropyl-6-(trifluoromethyl)-
N^3,N^3-Di-n-propyl-2,4-dinitro-6-(trifluoromethyl)-*m*-phenylenediamine
N^3,N^3-Di-N-propyl-2,4-dinitro-6-(trifluoromethyl)-*m*-phenylenediamine
N^3,N^3-Dipropyl-2,4-dinitro-6-(trifluoromethyl)-*m*-phenylenediamine
USB 3153
1,3-Diamino-N^1,N^1-dipropyl-2,6-dinitro-4-(trifluoromethyl)benzene
2,4-Dinitro-N^3,N^3-dipropyl-6-(trifluoromethyl)-1,3-benzenediamine (ACN)

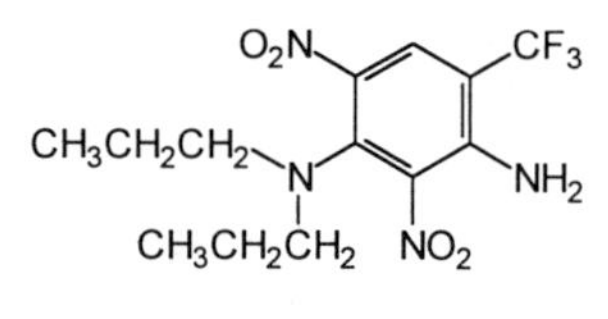

Merck Index No. 7773
MP: 124°, 124 - 125°
BP:
Density: 1.47^{25}
Solubility: 0.03 mg/l water 25°, 205 g/l acetone, 45 g/l acetonitrile, 74 g/l benzene, 93 g/l chloroform, 7 g/l ethanol, 20 g/l hexane, 37 g/l xylene
Octanol/water PC: 12672 +- 2270
LD_{50}: rat orl 15380 mg/kg; rat orl >5000 mg/kg; rat skn >2000 mg/kg; mus orl >15000 mg/kg; mus orl 15,380 mg/kg.
CFR: 40CFR180. temp 1975

CAS RN 41198-08-7
$C_{11}H_{15}BrClO_3PS$

Profenofos (ACN)
Phosphorothioic acid, O-(4-bromo-2-chlorophenyl) O-ethyl S-propyl ester
Curacron
CGA 15324
O-(4-Bromo-2-chlorophenyl) O-ethyl S-propyl phosphorothioate (ACN)
O-(4-Bromo-2-chlorophenyl)-O-ethyl-S-propyl phosphorothioate
Phosphorothioic acid, O-(4-bromo-2-chlorophenyl)-O-ethyl-S-propyl ester
Polycron
Profenophos
Selecron

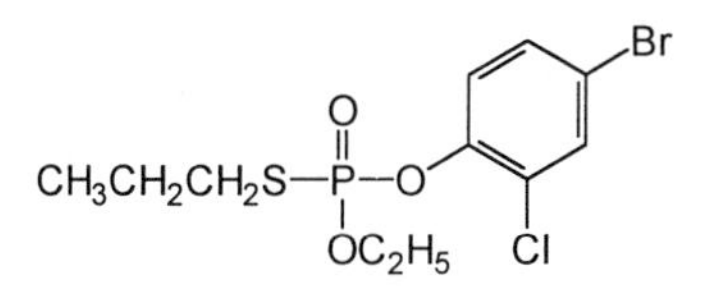

Merck Index No.
MP:
BP: 110° $^{0.001}$
Density: 1.455^{20}
Solubility: 20 mg/l water 20°, misc. most org. solvents
Octanol/water PC:
LD_{50}: rat orl 358 mg/kg; rat ihl LC_{50} 3000 mg/m^3/4H; rat skn 1610 mg/kg; rat ipr 520 mg/kg; rat scu 3200 mg/kg; mus orl 162 mg/kg; mus ipr 116 mg/kg; mus scu 2000 mg/kg; rbt orl 700 mg/kg; rbt skn 192 mg/kg; ckn orl 1.9 mg/kg.
CFR: 40CFR180.348

CAS RN 26399-36-0
$C_{14}H_{16}F_3N_3O_4$

Profluralin (ACN)
Benzenamine, N-(cyclopropylmethyl)-2,6-dinitro-N-propyl-4-(trifluoromethyl)-
p-Toluidine, N-(cyclopropylmethyl)-α,α,α-trifluoro-2,6-dinitro-N-propyl-
p-Toluidine, N-(cyclopropylmethyl)-2,6-dinitro-N-propyl-α,α,α -trifluoro-
N-(Cyclopropylmethyl)-α,α,α-trifluoro-2,6-dinitro-N-propyl-*p*-toluidine(ACN)
CG-10832
CGA (VAN)
CGA 10332
CGA 10832
Er5461
ER 5461
GA-10832
Pregard
SGA 10832
Tolban

Merck Index No.
7781
MP: 33 - 36°
BP:
Density:
Solubility: 0.1 mg/l water 20°, readily sol. most org. solvents
Octanol/water PC:
LD_{50}: rat orl 1808 mg/kg; rat orl 10000 mg/kg; rat ihl LCLo >3970 mg/m^3/1H; rat skn >3170 mg/kg; rbt skn 13754 mg/kg.
CFR: 40CFR180.404

CAS RN 7287-19-6
$C_{10}H_{19}N_5S$

Prometryn (ACN)
1,3,5-Triazine-2,4-diamine, N,N'-bis(1-methylethyl)-6-(methylthio)-
s-Triazine, 2,4-bis(isopropylamino)-6-(methylthio)-
s-Triazine, 4,6-bis(isopropylamino)-2-(methylmercapto)-
A 1114
Caparol
G 34161
Gesagard
Gesagard 50
Gesagarde 50 Wp
Mercasin
Mercazin
Merkazin
N,N'-Bis(1-methylethyl-6-(methylthio)-1,3,5-triazine-2,4-diamine
Polisin
Primatol Q
Prometrex
Prometrin
Prometryne
Selectin
Selectin 50
Selektin
Sesagard
Uvon
2-(Methylmercapto)-4,6-bis(isopropylamino)-*s*-triazine
2-(Methylthio)-4,6-bis(isopropylamino)-*s*-triazine
2,4-Bis(isopropylamino)-6-(methylmercapto)-*s*-triazine
2,4-Bis(isopropylamino)-6-(methylthio)-*s*-triazine (ACN)
2,4-Bis(isopropylamino)-6-(methylthio)-1,3,5-triazine
2,4-Bis(isopropylamino)-6-methylmercapto-*s*-triazine
2,4-Bis(Isopropylamino)-6-methylthio-*s*-triazine

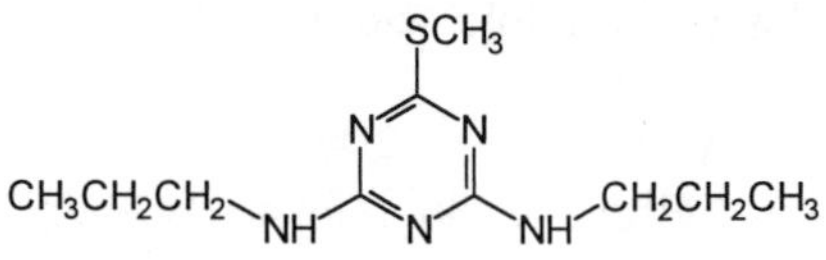

Merck Index No. 7800
MP: 118 - 120°
BP:
Density: 1.157^{20}
Solubility: 48 mg/l water 20°, 240 g/l acetone, 160 g/l methanol, 300 g/l dichloromethane, 5.5 g/l hexane, 170 g/l toluene.
Octanol/water PC: 2190
LD_{50}: rat orl 1800 mg/kg; rat orl 5235 mg/kg; rat orl 3750 mg/kg; rat skn >3100 mg/kg; mus orl 2138 mg/kg; rbt skn >3100 mg/kg; mam unr 1800 mg/kg.
CFR: 40CFR180.222

CAS RN 23950-58-5
$C_{12}H_{11}Cl_2NO$

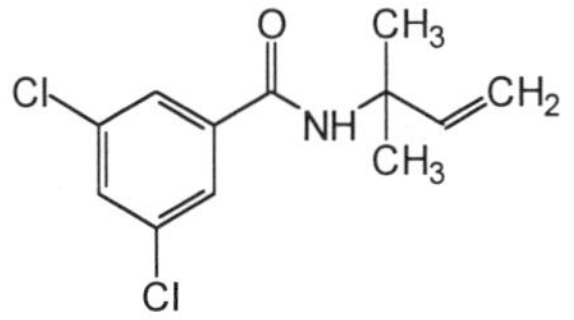

Propyzamide
Benzamide, 3,5-dichloro-N-(1,1-dimethyl-2-propynyl)-
Kerb
Kerb 50W
N-(1,1-Dimethyl-2-propynyl)-3,5-dichlorobenzamide
N-(1,1-Dimethylpropynyl)-3,5-dichlorobenzamide
Promamide
Pronamide
RH 315
3,5-Dichloro-N-(1,1-dimethyl-2-propynyl)benzamide (ACN)
3,5-Dichloro-N-(1,1-dimethylpropynyl)benzamide

Merck Index No. 7886
MP: 155 - 156°
BP:
Density:
Solubility: 15 mg/l water 25°, 150 g/l methanol, 150 g/l isopropanol, 200 g/l cyclohexanone, 200 g/l methyl ethyl ketone, 330 g/l dimethylsulfoxide, mod. sol. benzene, xylene, carbon tetrachloride, sl. sol. petroleum ether
Octanol/water PC: 1234-1903 ± 540
LD_{50}: rat orl 8350 mg/kg; rat orl 5620 mg/kg; rat orl 3350 mg/kg; dog orl 10000 mg/kg; rbt skn LDLo >3160 mg/kg; dck orl >14000 mg/kg.
CFR: 40CFR180.317

CAS RN 1918-16-7
$C_{11}H_{14}ClNO$

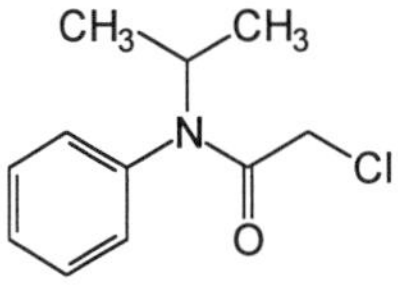

Propachlor
Acetamide, 2-chloro-N-(1-methylethyl)-N-phenyl-
Acetanilide, 2-chloro-N-isopropyl-
α-Chloro-N-isopropylacetanilide
Bexton
Bexton 4L
CIPA
CP 31393
N-Isopropyl-α-chloroacetanilide
N-Isopropyl-2-chloroacetanilide
Niticid
Propachlore
Ramrod
Ramrod 65
Satecid
2-Chloro-N-(1-methylethyl)-N-phenylacetamide
2-Chloro-N-isopropyl-N-phenylacetamide
2-Chloro-N-isopropylacetanilide (ACN)
2'-Chloro-N-isopropylacetanilide

Merck Index No. 7805
MP: 77, 67 - 76° (technical)
BP: 110° $^{0.03}$
Density: 1.242^{25}
Solubility: 613 mg/l water 25°, 448 g/kg acetone, 737 g/kg benzene, 342 g/kg toluene, 408 g/kg ethanol, 239 g/kg xylene, 602 g/kg chloroform, 174 g/kg carbon tetrachloride, 219 g/kg diethyl ether, sl. sol. aliphatic hydrocarbons
Octanol/water PC:
LD_{50}: rat orl 1800 mg/kg, rat orl 710 mg/kg; rat skn LD >2000 mg/kg; rat unr 1056 mg/kg; mus orl 290 mg/kg; mus unr 306 mg/kg; rbt orl 392 mg/kg; rbt skn 380 mg/kg; qal orl 91 mg/kg; dck orl 512 mg/kg; mam unr 800 mg/kg.
CFR: 40CFR180.211

CAS RN 709-98-8
$C_9H_9Cl_2NO$

Propanil
Propanamide, N-(3,4-dichlorophenyl)-
Propionanilide, 3',4'-dichloro-
BAY 30130
Chem rice
Crystal propanil-4
Dichloropropionanilide
Dipram
DCPA (VAN)
DPA (VAN)
FW 734
Grascide
Montrose propanil
N-(3,4-Dichlorophenyl)propanamide
N-(3,4-Dichlorophenyl)propionamide
Prop-Job
Propanex
Propanid
Propanide
Propionic acid 3,4-dichloroanilide
Riselect
Rogue
Rosanil
S 10165
Stam
Stam lv 10
Stam m-4
Stam F 34
Stam LV 10
Strel
Surcopur
Surpur
Synpran N
STAM
Vertac
3,4-Dichloropropionanilide
3',4'-Dichloropropionanilide (ACN)

Merck Index No. 7814
MP: 92 - 93°, 88 - 91° (technical), 91 - 93°.
BP:
Density: 1.25^{25}
Solubility: 225 mg/ml water 25°, 70 g/l benzene, 1700 g/l acetone, 1100 g/l ethanol, 52 g/l isophorone, 350 g/l cyclohexanone, 250 g/l methyl ethyl ketone, readily sol. isopropanol, toluene, xylene
Octanol/water PC: 193
LD_{50}: rat orl 1285-1483 mg/kg; rat orl 1383 mg/kg; rat orl 367 mg/kg; rat skn >5000 mg/kg; rat unr 1300 mg/kg; mus orl 360 mg/kg; mus orl 4000 mg/kg; dog orl 1217 mg/kg; rbt skn LD >2000 mg/kg; rbt skn 7080 mg/kg; rbt unr 500 mg/kg; dck orl 375 mg/kg; mam orl 2527 mg/kg.
CFR: 40CFR180.274

CAS RN 2312-35-8
$C_{19}H_{26}O_4S$

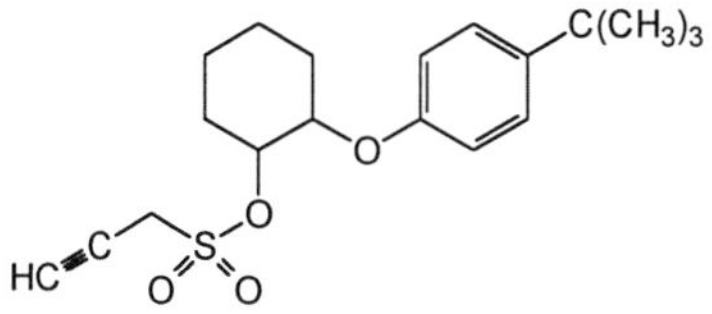

Propargite (ACN)
Sulfurous acid, 2-[4-(1,1-dimethylethyl)phenoxy]cyclohexyl 2-propynyl ester
Sulfurous acid, 2-(*p-tert*-butylphenoxy)cyclohexyl 2-propynyl ester (8CI)
2-(*p-tert*-butylphenoxy)cyclohexyl 2'-propynyl sulfite
2-(*p-tert*-Butylphenoxy)cyclohexyl propargyl sulfite
2-(*p-tert*-Butylphenoxy)cyclohexyl 2-propynyl sulfite (ACN)
BPPS
Comite
Cyclosulfyne
D 014
ENT 27226
Naugatuck D 014
Omite
Omite 57E
Omite 85E
Propargil
Uniroyal D 014

Merck Index No. 7818
MP:
BP:
Density: 1.085 - 1.115^{25}
Solubility: 0.5 mg/l water 25°, 10.5 ppm water, misc. most org. solvents, e.g. acetone, benzene, ethanol, methanol, hexane, heptane
Octanol/water PC: 5314
LD_{50}: rat orl 2200 mg/kg, rat orl 1480 mg/kg; rat orl 1480 mg/kg; rat ihl LC_{50} >940 mg/m^3/4H; rat skn 250 mg/kg; mus unr 780 mg/kg; rbt skn >3400 mg/kg.
40CFR180.259

CAS RN 139-40-2
$C_9H_{16}ClN_5$

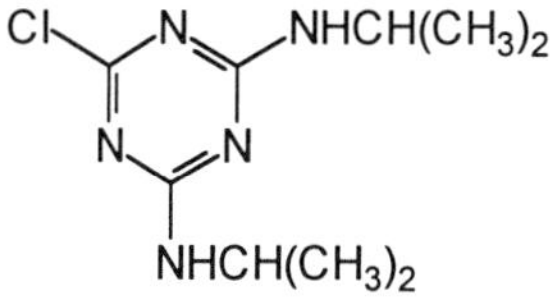

Propazine (ACN)
1,3,5-Triazine-2,4-diamine, 6-chloro-N,N'-bis(1-methylethyl)- (9CI)
s-Triazine, 2-chloro-4,6-bis(isopropylamino)- (8CI)
component of Milocep
s-Triazine, 2,4-bis(isopropylamino)-6-chloro-
G 30028
Geigy 30,028
Gesamil
G30028
Milogard
Plantulin
Primatol P
Propasin
Propazin (VAN)
Propazine (herbicide)
Prozinex
2-Chlor-4,6-bis(isopropylamino)-*s*-triazine
2-Chloro-4,6-bis(isopropylamino)-*s*-triazine (ACN)
2-Chloro-4,6-bis(isopropylamino)-1,3,5-triazine
2-Chloro-4,6-bis(isopropylamino)*s*-triazine

Merck Index No. 7822
MP: 212 - 214°, 213°
BP:
Density: 1.162^{20}
Solubility: 5.0 mg/l water 20°, 8.6 ppm water 20°, 6.2 g/kg benzene, 6.2 g/kg toluene, 5.0 g/kg diethyl ether, 2,5 g/kg carbon tetrachloride
Octanol/water PC:
LD_{50}: rat orl >7000 mg/kg; rat orl >5000 mg/kg; rat orl 3840 mg/kg; rat skn >3100 mg/kg; rat scu 395 mg/kg; mus orl 3180 mg/kg; rbt skn >10200 mg/kg; gpg orl 1200 mg/kg; mam unr 6000 mg/kg.
40CFR180.243

CAS RN 31218-83-4
$C_{10}H_{20}NO_4PS$

Propetamphos (ACN)
2-Butenoic acid, 3-[[(ethylamino)methoxyphosphinothioyl]oxy]-, 1-methylethyl ester, (E)- (E)-1-Methylethyl 3-[[(ethylamino)methoxyphosphinothioyl]oxy]-2-butenoate (ACN)
Isopropyl 3-hydroxycronate, O-methyl ethylphosphoramidothioate
Isopropyl-3-hydroxycronate, O-methylethyl phosphoramidothioate
O-2-Isopropoxycarbonyl-1-methylvinyl O-methylethylphosphoramidothioate
Safrotin
Safrotin 4EC
TSAR
1-Methylethyl 3-[[(ethylamino)methoxyphosphinothioyl]oxy]-2-butenoate
1-Methylethyl-(E)-3[[(ethylamino)methoxyphosphinothioyl]oxy]-2-butenoate

S
CH3 O CH3
CH3O—P—O O CH3
NHCH2CH3

Merck Index No. 7827
MP:
BP: 87 - 89° $^{0.005}$
Density: 1.1294^{20}
Solubility: 110 mg/l water 24°, readily misc. with acetone, methanol, ethanol, hexane, diethyl ether, dimethylsulfoxide, chloroform, xylene
Octanol/water PC:
LD_{50}: rat orl 75 mg/kg; rat orl 119 mg/kg; rat orl 82 mg/kg; rat skn 564 mg/kg; mam orl 130 mg/kg; mam skn 4000 mg/kg.
40CFR185.5100

CAS RN 75-56-9
C_3H_6O

Propylene oxide (ACN)(8CI)
Oxirane, methyl- (9CI)
Epoxy propane
Epoxypropane
Ethylene oxide, methyl-
Methyl ethylene oxide
Methyloxirane
NCI-C50099
Oxyde de propylene (French)
Propane, epoxy-
Propane, 1,2-epoxy-
Propene oxide
Propylene epoxide
Propylene oxide, inhibited
1,2-Epoxypropane
1,2-Propylene oxide
2,3-Epoxypropane
3-Methyl-1,2-epoxypropane

O
CH3

Merck Index No. 7869
MP: -112.13°
BP: 34.23°
Density: 0.859^{0}
Solubility: 40.5% w/w water 20°, misc. alcohol, ether.
Octanol/water PC:
LD_{50}: man ihl LCLo 1400 mg/m^3/10M; rat orl 380 mg/kg; rat orl 1140 mg/kg; rat ihl LCLo 4000 ppm/4H; rat ipr 150 mg/kg; mus orl 440 mg/kg; mus ihl LC_{50} 1740 ppm/4H; mus ipr 175 mg/kg; dog ihl LCLo 2005 ppm/4H; rbt skn 1245 mg/kg; gpg orl 660 mg/kg; gpg ihl LCLo 4000 ppm/4H; mam orl 440 mg/kg.
40CFR185.5150

CAS RN 114-26-1

$C_{11}H_{15}NO_3$

Propoxur

Phenol, 2-(1-methylethoxy)-, methylcarbamate (9CI)
Carbamic acid, methyl-, o-isopropoxyphenyl ester (8CI)
o-IPMC
o-Isopropoxyphenyl methylcarbamate (ACN)
o-Isopropoxyphenyl N-methylcarbamate
Aprocarb
Arprocarb
Bay 9010 Baygon
Bayer B 5122
Bayer 39007
Baygon
Blattanex
Blattosep
Bolfo
Boygon
Brygou
BAY 39007
BAY 5122
BAY 9010
Chemagro 9010
Dalf dust
DDVP
ENT 25,671
Invisi-Gard
Isocarb
IPMC
N-Methyl-2-isopropoxyphenylcarbamate
OMS 33
OMS33
Phenol, o-isopropoxy-, methylcarbamate
Propotox
Propoxure
PHC
Sendran
Suncide
Tendex
Tugon fliegenkugel
Unden
Unden (pesticide) (VAN)

Merck Index No. 7849
MP: 84 - 87°, 91.5°
BP: dec.
Density: 1.12^{20}
Solubility: 2 g/l water 20°, >200 g/l alcohols, acetone, methyl ethyl ketone, cyclohexanone, dichloromethane, 100 g/l chloroform, toluene
Octanol/water PC:
LD_{50}: wmn orl LDLo 24 mg/kg; rat orl 70 mg/kg; rat orl 95 mg/kg; rat orl 104 mg/kg; rat orl 83 mg/kg; rat orl 86 mg/kg; rat ihl LC_{50} 1440 mg/m^3/1H; rat skn 800 mg/kg; rat ipr 30 mg/kg; rat scu 56 mg/kg; rat ivn 11 mg/kg; rat ims 53 mg/kg; rat unr 100 mg/kg; mus orl 100-109 mg/kg; mus orl 23.5 mg/kg; mus skn >1360 mg/kg; mus ipr 12 mg/kg; mus scu 11.4 mg/kg; gpg orl 40 mg/kg; ham ipr 500 mg/kg; pgn orl 7.5 mg/kg; ckn orl 46.5 mg/kg; qal orl 28 mg/kg; dck orl 9.58 mg/kg; mam orl 90 mg/kg; mam skn 800 mg/kg; mam unr 100 mg/kg; bwd orl 3.8 mg/kg.
40CFR180. (pend. 1978)

CAS RN 72-54-8
$C_{14}H_{10}Cl_4$

_p,p'_-DDD
Benzene, 1,1'-(2,2-dichloroethylidene)bis[4-chloro- (9CI)
Ethane, 1,1-dichloro-2,2-bis(_p_-chlorophenyl)- (8CI)
(Dichlorodiphenyl)dichloroethane
_p,p'_-(Dichlorodiphenyl)dichloroethane
_p,p'_-TDE
Dichloro diphenyl dichloroethane (ACN)
Dichlorodiphenyldichloroethane
Dilene
DDD
DDD-_p,p'_
Ethane, 1,1-bis(_p_-chlorophenyl)-2,2-dichloro-
ENT 4,225
ME1700
NCI-C00475
Rhothane
Tetrachlorodiphenylethane
TDE
1,1-Bis(4-chlorophenyl)-2,2-dichloroethane
1,1-Dichloor-2,2-bis(4-chloor fenyl)-ethaan (Dutch)
1,1-Dichlor-2,2-bis(4-chlor-phenyl)-aethan (German)
1,1-Dichloro-2,2-bis(_p_-chlorophenyl)ethane
1,1-Dichloro-2,2-bis(4-chlorophenyl)-ethane (French)

Cl Cl CHCl2

2,2-Bis(p-chlorophenyl)-1,1-dichloroethane
2,2-Bis(4-chlorophenyl)-1,1-dichloroethane
4,4'-DDD

Merck Index No. 3049
MP: 109 - 110°
BP:
Density:
Sol:
Octanol/water PC:
LD_{50}: rat orl 113 mg/kg; rat orl >4000 mg/kg; mus orl LDLo 600 mg/kg; rbt skn 1200 mg/kg; ham orl >5000 mg/kg.
40CFR180.187 (removed 12/24/86)

CAS RN 122-42-9
$C_{10}H_{13}NO_2$

Propham
Agermin
Ban-Hoe
Beet-Kleen
Birgin
Collavin
IFC
IFK
INPC
IPC
IPPC
ISO.PPC
1-methylethyi phenylcarbamate
isopropyl phenylcarbamate
isopropyl carbanilate

NH O CH3 O CH3

Merck Index No. 7828
MP: 87 - 88°
BP: sublimes
Density: 1.09^{20}
Solubility: 250 mg/l water 20, sol. esters, alcohols, acetone, benzene, cyclohexane, xylene
Octanol/water PC:
LD_{50}: hmn orl 714 mg/kg; rat orl 1000 mg/kg; rat orl 5000 mg/kg; rat skn >5000 mg/kg; mus orl 2160 mg/kg; mus orl 3000 mg/kg; mus ipr 200 mg/kg; qal unr >2000 mg/kg; mam unr 1000 mg/kg.
40CFR180.319

CAS RN 1698-60-8
$C_{10}H_8ClN_3O$

Pyrazon (ACN)
Chloridazon
3(2H)-Pyridazinone, 5-amino-4-chloro-2-phenyl- (8CI9CI)
Blurex
Burex
BAS 11916H
Dazon
HS 119-1
HS119-1
Phenazon (VAN)
Phenazone (VAN)
Phenazone (herbicide) (VAN)
Phenosane
Piramin
Pyramin
Pyramin (herbicide)
Pyramin rb
Pyramine (VAN)
Pyrazone
Pyrazonl
PAC
PAC (pesticide)
PCA (VAN)
1-Fenyl-4-amino-5-chlor-6-pyridazinon (Czech)
1-Phenyl-4-amino-5-chloro-6-pyridazone
1-Phenyl-4-amino-5-chloropyridaz-6-one
1-Phenyl-4-amino-5-chloropyridazin-6-one
1-Phenyl-4-amino-5-chloropyridazone-6
1-Phenyl-4-amino-5-chlorpyridaz-6-one
5-Amino-4-chloro-α-phenyl-3-(2H)-pyridazinone
5-Amino-4-chloro-2-phenyl-3(2H)-pyradazinone
5-Amino-4-chloro-2-phenyl-3(2H)-pyridazinone (ACN)
5-Amino-4-chloro-2-phenylpyridazin-3-one
5-Amino-4-chloro-2,3-dihydro-3-oxo-2-phenylpyridazine

Cl, O, H_2N, N, N

Merck Index No.
MP: 205 - 206° (dec), 198 - 202° (technical)
BP:
Density:
Solubility: 400 mg/l water 20°, 34 g/kg methanol, 28 g/kg acetone, 6 g/kg ethyl acetate, 3.3 g.kg dichloromethane, 2.1 g/kg chloroform, 0.7 g/kg benzene, 0.7 g/kg diethyl ether, >0.1 g/kg cyclohexane
Octanol/water PC: 158
LD_{50}: rat orl 647 mg/kg; rat orl 3830 mg/kg; rat orl 2140 mg/kg; rat skn >2500 mg/kg; rat ipr 600 mg/kg; mus orl 1000 mg/kg; mus orl 2500 mg/kg; mus ipr 410 mg/kg; rbt orl 2000 mg/kg; rbt skn 2500 mg/kg; gpg orl 760 mg/kg; mam unr 2000 mg/kg.
40CFR180.316

CAS RN 5512-33-9
$C_{19}H_{23}ClN_2O_2S$

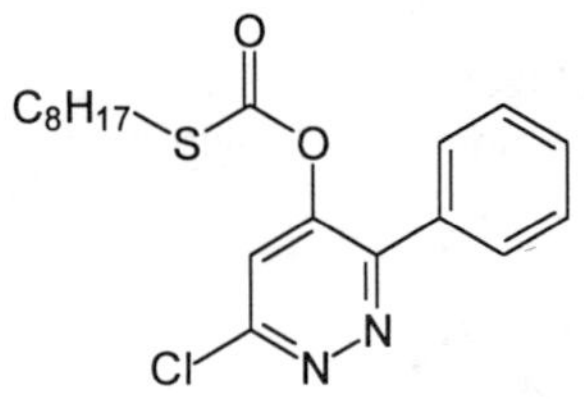

Pyridate
O-(6-Chloro-3-phenyl-4-pyridazinyl) S-octylcarbonothioate
6-Chloro-3-phenyl-4-pyridazinyl S-octylthiocarbonate
CL 11344
Lentagran
Tough

Merck Index No.
MP: 27°
BP: 220° $^{0.1}$
Density: 1.16^{20} SG
Solubility: 1.5 mg/l water 20°. Soluble in most organic solvents.
Octanol/water PC:
LD_{50}: rat orl 1960 mg/kg; rat orl 2400 mg/kg; rat ihl LC_{50} >4370 mg/m^3/4H; rbt scu >3450 mg/kg.
40CFR180. Temp (1988)

CAS RN 7784-46-5
$AsHO_2.Na$

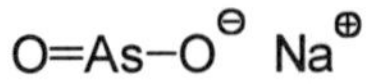

Sodium arsenite (ACN)
Arsenenous acid, sodium salt (9CI)
Arsenious acid, sodium salt (8CI)
Arsenious acid, monosodium salt
Arsenite de sodium (French)
Atlas
Atlas 'a'
Atlas 'A'
Atlas a
Atlas A
Chem pels C
Chem-Sen 56
Kill-All
Penite
Prodalumnol
Prodalumnol double
Sodium arsenite ($NaAsO_2$)
Sodium arsenite liquid (DOT)
Sodium arsenite, (liquid)
Sodium metaarsenite

Merck Index No. 8523
MP:
BP:
Density:
Solubility: sol. water, sl. sol. alcohol
Octanol/water PC:
LD_{50}: chd orl TDL0 1 mg/kg; chd orl LDLo 2 mg/kg; rat orl 41 mg/kg; rat skn 150 mg/kg; rat ipr LDLo 7 mg/kg; rat ivn LDLo 6 mg/kg; mus ipr 1.17 mg/kg; mus ims 14 mg/kg; rbt orl LDLo 12 mg/kg; rbt ivn 7.6 mg/kg; mam orl 10 mg/kg.
40CFR180.335

CAS RN 10453-86-8
$C_{22}H_{26}O_3$

(-)-*cis*-Resmethrin
Cyclopropanecarboxylicacid,2,2-dimethyl-3-(2-methyl-1-propenyl)-,[5-(phenylmethyl)-3-furanyl]methylester(9CI)
Cyclopropanecarboxylic acid, 2,2-dimethyl-3-(2-methylpropenyl)-, (5-benzyl-3-furyl) methylester (8CI)
(+)-*trans, cis*-Resmethrin
(5-Benzyl-3-furyl)methyl chrysanthemate
(5-Benzyl-3-furyl)methyl *cis,trans*-(.+-.)-2,2-dimethyl-3-(2-methylpropenyl)cyclopropane carboxylate)
(5-Benzyl-3-furyl)methyl 2,2-dimethyl-3-(2-methylpropenyl)-cyclopropanecarboxylate
(5-Benzyl-3-furyl)methyl 2,2-dimethyl-3-(2-methylpropenyl)cyclopropanecarboxylate
(5-Benzyl-3-furyl)methyl 2,2-dimethyl-3-(2-methylpropenyl)cyclopropanecarboxylate (approx.70% *trans*, 30% *cis* isomers)
(5-Benzyl-3-furyl)methyl-2,2-dimethyl-3-(2-methylpropenyl)cyclopropane carboxylate
(5-Benzyl-3-furyl)methyl-2,2-dimethyl-3-(2-methylpropenyl)cyclopropanecarboxylate
d-*trans*-5-Phenylmethyl)-3-Furanyl methyl 2,3-dimethyl-3-(2-methyl-1-propenyl) cyclopropane carboxylate
[5-(Phenylmethyl)-3-furanyl]methyl 2,2-dimethyl-3-(2-methyl-1-propenyl)cyclopropane carboxylate
Benzofuroline
Benzyfuroline
Bioresmethrin
Chryson
Chrysron
Cismethrin
D-Resmethrin
ENT 27474
FMC 17370
NIA 17370
NIA-17370
NRDC 104 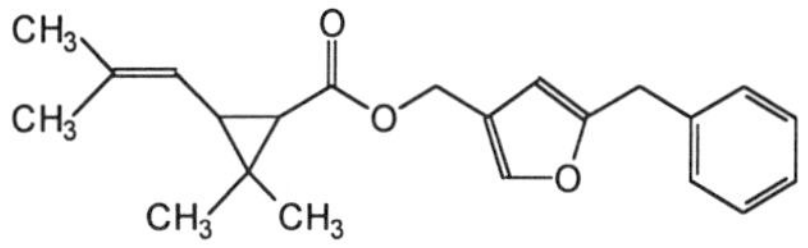
Cyclopropanecarboxylic acid, 2,2-dimethyl-3-(2-methylpropenyl)-, (5-benzyl-3-furyl) methyl ester, (+)-(Z,E)-
Cyclopropanecarboxylic acid, 2,2-dimethyl-3-(2-methylpropenyl)-, (5-benzyl-3-furyl) methyl ester, (-)-(Z)-
Cyclopropanecarboxylic acid, 2,2-dimethyl-3-(2-methylpropenyl)-, (5-benzyl-3-furyl)methyl ester, cis- mixed with *trans*-2,2-dimethyl-3-(2-methylpropenyl)cyclopropanecarboxylic acid(5-benzyl-3-furyl)methyl ester
Cyclopropanecarboxylic acid, 2,2-dimethyl-3-(2-methylpropenyl)-, [4-(2-benzylfuryl) methyl]ester
For-Syn
Penick SBP 1382
Penick 1382
Pyrethroid SBP-1382
Resmethrin (ACN)
Resmethrin racemic mixture
Resmethrin(SBP-1382)
Resmethrin, (+)-, *trans,cis-*
Resmetrina (Portuguese)
Synthrin

Merck Index No.
MP: 71.5 - 83°
BP:
Density:
Solubility:
Octanol/water PC:
LD_{50}: rat orl 1244 mg/kg; rat ihl LC >420 mg/m^3/4H; rat skn 3040 mg/kg; rat ipr LDLo 19200 mg/kg; rat scu >5000 mg/kg; rat ivn LDLo 160 mg/kg; mus orl 300 mg/kg; mus skn >5000 mg/kg; mus ipr >1000 mg/kg; mus scu >2000 mg/kg; dog ihl LC >420 mg/m^3/4H; dog ivn LDLo 250 mg/kg; rbt skn 2500 mg/kg; rbt ivn LDLo 250 mg/kg; mam orl >3375 mg/kg; mam ipr >1000 mg/kg; bwd orl 75 mg/kg.
40CFR185.5300

CAS RN 1134-23-2
$C_{11}H_{21}NOS$

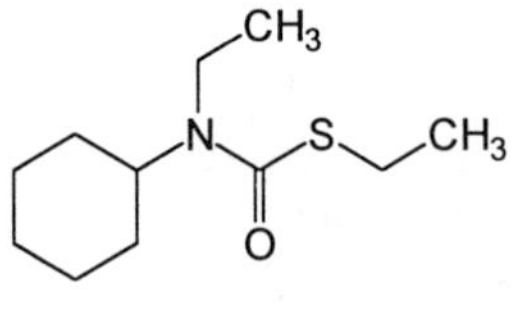

Cycloate
Carbamothioic acid, cyclohexylethyl-, S-ethyl ester (9CI)
Cyclohexanecarbamic acid, N-ethylthio-, S-ethyl ester (8CI)
Carbamic acid, cyclohexylethylthio-, S-ethyl ester
Eurex
Hexylthiocarbam
R 2063
Ro-Neet
Ronit
RoNeet
S-Ethyl (cyclohexyl)ethylthiocarbamate
S-Ethyl cyclohexylethylcarbamothioate
S-Ethyl cyclohexylethylthiocarbamate (ACN)
S-Ethyl ethylcyclohexylthiocarbamate
S-Ethyl [(N-ethyl)N-cyclohexyl]thiocarbamate
S-Ethyl N-cyclohexyl-N-ethyl(thiocarbamate)
S-Ethyl N-ethylcyclohexanecarbamothioate

Merck Index No.
MP: 11.5°
BP: 145 - 146° ^{10}
Density: 1.0156^{30}
Solubility: 75 mg/l water 20°, misc. most org. solvents, e.g. acetone, benzene, methanol, ethanol, isopropanol, xylene, kerosene etc.
Octanol/water PC: 13000
LD_{50}: rat orl 1678 mg/kg; rat orl 2000-3190 mg/kg; rat orl 3160-4100 mg/kg; rat skn 2467 mg/kg; rat unr 2325 mg/kg; mus orl 1275 mg/kg; mus unr 2300 mg/kg; rbt skn 3000 mg/kg; gpg orl 1600 mg/kg; qal unr >2000 mg/kg; mam unr 2300 mg/kg.
40CFR180.212

CAS RN 7778-43-0
$AsH_3O_4.2Na$

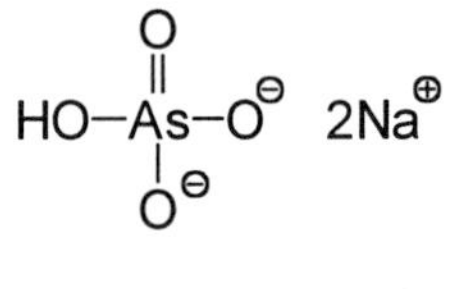

Sodium arsenate (ACN)
Arsenic acid (H_3AsO_4), disodium salt (8CI9CI)
Arsenic acid, disodium salt
Disodium arsenate
Disodium arsenic acid
Disodium hydrogen arsenate
Disodium hydrogen orthoarsenate
Disodium monohydrogen arsenate
Sodium acid arsenate
Sodium arsenate (Na_2HAsO_4)
Sodium arsenate dibasic, anhydrous
Sodium arsenate, exsiccated

Merck Index No. 8522
MP: 57°
BP:
Density: 1.87
Solubility: sol. in 1.3 parts water, sol. glycerol, sl. sol alcohol
Octanol/water PC:
LD_{50}: rat ipr LDLo 34.72 mg/kg.
40CFR180.196 (revoked 5/4/88)

CAS RN 299-84-3
$C_8H_8Cl_3O_3PS$

Ronnel (USAN)
Phosphorothioic acid, O,O-dimethyl O-(2,4,5-trichlorophenyl) ester (8CI9CI)
Blitex
Dermafos
Dimethyl (2,4,5-trichlorophenyl) phosphorothionate
Dow ET 14
Dow ET 57
Ectoral
Etrolene
ENT 23,284
ET 14
ET 57 (VAN)
Fenchloorfos (Dutch)
Fenclofos
Fenclofosum
Gesektin K
Korlan
Korlane
Moor Man's Medicated Rid-Ezy
Moorman's Medicated Rid-Ezy
Nanchor
Nanker
Nankor
O-(2,4,5-Trichloor-fenyl)-O,O-dimethyl-monothiofosfaat (Dutch)
O-(2,4,5-Trichlor-phenyl)-O,O-dimethyl-monothiophosphat (German)
O-(2,4,5-Tricloro-fenil)-O,O-dimetil-monotiofosfato (Italian)
O,O-Dimethyl O-(2,4,5-trichlorophenyl) phosphorothioate (ACN)
O,O-Dimethyl O-(2,4,5-trichlorophenyl) thiophosphate
OMS 123
Phenchlorfos
Phenol, 2,4,5-trichloro-, O-ester with O,O-dimethyl phosphorothioate
Phosphorothionic acid O,O-dimethyl O-(2,4,5-trichlorophenyl) ester
Remelt
Rovan
Thiophosphate de O,o-dimethyle et de o-(2,4,5-trichlorophenyle) (French)
Trichlorometafos

Cl
Cl
S
CH_3O—P—O
OCH_3
Cl

Merck Index No. 8239
MP: 41°
BP:
Density:
Solubility: 40 mg/l water 25°, sol. acetone, carbon tetrachloride, diethyl ether, dichloromethane, toluene, kerosene
Octanol/water PC:
LD_{50}: rat orl 625 mg/kg; rat orl 1250 mg/kg; rat orl 2630 mg/kg; rat skn 2000 mg/kg; rat ipr 2823 mg/kg; rat scu 1380 mg/kg; mus orl 2000 mg/kg; mus ipr 118 mg/kg; dog orl 500 mg/kg; rbt orl 420 mg/kg; rbt skn 1000 mg/kg; gpg orl 1400 mg/kg; gpg skn 2000 mg/kg; ckn orl 6375 mg/kg; dck orl 3500 mg/kg; trk orl 500 mg/kg; dom orl 1000 mg/kg; mam unr 400 mg/kg; bwd orl 80 mg/kg.
40CFR180.177

CAS RN 299-86-5
$C_{12}H_{19}ClNO_3P$

Crufomate (USAN)
Ruelene
Phosphoramidic acid, methyl-, 2-chloro-4-(1,1-dimethylethyl)phenyl methyl ester (9CI)
Phosphoramidic acid, methyl-, 4-*tert*-butyl-2-chlorophenyl methyl ester (8CI)
o-(4-*tert*-Butyl-2-chlor-phenyl)-O-methyl-phosphorsaeure-N-methylamid (German)
Amidofos
Amidophos
Crufomat
Crufomato
Crufomatum
Cruformate
Dowco 132
ENT-25602
Montrel 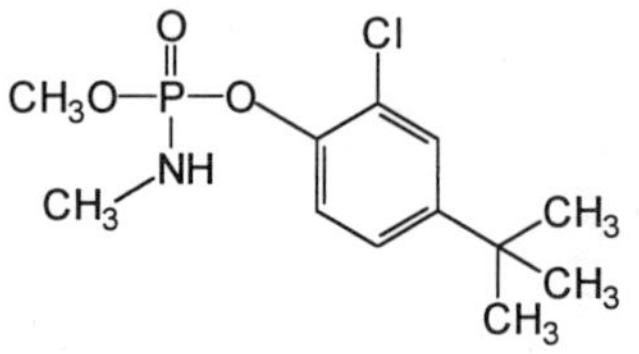
Methylphosphoramidic acid, 2-chloro-4-(1,1-dimethylethyl)phenyl methyl ester
Methylphosphoramidic acid, 4-*tert*-butyl-2-chlorophenyl methyl ester
N,O-Dimethyl-O-(4-*tert*-butyl-2-chlorophenyl)phosphoramidate
O-(4-*tert* Butyl-2-chloor-fenyl)-O-methyl-fosforzuur-N-methyl-amide (Dutch)
O-(4-*Terz*.-butil-2-cloro-fenil)-O-metil-fosforammide (Italian)
O-Methyl O-2-chloro-4-*tert*-butylphenyl N-methylamidophosphate
Phenol, 4-*tert*-butyl-2-chloro-, ester with methyl methylphosphoramidate
Phosphoramidic acid, methyl-, 2-Chloro-4-(1,1-dimethyethyl)phenyl methyl ester
Phosphoramidic acid, methyl-4-*tert*-butyl-2-chlorophenyl methyl ester
Ruelene drench
Rulene
2-Chloro-4-(1,1-dimethylethyl)phenyl methyl methylphosphoramidate
4-*tert*-Butyl-2-chlorophenyl methyl methylphosphoramidate (ACN)
4-*tert*-Butyl-2-chlorophenyl O-methyl methylphosphoroamidate
4-*tert*-Butyl-2-chlorophenylmethyl methylphosphoramidite
4-*Tert*. butyl 2-chlorophenyl methylphosphoramidate de methyle (French)

Merck Index No. 2607
MP: 60 - 60.5°
BP: 117 - 118° $^{0.01}$
Density:
Solubility: insol. water, sol. acetone, benzene, carbon tetrachloride, insol. petroleum ether
Octanol/water PC:
LD_{50}: rat orl 635 mg/kg; rat orl 460 mg/kg; rat ihl LCLo 12 mg/m^3/4H; rat unr 770 mg/kg; dog orl 1000 mg/kg; rbt orl 400 mg/kg; rbt skn 2000 mg/kg; gpg orl 1000 mg/kg; gpg scu LDLo 50 mg/kg; mam orl 251 mg/kg; bwd orl 100 mg/kg.
40CFR180.285

CAS RN 152-16-9
$C_8H_{24}N_4O_3P_2$

Schradan
Diphosphoramide, octamethyl- (9CI)
Pyrophosphoramide, octamethyl- (8CI)
Bis(bisdimethylamino)phosphorous anhydride
Bis(dimethylamino)phosphonous anhydride
Bis(dimethylamino)phosphoric anhydride
Bis(dimethylaminophosphorous) anhydride
Bis-N,N,N',N'-tetramethylphosphorodiamidic anhydride
Bis[bis(dimethylamino)phosphonous]anhydride
ENT-17291
Lethalaire g-59
Octamethyl pyrophosphoramide
Octamethyl pyrophosphortetramide
Octamethyl tetramido pyrophosphate
Octamethyl-difosforzuur-tetramide (Dutch)
Octamethyl-diphosphorsaeure-tetramid (German)
Octamethyldiphosphoramide
Octamethylpyrophosphoramide (ACN)
Octamethylpyrophosphoric acid amide
Octamethylpyrophosphoric acid tetramide
Ompacide
Ompatox
Ompax
Ottometil-pirofosforammide (Italian)
OMPA
Pestox
Pestox III
Pestox 3
Pestox-III
Pestox-3
Pyrophosphoric acid, octamethyltetraamide
Pyrophosphoryltetrakisdimethylamide
Schradane (French)
Schraden
Tetrakisdimethylaminophosphonous anhydride

$(CH_3)_2N-P(=O)(N(CH_3)_2)-O-P(=O)(N(CH_3)_2)-N(CH_3)_2$

Merck Index No. 8351
MP: 14 - 20°
BP: 120 - 125°$^{0.5}$, 154°^{2}
Density: 1.09^{25}
Solubility: misc. water, sol. ketones, nitriles, esters, aromatic hydrocarbons and alcohols, insol. higher aliphatic hydrocarbons
Octanol/water PC:
LD_{50}: man orl TDLo 0.643 mg/kg/30D-I; rat orl 5 mg/kg; rat orl 9.1 mg/kg; rat orl 42 mg/kg; rat ihl LCLo 8 mg/m^3/4H; rat skn 15 mg/kg; rat ipr 4.9 mg/kg; rat scu 9 mg/kg; rat unr 9 mg/kg; mus orl 26.74 mg/kg; mus ipr 10 mg/kg; mus scu 14 mg/kg; mus ivn 9.4 mg/kg; mus unr 28 mg/kg; dog ivn 5 mg/kg; rbt orl 25 mg/kg; rbt skn LDLo 20 mg/kg; rbt ivn 6 mg/kg; rbt ocu LDLo 5mg/kg; gpg ipr 10 mg/kg; bwd orl 11 mg/kg.
40CFR180.166 (revoked 1/15/86)

CAS RN 26259-45-0
$C_{10}H_{19}N_5O$

***sec*-Bumeton** 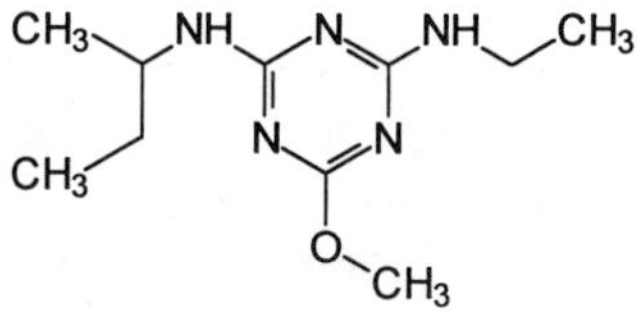
1,3,5-Triazine-2,4-diamine, N-ethyl-6-methoxy-N'-(1-methylpropyl)- (9CI)
s-Triazine,2-(*sec*-butylamino)-4-(ethylamino)-6-methoxy- (8CI)
Etazine
Ezitan
Geigy G.S. 14254
GS 14254
Isobumeton
N-Ethyl-6-methoxy-N-(1-methylpropyl)-1,3,5-triazine-2,4-diamine
N-Ethyl-6-methoxy-N'-(1-methylpropyl)-1,3,5-triazine-2,4-diamine
Secbumeton (ACN)
Sumitol
Sumitol 80W
Terbut
2-(*sec*-Butylamino)-4-(ethylamino)-6-methoxy-*s*-triazine
2-Methoxy-4-(*sec*-butylamino)-6-(aethylamino)-*s*-triazin (German)

Merck Index No.
MP: 86 - 88°
BP:
Density:
Solubility: sp. sol. water
Octanol/water PC:
LD_{50}: rat orl 1000 mg/kg; rbt skn 1910 mg/kg.
40CFR180.323 (revoked 5/4/88)

CAS RN 117-18-0
$C_6HCl_4NO_2$

Tecnazene
TCNB 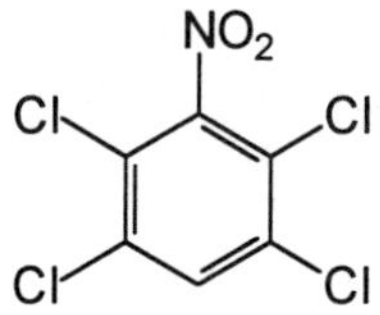
Benzene, 1,2,4,5-tetrachloro-3-nitro- (8CI9CI)
Benzene, 1-nitro-2,3,5,6-tetrachloro-
Benzene, 3-nitro-1,2,4,5-tetrachloro-
Chipman 3,142
Folosan (VAN)
Folosan DB-905
Fusarex
Myfusan
Tecnazen
Tecnazene Fumite
Tetrachloronitrobenzene (VAN)
1,2,4,5-Tetrachloro-3-nitrobenzene (ACN)
2,3,5,6-Tetrachlor-3-nitrobenzol (German)
2,3,5,6-Tetrachloro-1-nitrobenzene
2,3,5,6-Tetrachloronitrobenzene

Merck Index No.
MP: 99°
BP: 304° (dec)
Density: 1.744^{25}
Solubility: 0.44 mg/l water 20°, 40 g/l ethanol, readily sol. benzene, carbon disulfide, chloroform, ketones, aromatic and chlorinated hydrocarbons
Octanol/water PC:
LD_{50}: rat orl 7500 mg/kg; rat orl 2047 mg/kg; rat orl 1256 mg/kg; rat unr 250 mg/kg.
40CFR180.203

CAS RN 136-78-7
$C_8H_8Cl_2O_5S.Na$

Sesone
Ethanol, 2-(2,4-dichlorophenoxy)-, hydrogen sulfate, sodium salt (9CI)
Ethanol, 2-(2,4-dichlorophenoxy)-, hydrogen sulfate sodium salt (8CI)
(2,4-Dichlorophenoxy)ethyl sulfate, sodium salt
Crag herbicide
Crag Herbicide 1
Crag Sesone
Crag SES
Crag 1
Disul
Disul-sodium
Disul-Na
Experimental Herbicide 1
Natrium-2,4-dichlorphenoxyaethylsulfat (German)
Seson
Sodium 2-(2,4-dichlorophenoxy)ethyl sulfate
Sodium 2,4-dichlorophenoxyethyl sulfate
Sodium 2,4-dichlorophenoxyethyl sulphate
Sodium 2,4-dichlorophenoxythyl sulfate
Sodium 2,4-dichlorophenyl cellosolve sulfate
SES
SES-T
2-(2,4-Dichlorophenoxy)ethanol hydrogen sulfate sodium salt
2-(2,4-Dichlorophenoxy)ethyl sulfate, sodium salt (ACN)
2-(2,4-Dichlorophenoxy)ethyl sulfate,sodium salt
2,4-Des sodium
2,4-Des-na
2,4-Des-natrium (German)
2,4-Dichlorophenoxyethyl hydrogen sulfate, sodium salt
2,4-Dichlorophenoxyethyl sulfate sodium
2,4-Dichlorophenoxyethylsulfuric acid, sodium salt
2,4-DES na
2,4-DES sodium

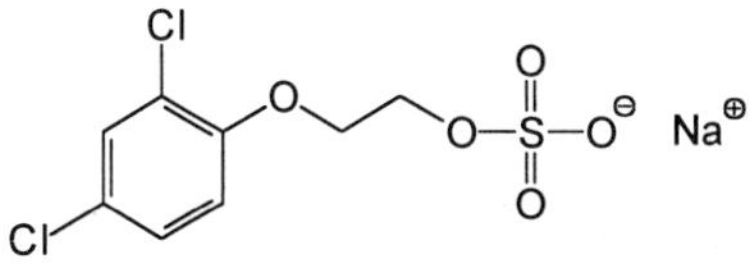

Merck Index No. 3372
MP: 245° (dec)
BP:
Density:
Solubility: 25.5% water 25°, sol. most org. solvents, insol. methanol.
Octanol/water PC:
LD_{50}: rat orl 480 mg/kg; rat orl 730 mg/kg; mam orl 1230 mg/kg.
40CFR180.102

CAS RN 93-72-1
$C_9H_7Cl_3O_3$

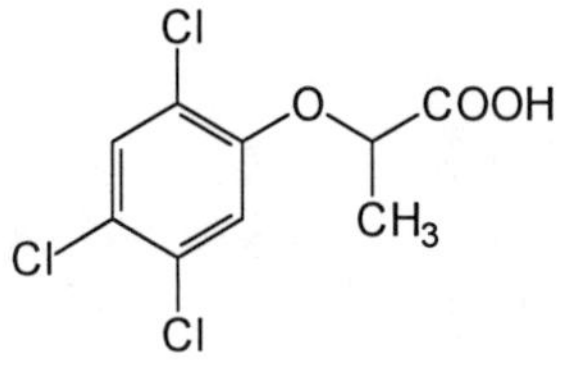

Silvex (ACN)
Propanoic acid, 2-(2,4,5-trichlorophenoxy)- (9CI)
Propionic acid, 2-(2,4,5-trichlorophenoxy)- (8CI)
α-(2,4,5-Trichlorophenoxy)propionic acid
(2,4,5-Trichlorophenoxy)-α-propionic acid
(2,4,5-Trichlorophenoxy)propionic acid (VAN)
Acide 2-(2,4,5-trichloro-phenoxy) propionique (French)
Acido 2-(2,4,5-tricloro-fenossi)-propionico (Italian)
Aqa-Vex
Aqua-Vex
Color-Set
Ded-Weed
Double Strength Kuron
Farmco TP-70-2
Fenoprop
Fenormone
Fruitone T
Herbicides, silvex
Kuran
Kurosal
Kurosal G
Kurosal SL
Propon
Silvex, acid
Silvi-Rhap
Sta-fast
Weed-B-gon
2-(2,4,5-Trichloor-fenoxy)-propionzuur (Dutch)
2-(2,4,5-Trichlor-phenoxy)-propionsaeure (German)
2-(2,4,5-Trichlorophenoxy)propanoic acid
2-(2,4,5-Trichlorophenoxy)propionic acid (ACN)
2-(2,4,5-TP)
2-Phenoxypropionic acid, 2,4,5-trichloro-
2,4,5-TCPPA
2,4,5-TP
2,4,5-TP acid

Cl
O
COOH
CH3
Cl
Cl

Merck Index No. 8483
MP: 181.6°
BP:
Density:
Solubility: 0.014% water 25°, 15.2% acetone, 0.16% benzene, 0.024% carbon tetrachloride, 7.13% diethyl ether, 0.017% heptane, 10.5% methanol
Octanol/water PC:
LD_{50}: rat orl 650 mg/kg; mus orl 276 mg/kg; mus ipr LD >250 mg/kg; rbt skn >3200 mg/kg; mam orl 650 mg/kg; mam unr 650 mg/kg.
40CFR180.340

CAS RN 122-34-9
$C_7H_{12}ClN_5$

Simazine
1,3,5-Triazine-2,4-diamine, 6-chloro-N,N'-diethyl- (9CI)
s-Triazine, 2-chloro-4,6-bis(ethylamino)- (8CI)
s-Triazine, 2,4-bis(ethylamino)-6-chloro-
A 2079
Aktinit S
Amizine
Aquazine
Batazina
Bitemol
Bitemol S 50
Bitemol S-50
Cat (herbicide)
Cekusan
Cekuzina-S
CAT
CAT (herbicide)
CDT (VAN)
CET
DCT
ENT-51142
Framed
G 27692
G-27692
Geigy 27,692
Gesapun
Gesatop
Gesatop-50
H 1803
Herbazin
Herbazin 50
Premazine
Primatol S
Princep
Printop
Radocon
Radokor

Merck Index No. 8485
MP: 225 - 227°, 226 - 227°
BP:
Density: 1.302^{20}
Solubility: 5 mg/l water 20°, 400 mg/l methanol, 900 mg/l chloroform, 300 mg/l diethyl ether, 2 mg/l petroleum ether, 3 mg/l pentane
Octanol/water PC: 91.2
LD_{50}: rat orl 971 mg/kg; rat orl >5000 mg/kg; rat ihl LC_{50} 9800 mg/m^3/4H; rat skn >5000 mg/kg; mus orl 5000 mg/kg; mus ivn 100 mg/kg; mus unr 1390 mg/kg; rbt skn >8160 mg/kg; mam orl 2014 mg/kg; mam unr 1400 mg/kg.
40CFR180.213

CAS RN 26628-22-8
N_3Na

$N{\equiv}N{=}N^{\ominus}$ $Na^{\oplus}$

Sodium azide (DOT)(8CI)
Sodium azide ($Na(N_3)$) (9CI)
Azide
Azium
Azoture de sodium (French)
Kazoe
Natriumazid (German)
Natriummazide (Dutch)
NCI-C06462
NSC 3072
Smite
Sodium azide(DOT)
Sodium azide (NaN_3)
Sodium, azoture de (French)
Sodium, azoturo di (Italian)
U-3886

Merck Index No. 8526
MP:
BP:
Density: 1.846
Solubility: 40.16% water 10°, 41.70% water 17°, sol. NH3, sl. sol. alcohol, insol. ether
Octanol/water PC:
LD_{50}: wmn orl LDLo 14 mg/kg; wmn orl TDLo 3 mg/kg; hmn orl TDLo 0.71 mg/kg; man orl LDLo 143 mg/kg; rat orl 27 mg/kg; rat orl 45 mg/kg; rat ipr LDLo 30 mg/kg; rat scu LDLo 35 mg/kg; mus orl 27 mg/kg; mus ipr 28 mg/kg; mus scu LDLo 17 mg/kg; mus ivn 19 mg/kg; mus unr 27 mg/kg; mky ivn LDLo 12 mg/kg; rbt skn 20 mg/kg; rbt scu LDLo 17 mg/kg; bwd orl 23.7 mg/kg.
40CFR180. temp (1975)

CAS RN 128-04-1
$C_3H_7NS_2.Na$

CH3 N S S Na⊕

Sodium dimethyldithiocarbamate (ACN)
Carbamodithioic acid, dimethyl-, sodium salt (9CI)
Carbamic acid, dimethyldithio-, sodium salt (8CI)
Aceto SDD 40
Alcobam NM
Brogdex 555
Dibam
Dibam A
Dimethyldithiocarbamate sodium salt
Dimethyldithiocarbamic acid, sodium salt
DDC
DMDK
Methyl namate
Methyl Namate
N,N-Dimethyldithiocarbamate sodium salt
N,N-Dimethyldithiocarbamic acid, sodium salt
NCI-C02835
Sharstop 204
Sodium dimethylaminecarbodithioate
Sodium dimethylaminocarbodithioate
Sodium N,N-dimethyldithiocarbamate
Sta-fresh 615
Steriseal liquid #40SDDCVulnopol NM
Wing Stop B

Merck Index No.
MP:
BP:
Density:
Solubility:
Octanol/water PC:
LD_{50}: rat orl 1000 mg/kg; rat ipr 1000 mg/kg; mus orl 1500 mg/kg; mus ipr 573 mg/kg.
40CFR180.152

CAS RN 38827-66-6
$C_{21}H_{16}O_2.Na$

Sonar
2-Naphthalenol, 1,1'-methylenebis-, monosodium salt (9CI)
Squaxin
1,1'-Methylenedi-2-naphthol, monosodium salt (ACN)

HO
OH

Merck Index No.
MP: 132°
BP:
Density:
Solubility:
Octanol/water PC:
LD_{50}:
40CFR180.420

CAS RN 57-92-1
$C_{21}H_{39}N_7O_{12}$

Streptomycin (ACN)(8CI)
D-Streptamine,
O-2-deoxy-2-(methylamino)-α-L-glucopyranosy
l-(1→2)-O-5-deoxy-3-C-formyl-α-L-
lyxofuranosyl-
(1→4)-N,N'-bis(aminoiminomethyl)- (9CI)
Agri-strep
Agri-Strep
Agrimycin
Agrimycin 17
Chemform
Gerox
Hokko-Mycin
N-Methyl-L-glucosamidinostreptosidostreptidine
Neodiestreptopab
NSC 14083
NSC-14083
Strepcen
Streptomycin A
Streptomycine
Streptomycinum
Streptomyzin (German)

Merck Index No. 8786
MP:
BP:
Density:

HO HO HOCH2 O NHCH3 O CH3 O HO CHO NH NH2 NH OH HO OH NH NH2 NH

Solubility: (sesquisulfate) >20 g/l water 28°, 0.85 g/l methanol, 0.3 g/l ethanol, 0.01 g/l isopropanol, 0.035 g/l carbon tetrachloride, 0.035 g/l diethyl ether, 0.015 g/l petroleum ether
Octanol/water PC:
LD_{50}: hmn orl TDLo 400 mg/kg/28D-I; hmn ipr TDLo 143 mg/kg; hmn par TDLo 28 mg/kg/D; rat orl >10000 mg/kg; rat orl 9000 mg/kg; rat scu 600 mg/kg; rat ivn LDLo 175 mg/kg; mus ipr 1250 mg/kg; mus scu 400 mg/kg; mus scu 325 mg/kg; mus scu 520 mg/kg; mus ivn 90.2 mg/kg; mus unr 90 mg/kg; dog scu LDLo 300 mg/kg; mky scu LDLo 400 mg/kg; cat orl LDLo 2000 mg/kg; cat scu LDLo 600 mg/kg; cat ivn LDLo 150 mg/kg; rbt scu LDLo 600 mg/kg; rbt ivn LDLo 175 mg/kg; gpg scu LDLo 600 mg/kg.
40CFR180.245

CAS RN 8001-50-1
W99

Structure information not available for this compound.

Strobane (9CI)
Compound 3961
Dichloricide Aerosol
Dichloricide Mothproofer
ENT 19,442
Insecticide 3960-x14
Strobane(R)
Terpene polychlorinate
Terpene polychlorinates
Terpene polychlorinates (65% or 66% chlorine)consists of chlorinated camphene, pinene, and related polychlorinates
Terpene polychlorinates* consists of chlorinated camphene, pinene, and related polychlorinated *(65% or 66% chlorine)
3960-X14

Merck Index No. 8796
MP:
BP:
Density: 1.6267
Solubility:
Octanol/water PC:
LD_{50}: rat orl 200 mg/kg; rat unr 250 mg/kg; mus orl 200 mg/kg; dog orl 200 mg/kg.
40CFR180.164 (removed 1/15/86)

CAS RN 7446-09-5
O_2S

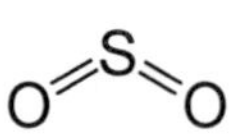

Sulfur dioxide (ACN)(DOT)(8CI9CI)
Fermenicide liquid
Fermenicide powder
Fermenticide liquid
Siarki dwutlenek (Polish)
Sulfur oxide
Sulfur oxide (SO_2)
Sulfur oxide (S_2O_4)
Sulfurous acid anhydride
Sulfurous anhydride
Sulfurous oxide
Sulphur dioxide

Merck Index No. 8950
MP: -72°
BP: -10°
Density: 1.5
Solubility: 17.7% water 0°, 11.9% water 15°, 8.5% water 25°, 6.4% water 35°, 25% ethanol, 32% methanol, sol. chloroform, diethyl ether
Octanol/water PC:
LD_{50}: hmn ihl LCLo 1000 ppm/10M; hmn ihl TCLo 3 ppm/5D; hmn ihl TCLo 12 ppm/1H; hmn ihl LCLo 3000 ppm/5M; rat ihl LC_{50} 2520 ppm/1H; mus ihl LC_{50} 3000 ppm/30M; gpg ihl LCLo 1039 ppm/24H; frg ihl LCLo 1 pph/15M; mam ihl LCLo 3000 ppm/5M.
40CFR180.444

CAS RN 80-00-2
$C_{12}H_9ClO_2S$

Sulphenone
Benzene, 1-chloro-4-(phenylsulfonyl)- (9CI)
Sulfone, *p*-chlorophenyl phenyl (8CI)
p-Chlorophenyl phenyl sulfone (ACN)
p-Monochlorophenyl phenyl sulfone
r-242-b
Compound r-242
ENT 17,941
ENT-1741
R 242
Sulfenon
Sulfenone
Trifenson
4-Chlorodiphenyl sulfone
4-Chlorophenyl phenyl sulfone

Merck Index No. 8969
MP: 94°
BP:
Density:
Solubility: Insoluble in water, 744 g/l acetone, 656 g/l dioxane, 210 g/l isopropanol, 4 g/l hexane, 444 g/l benzene, 294 g/l toluene, 182 g/l xylene, 49 g/l carbon tetrachloride. Slightly soluble in petroleum oils.
Octanol/water PC:
LD_{50}: rat orl 1400 mg/kg; rat orl 2700 mg/kg; rat ipr LDLo 500 mg/kg; mus orl 2700 mg/kg; mus ipr 1000 mg/kg; rbt skn >1000 mg/kg;
40CFR180.112 (revoked 5/4/88)

CAS RN 2008-41-5
$C_{11}H_{23}NOS$

Butylate
Sutan
Carbamothioic acid, bis(2-methylpropyl)-, S-ethyl ester (9CI)
Carbamic acid, diisobutylthio-, S-ethyl ester (8CI)
Butilate
Ethyl N,N-diisobutylthiocarbamate
Ethyl N,N-diisobutylthiolcarbamate
R 1910
R1910
S-Ethyl bis(2-methylpropyl)carbamothioate
S-Ethyl diisobutyl thiocarbamate
S-Ethyl diisobutylthiocarbamate (ACN)
S-Ethyl N,N-diisobutyl thiolcarbamate
S-Ethyl N,N-diisobutylthiocarbamate
Stauffer R-1,910

C_2H_5S—C(=O)—N($CH_2CH(CH_3)_2$)$_2$

Merck Index No. 1546
MP:
BP: 137.5 - 138° ^{21}
Density: 0.9402^{25}, 0.9417
Solubility: 45 mg/l water 22°, misc. acetone, ethanol, xylene, methyl isobutyl ketone, kerosene
Octanol/water PC: 14000
LD_{50}: rat orl 4000 mg/kg; rat orl 4659 mg/kg; rat orl 5431 mg/kg; mus ipr 365 mg/kg; rbt skn >5000 mg/kg.
40CFR180.

CAS RN 93-76-5
$C_8H_5Cl_3O_3$

2,4,5T 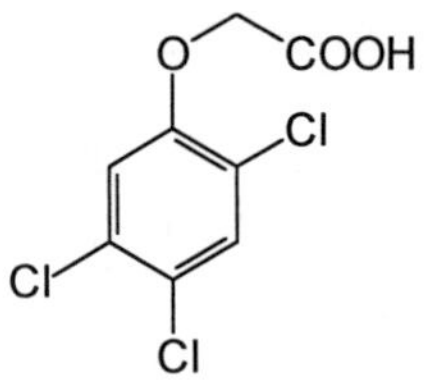
Acetic acid, (2,4,5-trichlorophenoxy)- (8CI9CI)
(Trichlorophenoxy)acetic acid
(2,4,5-Trichloor-fenoxy)-azijnzuur (Dutch)
(2,4,5-Trichlor-phenoxy)-essigsaeure (German)
(2,4,5-Trichlorophenoxy)acetic acid (ACN)
(2,4,5-Trichlorophenoxyacetic) acid
Acide 2,4,5-trichloro phenoxyacetique (French)
Acido (2,4,5-tricloro-fenossi)-acetico (Italian)
Brush rhap
Brush-off 445 low volatile brush killer
Brush-Khap
Brushtox
BCF-Bushkiller
Co-Op Concentrated 2,4,5-T
Dacamine
Dacamine 4T
Debroussaillant Concentre
Debroussaillant Super Concentre
Decamine 4T
Ded-weed brush killer
Ded-weed lv-6 brush kil and t-5 brush kil
Ded-Weed
Dinoxol
Envert-T
Estercide t-2 and t-245
Esteron
Esteron Brush Killer
Esteron 245
Fence rider
Forron
Forst U 46
Fortex (VAN)
Fruitone A
Inverton 245
Line rider
Phenoxyacetic acid, 2,4,5-trichloro-
Phortox
Reddon
Reddox
Spontox

O COOH Cl Cl Cl

Merck Index No. 8999
MP: 153°
BP:
Density: 1.80^2
Solubility: 238 mg/l water 20°, sol. alcohol
Octanol/water PC:

LD_{50}: rat orl 500 mg/kg; rat orl 300 mg/kg; rat skn 1535 mg/kg; rat unr 500 mg/kg; mus orl 242 mg/kg; mus orl 389 mg/kg; dog orl 100 mg/kg; gpg orl 381 mg/kg; ham orl 425 mg/kg; ckn orl 310 mg/kg; mam orl 500 mg/kg.
40CFR180. (pend)

CAS RN 28300-74-5
$C_8H_4O_{12}Sb_2.3H_2O.2K$

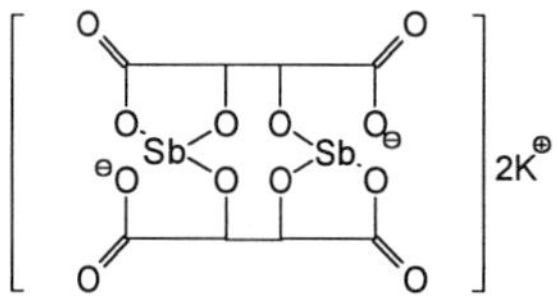

Tartar emetic
Antimonate(2-), bis[μ-tartrato(4-)]di-, dipotassium, trihydrate (8CI)
Antimonate(1-), aqua[tartrato(4-)]-, potassium, hemihydrate, dimer
Antimonate(1-), oxo(tartrato)-, potassium hemihydrate, dimer
Antimonate(2)-, bis[μ-tartrato(4-)]di-, dipotassium, trihydrate
Antimony potassium tartrate (VAN)
Antimonyl potassium tartrate, hemihydrate
Emetique (French)
ENT 50,434
Antimonate(2-), bis[μ.-[2,3-dihydroxybutanedioato(4-)-O1,O2:O3,O4]]di-, dipotassium, trihydrate, stereoisomer
Potassium antimony tartrate
Potassium antimonyl tartrate
Potassium antimonyl tartrate, hydrate
Potassium antimonyl D-tartrate
Potassium antimonyltartrate
Potassium di-μ-tartratodiantimonate(III), trihydrate
Tartarized antimony
Tarter emitic
Tartox
Tartrate antimonio-potassique (French)
Tartrated antimony

Merck Index No. 732
MP:
BP:
Density: 2.6
Solubility: 1 g in 12 ml. water, 3 ml boiling water, 1 g in 15 ml glycerol, insol. alcohol
Octanol/water PC:
LD_{50}: hmn orl LDLo 2 mg/kg; hmn ivn TDLo 1.392 mg/kg; man ivn LDLo 12 mg/kg/1W-I; man ivn 249 mg/kg/9D-I; rat orl 115 mg/kg; rat ipr 11 mg/kg; rat ims 33mg/kg; mus orl LDLo 600 mg/kg; mus ipr 33 mg/kg; mus scu 55 mg/kg; mus ivn 65 mg/kg; mus ivn 45 mg/kg; rbt orl 115 mg/kg; rbt ivn 12 mg/kg; gpg irp 15 mg/kg; gpg ims LDLo 55 mg/kg.
40CFR180.179

CAS RN 76-03-9
$C_2HCl_3O_2$

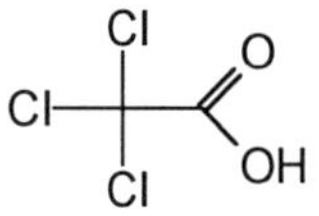

Trichloroacetic acid (ACN)
Acetic acid, trichloro- (8CI9CI)
Aceto caustin
Acide trichloracetique (French)
Acido tricloroacetico (Italian)
Amchem Grass Killer
Konesta
Trichloorazijnzuur (Dutch)
Trichloracetic acid
Trichloressigsaeure (German)
Trichloroacetic acid solid (DOT)
Trichloroacetic acid solution (DOT)
Trichloroethanoic aci
TCA
Varitox

Merck Index No. 9539
MP: 57 - 58°
BP: 196 - 197°
Density: 1.629^{61}
Solubility: Soluble in 0.1 part water, very soluble in alcohol, ether.
Octanol/water PC:
LD_{50}: rat orl 5000 mg/kg; mus ipr LDLo 500 mg/kg; mus scu 270 mg/kg.
40CFR180.310

CAS RN 34014-18-1
$C_9H_{16}N_4OS$

Tebuthiuron (ACN)
Urea, N-[5-(1,1-dimethylethyl)-1,3,4-thiadiazol-2-yl]-N,N'-dimethyl- (9CI)
Urea, 1-(5-*tert*-butyl-1,3,4-thiadiazol-2-yl)-1,3-dimethyl- (8CI)
N-(5-(1,1-Dimethylaethyl)-1,3,4-thiadiazol-2-yl)-N,N'-dimethylharnstoff (German)
N-[5-(1,1-Dimethylethyl)-1,3,4-thiadiazol-2-yl]-N,N'-dimethylurea (ACN)
N,N'-Dimethyl-N-[5-(1,1-dimethylethyl)-1,3,4-thiadiazol-2-yl]urea
Urea, 1-[5-(*tert*-butyl)-1,3,4-thiadiazol-2-yl]-1,3-dimethyl-
Urea, 1,2-dimethyl-1-[5-(1,1-dimethylethyl)-1,3,4-thiadiazol-2-yl]-
1-(5-*tert*-Butyl-1,3,4-thiadiazol-2-yl)-1,3-dimethylurea
1-(5-*tert*-Butyl-1,3,4-thiadiazol-2yl)-3-dimethylharnstoff (German)
Perfmid
Preflan
Prefmid
Spike
SPIKE
Tebulan
Tiurolan
Brulan
E-103
EL 103
EL-103

$(CH_3)_3C$ N—N O S N NHCH$_3$ CH$_3$

Merck Index No. 9053
MP: 161.5 - 164° (dec), 160 - 163
BP:
Density:
Solubility: 2.5 g/l water 25°, 3.7 g/l benzene, 6.1 g/l hexane, 60 g/l 2-methoxyethanol, 60 g/l acetonitrile, 70 g/l acetone, 170 g/l methanol, 250 g/l chloroform
Octanol/water PC: 61 (pH 7 25)
LD_{50}: rat orl 644 mg/kg; rat ihl LCLo >3696 mg/m^3; rat skn >5000 mg/kg; rat ipr 480 mg/kg; rat scu 500 mg/kg; mus orl 579 mg/kg; mus ipr 505 mg/kg; mus scu 545 mg/kg; dog orl >500 mg/kg; cat orl >200 mg/kg; rbt orl 286 mg/kg; ckn orl >500 mg/kg; qal orl >500 mg/kg; dck orl >500 mg/kg.
40CFR180.390

CAS RN 5902-51-2
$C_9H_{13}ClN_2O_2$

Terbacil (ACN)
2,4(1H,3H)-Pyrimidinedione, 5-chloro-3-(1,1-dimethylethyl)-6-methyl- (9CI)
5-Chloro-3-(1,1-dimethylethyl)-6-methyl-2,4(1H,3H)-pyrimidinedione
Uracil, 3-*tert*-butyl-5-chloro-6-methyl- (8CI)
Compound 732
Du Pont 732
Experimental Herbicide 732
Sinbar
Turbacil
3-*tert*-Butyl-5-chloro-6-methyluracil (ACN)
5-Chloro-3-*tert*-butyl-6-methyluracil

H CH$_3$ N O Cl N C(CH$_3$)$_3$ O

Merck Index No. 9085
MP: 175 - 177°
BP: sublimes < 175°
Density: 1.34^{25}
Solubility: 710 mg/l water 25°, 337 g/kg dimethylformamide, 220 g/kg cyclohexanone, 121 g/kg methyl isobutyl ketone, 88 g/kg butyl acetate, 65 g/kg xylene, sp. sol. mineral oils and aliphatic hydrocarbons, readily sol. in strong aq. alkali
Octanol/water PC:
LD_{50}: rat orl >5000 mg/kg; rat par 7500 mg/kg; mam unr 5000 mg/kg.
40CFR180.209

CAS RN 107-49-3
$C_8H_{20}O_7P_2$

TEPP
Diphosphoric acid, tetraethyl ester (9CI)
Pyrophosphoric acid, tetraethyl ester (8CI)
Bis[O,O-diethylphosphoric anhydride]
Bladan (VAN)
Ethyl pyrophosphate
Ethyl pyrophosphate ($Et_4P_2O_7$)
Ethyl pyrophosphate, tetra-
ENT 18,771
Fosvex
Grisol
Hexamite
HEPT
Insecticides contg. tetraethylpyrophosphate
Killax
Kilmite 40
Lethalaire g-52
Lirohex
Mortopal
Nifos
Nifost
O,O,O,O-Tetraaethyl-diphosphat, bis(O,O-diaethylphosphorsaeure-anhydrid (German)
O,O,O,O-Tetraethyl-difosfaat (Dutch)
O,O,O,O-Tetraetil-pirofosfato (Italian)
Pyrophosphate de tetraethyle (French)
Pyrophosphoric acid tetraethyl ester (liquid mixture)
Pyrophosphoric acid, tetraethyl ester (dry mixture)
Tetraethyl diphosphate
Tetraethyl pyrofosfaat(belgian)
Tetrastigmine
Tetron
Tetron-100
TEP
Vapotone

Merck Index No. 9138
MP: 170 - 213° (dec)
BP: $82^{\circ\,0.05}$, $124^{\circ\,1.0}$, $138^{\circ\,2.3}$
Density:
Solubility: misc. water (hydrolyzed), misc, acetone, ethanol, methanol, benzene, chloroform, carbon tetrachloride, glycerol, ethylene glycol, propylene glycol, toluene, xylene, not misc. petroleum ether, kerosene
Octanol/water PC:
LD_{50}: hmn orl LDLo 1.429 mg/kg; hmn orl TDLo 0.309 mg/kg; hmn ims 0.286 mg/kg; hmn par TDLo 0.071 mg/kg; rat orl 0.500 mg/kg; rat orl 1.1 mg/kg; rat skn 2.4 mg/kg; rat ipr 0.650 mg/kg; rat scu 0.279 mg/kg; rat ivn 0.30 mg/kg; rat ims 1.8 mg/kg; rat unr 1.12 mg/kg; mus orl 3 mg/kg; mus skn 8 mg/kg; mus ipr 0.83 mg/kg; mus scu 0.50 mg/kg; mus ivn 0.20 mg/kg; rbt scu 2 mg/kg; rbt ivn LDLo 0.30 mg/kg; rbt unr 1 mg/kg; gpg orl 2.3 mg/kg; pgn ivn 0.18 mg/kg; dck orl 3.56 mg/kg; dck skn 64 mg/kg; frg par 34 mg/kg; bwd orl 1.3 mg/kg.
40CFR180.347

CAS RN 5915-41-3
$C_9H_{16}ClN_5$

C_2H_5NH, N, $NHC(CH_3)_3$, N, N, Cl

Terbuthylazine (ACN)
1,3,5-Triazine-2,4-diamine, 6-chloro-N-(1,1-dimethylethyl)-N'-ethyl- (9CI)
s-Triazine, 2-(*tert*-butylamino)-4-chloro-6-(ethylamino)- (8CI)
G 13529
Gardoprim
GS 13529
Primatol M
Primatol-M80
Sorgoprim
Terbutylethylazine
2-(*tert*-Butylamino)-4-chloro-6-(ethylamino)-*s*-triazine (ACN)
2-(*tert*-Butylamino)-4-chloro-6-(ethylamino)-1,3,5-tria zine
2-(*tert*-Butylamino)-6-chloro-4-(ethylamino)-*s*-triazine
2-Chloro-4-(ethylamino)-6-(*tert*-butylamino)-*s*-triazine
6-Chloro-N-(1,1-dimethylethyl)-N'-ethyl-1,3,5-triazine -2,4-diamine

Merck Index No.
MP: 177 - 179°
BP:
Density: 1.188^{20}
Solubility: 8.5 mg/l water 20°, 100 g/l dimethylformamide, 40 g/l ethyl acetate, 10 g/l isopropanol, 10 g/l tetralin, 10 g/l xylene, 14.3 g/l 1-octanol
Octanol/water PC: 1096
LD_{50}: rat orl 2160 mg/kg; rat orl 1845 mg/kg; rat ihl LC_{50} >3510 mg/m^3; rat skn >3000 mg/kg; rat par 2160 mg/kg; rat unr 2500 mg/kg; rbt skn >3000 mg/kg.
40CFR180.333

CAS RN 2593-15-9
$C_5H_5Cl_3N_2OS$

Cl_3C, N, OC_2H_5, N—S

Etridiazole
Terrazole
1,2,4-Thiadiazole, 5-ethoxy-3-(trichloromethyl)- (8CI9CI)
Aaterra
Banrot
Echlomezol
Echlomezole
Ethazol
Ethazole
Ethazole (fungicide)
Etridiazol
ETCMTB
ETMT
Koban
MF-344
Olin Mathieson 2,424
OM 2425
Pansoil
Terrachlor-super X
Terracoat
Terracoat 121
Terracoat L21
Truban
3-(Trichloromethyl)-5-ethoxy-1,2,4-thiadiazole
5-Aethoxy-3-trichlormethyl-1,2,4-thiadiazol(German)
5-Ethoxy-3-(trichloromethyl)-1,2,4-thiadiazole(ACN)
5-Ethoxy-3-trichloromethyl-1,2,4-thiadiazol

Merck Index No.
MP: 19.9°
BP: 95° [1]
Density: 1.503^{25}
Solubility: 50 mg/l water 25°, readily sol. acetone, cyclohexanone, chloroform, carbon tetrachloride, benzene, xylene, ethanol
Octanol/water PC: 300-400
LD_{50}: rat orl 1077 mg/kg; rat orl 1100 mg/kg; mus orl 2000 mg/kg; dog orl >5000 mg/kg; rbt orl 779 mg/kg; rbt skn 1700 mg/kg.
40CFR180.370

CAS RN 116-29-0
$C_{12}H_6Cl_4O_2S$

Tetradifon (ACN)
Benzene, 1,2,4-trichloro-5-[(4-chlorophenyl)sulfonyl]- (9CI)
Sulfone, *p*-chlorophenyl 2,4,5-trichlorophenyl (8CI)
p-Chlorophenyl 2,4,5-trichlorophenyl sulfone
Akaritox
Aredion
Duphar
ENT 23,737
FMC 5488
Mition
NIA 5488
NIA-5488
Polacaritox
Roztoczol
Roztoczol extra
Roztozol
Sulfone, 2,4,4',5-tetrachlorodiphenyl
Tedion
Tedion V-18
Tetradichlone
Tetradiphon
Tetrafidon
V-18
2,4,4',5-Tetrachloor-difenyl-sulfon (Dutch)
2,4,4',5-Tetrachlor-diphenyl-sulfon (German)
2,4,4',5-Tetrachlorodiphenyl sulfone
2,4,4',5-Tetracloro-difenil-solfone (Italian)
2,4,5,4'-Tetrachlorodiphenyl sulfone
4-Chlorophenyl 2,4,5-trichlorophenyl sulfone (ACN)
4-Chlorophenyl, 2,4,5-trichlorophenyl sulfone
4'-Chlorophenol 2,4,5-trichlorophenyl sulfone
4'-Chlorophenyl 2,4,5-trichlorophenyl sulfone

Cl Cl Cl S Cl O O

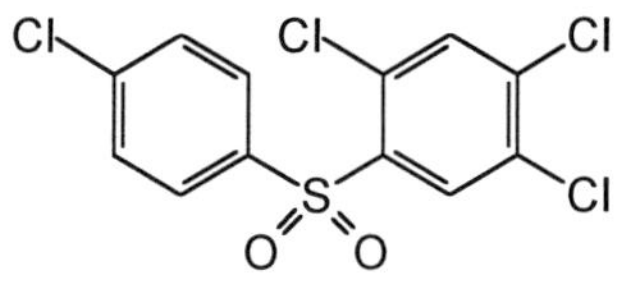

Merck Index No. 9132
MP: 148 - 149°, ≥144° (technical), 146.5 - 147.5°
BP:
Density: 1.151^{20}
Solubility: 0.05 mg/l water 10°, 0.08 mg/l water 20°, 0.34 mg/l water 50°, 82 g/l acetone, 148 g/l benzene, 255 g/l chloroform, 200 g/l cyclohexanone, 223 g/l dioxane, 10 g/l kerosene, 10 g/l methanol, 135 g/l toluene, 115 g/l xylene
Octanol/water PC:
LD_{50}: rat orl >14700 mg/kg, rat orl 556 mg/kg; rat unr 5000 mg/kg; dog orl 2000 mg/kg; rbt skn 10000 mg/kg.
40CFR180.174

CAS RN 513-92-8
C_2I_4

Tetraiodoethylene (ACN)
Ethene, tetraiodo- (9CI)
Ethylene, tetraiodo- (8CI)
Diiodoform
Ethylene periodide
Ethylene tetraiodide
Iodoethylene
Periodoethene
Periodoethylene
Tetraiodoethene

Merck Index No. 9151
MP: 187°
BP:
Density: 2.98
Solubility: insol. water, sol. benzene, toluene, carbon disulfide, chloroform, sl. sol. diethyl ether
Octanol/water PC:
LD_{50}: rat orl 3300 mg/kg; mus orl 3810 mg/kg; mus ivn 56 mg/kg.
40CFR180.162

CAS RN 148-79-8
$C_{10}H_7N_3S$

Thiabendazole (USAN)
1H-Benzimidazole, 2-(4-thiazolyl)- (9CI)
Benzimidazole, 2-(4-thiazolyl)- (8CI)
Apl-Luster
Arbotect
Bioguard
Bovizole
Eprofil
Equizole
Lombristop
Mertec
Mertect
Mertect 160
Metasol TK 100
Mintesol
Mintezol
Minzolum
Mycozol
MK 360
MK-360
Nemapan
Omnizole
Polival
Tbz
Tebuzate
Tecto
Tecto RPH
Tecto 10P
Tecto 40F
Tecto 60
Testo
Thiaben
Thiabendazol
Thiprazole
Tiabenda

Merck Index No. 9217
MP: 304 - 305° (sublimes >310°)
BP:
Density:
Solubility: <50 mg/l water pH 5-12, 250 mg/l water pH 2-5, 10 g/l water pH 2, 4.2 g/l acetone, 7.9 g/l ethanol, 2.1 g/l ethyl acetate, 39 g/l dimethylformamide, 0.23 g/l benzene, 0.08 g/l chloroform, 80 g/l dimethylsulfoxide, 9.3 g/l methanol
Octanol/water PC:
LD_{50}: man orl TDLo 47.619 mg/kg/1D-I; rat orl 2080 mg/kg; mus orl 1300 mg/kg; rbt orl 3580 mg/kg; ckn orl 4000 mg/kg; dom orl 400 mg/kg.
40CFR180.242

CAS RN 51707-55-2
$C_9H_8N_4OS$

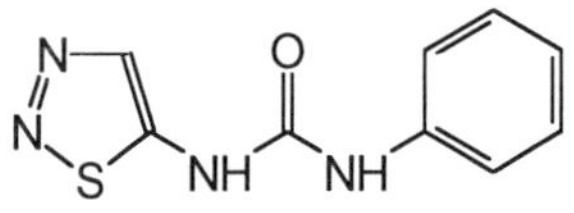

Thidiazuron (ACN)
Urea, N-phenyl-N'-1,2,3-thiadiazol-5-yl- (9CI)
Defolit
Dropp
N-Phenyl-N'-(1,2,3-thiadiazol-5-yl)urea
N-Phenyl-N'-(1,2,3-thiadiazol-5-ylurea
N-Phenyl-N'-1,2,3-thiadiazol-5-ylurea (ACN)
SN 49537
SN-49537
Thiadiazuron
Urea, 1-phenyl-3-(1,2,3-thiadazol-5-yl)-
1-Phenyl-3-(1,2,3-thiadiazol-5-yl)urea

Merck Index No.
MP: 210.5 - 212.5° (dec)
BP:
Density:
Solubility: 25 mg/l water 31°, >500 g/l dimethylsulfoxide, >500 g/l dimethylformamide, 21.5 g/l cyclohexanone, 4.5 g/l methanol, 8 g/l acetone, 0.8 g/l ethyl acetate, 0.013 g/l chloroform, 0.035 g/l benzene, 0.006 g/l hexane
Octanol/water PC: 59
LD_{50}: rat orl >4000 mg/kg; rat orl 5350 mg/kg; rat ihl LC_{50} >2300 mg/m^3/4H; rat ipr 4200 mg/kg; mus orl >5000 mg/kg; mus orl 3740 mg/kg; rbt orl 7100 mg/kg; rbt skn >1000 mg/kg; gpg orl 2813 mg/kg; qal orl >16000 mg/kg.
40CFR180.403

CAS RN 28249-77-6
$C_{12}H_{16}ClNOS$

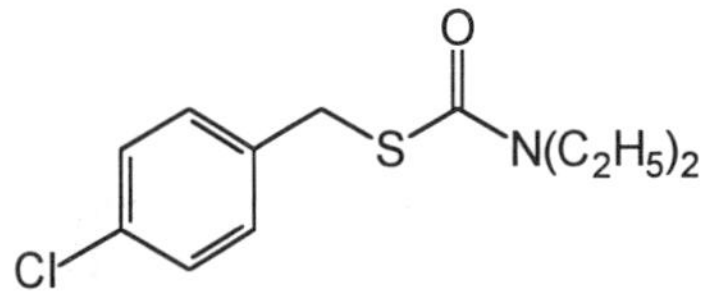

Thiobencarb (ACN)
Carbamothioic acid, diethyl-, S-[(4-chlorophenyl)methyl] ester (9CI)
Carbamic acid, diethylthio-, S-(*p*-chlorobenzyl) ester (8CI)
p-Chlorobenzyl diethylthiolcarbamate
p-Chlorobenzyl N,N-diethylthiolcarbamate
B 3015
Bencarb
Benthiocarb
Bolero
Carbamic acid, diethylthio-, S-(-chlorobenzyl)ester
Carbamothioic acid, diethyl-, S-[(4-chloro-phenyl) methyl)]
IMC 3950
IMC-3950
S-[(4-Chlorophenyl)methyl] diethylcarbamothioate (ACN)
S-4-Chlorobenzyl N,N-diethylthiocarbamate
Saturn
Saturno
5-(4-Chlorobenzyl) N,N-diethylthiol carbamate
5-(4-Chlorobenzyl)-N,N-diethylthiol carbamate

Merck Index No.
MP: 3.3°
BP: 126 - 129° $^{0.008}$
Density: 1.145 - 1.180^{20}
Solubility: 30 mg/l water 20°, readily sol. acetone, ethanol, xylene, methanol, benzene, hexane, acetonitrile
Octanol/water PC: 2630
LD_{50}: rat orl 1300 mg/kg; rat orl 920 mg/kg; rat ihl LC_{50} >7700 mg/m^3/4H; rat skn 2900 mg/kg; mus orl 560 mg/kg; mus skn 10000 mg/kg; mus ipr 1338 mg/kg; ckn orl 673 mg/kg.
40CFR180.401

CAS RN 59669-26-0
$C_{10}H_{18}N_4O_4S_3$

Thiodicarb
N,N'-[thiobis[[(methylimino)carbonyl]oxyl]]bis[ethanimidothioate] (ACN)
Dicarbasulf
Dimethyl Larvin
UC-51762

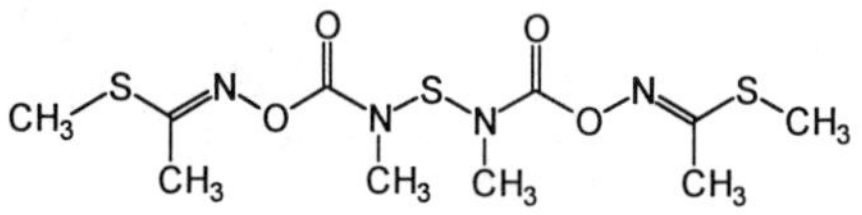

Merck Index No. 9258
MP: 173 - 174°
BP:
Density: 1.4^{20}
Solubility: 35 mg/l water 25°, 150 g/kg dichloromethane, 8 g/kg acetone, 5 g/kg methanol, 3 g/kg xylene
Octanol/water PC:
LD_{50}: rat orl 66 mg/kg (in water); rat orl 120 mg/kg (in corn oil); rat orl 160 mg/kg; rat ihl LC_{50} 520 mg/m^3/4H; rat skn >1600 mg/kg; dog orl 800 mg/kg; rbt orl 556 mg/kg; rbt skn 6310 mg/kg; brd orl 2023 mg/kg.
40CFR180.407

CAS RN 39196-18-4
$C_9H_{18}N_2O_2S$

Thiofanox (ACN)
2-Butanone, 3,3-dimethyl-1-(methylthio)-, O-[(methylamino)carbonyl]oxime (9CI)
Methyl ethyl ketone, 3,3-dimethyl-1-(methylthio)-, O-[(methylamino) carbonyl]oxime
Dacamox
Dacamox 15G
DS 15647
ENT 27851
Thiophanox
3,3-Dimethyl-1-(methylthio)-2-butanone O-(methylamino)carbomyl oxime
3,3-Dimethyl-1-(methylthio)-2-butanone O-[(methylamino)carbamoyl]oxime
3,3-Dimethyl-1-(methylthio)-2-butanone O-[(methylamino)carbonyl]oxime
3,3-Dimethyl-1-(methylthio)-2-butanone O-[(methylamino)carbonyl]oxime (ACN)
3,3-Dimethyl-1-(methylthio)butanone O-methylcarbamoyloxime

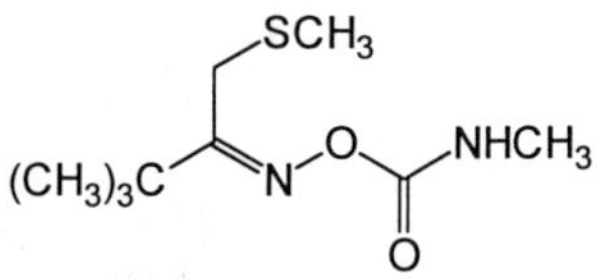

Merck Index No.
MP: 56.5 - 57.5°, 50 - 52°
BP:
Density:
Solubility:
Octanol/water PC:
LD_{50}: rat orl 8.50 mg/kg; rat ihl 70 mg/kg; rat skn 39 mg/kg; rbt skn 39 mg/kg; qal orl 1.2 mg/kg; dck orl 109 mg/kg.
40CFR180. (pend)

CAS RN 297-97-2
$C_8H_{13}N_2O_3PS$

Thionazin
Phosphorothioic acid, O,O-diethyl O-pyrazinyl ester (8CI9CI)
p-Nitrophenyldimethylthionophosphate
American Cyanamid 18133
AC 18133
ACC 18133
Cynem
Cynophos
CL 18133
Diethyl O-2-pyrazinyl phosphorothionate
Ethyl pyrazinyl phosphorothioate
Experimental nematocide 18,133
EN 18133
ENT 25,580
Nemafos
Nemafos 10 G
Nemaphos
Nematocide
Nematocide GR
Nematocide 18133
NCI-C02971
O,O-Diaethyl-O-(pyrazin-2yl)-monothiophosphat (German)
O,O-Diethyl O-(2-pyrazinyl) phosphorothioate
O,O-Diethyl O-(2-pyrazinyl) phosphothionate
O,O-Diethyl O-pyrazinyl phosphorothioate (ACN)
O,O-Diethyl O-pyrazinyl thiophosphate
O,O-Diethyl O-2-pyrazinyl phosphorodithoate
O,O-Diethyl O-2-pyrazinyl phosphorothioate
O,O-Diethyl O-2-pyrazinyl phosphothionate
O,O-Diethyl-O-2-pyrazinyl phosphorodithioate
Phosphorothioic acid, O,O-diethyl O-2-pyrazinyl ester
Pyrazinol, O-ester with O,O-diethyl phosphorothioate
Thionazine
Zinophos
Zynophos

S
C_2H_5O—P—O
OC_2H_5
N
N

Merck Index No. 9275
MP: -1.7°
BP: 80°
Density:
Solubility: sl. sol. water, misc. with most organic solvents
Octanol/water PC:
LD_{50}: rat orl 3.5 mg/kg; rat orl 6.4 mg/kg; rat skn 8 mg/kg; mus orl 5 mg/kg; rbt ocu 50 mg/kg; gpg skn 10 mg/kg; pgn orl 2.37 mg/kg; qal orl 3.16 mg/kg; dck orl 1.68 mg/kg; dck skn 7 mg/kg; bwd orl 2.42 mg/kg.
40CFR180.264 (revoked 5/4/88)

CAS RN 23564-05-8
$C_{12}H_{14}N_4O_4S_2$

Thiophanate-methyl (ACN)
Carbamic acid, [1,2-phenylenebis(iminocarbonothioyl)]bis-, dimethyl ester (9CI)
Allophanic acid, 4,4'-o-phenylenebis[3-thio-, dimethyl ester (8CI)
o-Bis(3-methoxycarbonyl-2-thioureido)benzene
Bis[(3-methoxycarbonyl)-2-thioureido]benzene
BAS 32500F
Cercobin methyl
Cercobin M
Dimethyl [(1,2-phenylene)bis(iminocarbonothioyl)]bis[carbamate]
Dimethyl [1,2-phenylenebis(iminocarbonothioyl)]biscarbamate
Dimethyl 4,4'-*o*-phenylene-bis[3-thioallophanate]
Dimethyl 4,4'-*o*-phenylenebis(3-thioallophanate)
Dimethyl 4,4'-*o*-phenylenebis[3-thioallophanate]
Enovit methyl
Enovit M
Enovit-Supper
Fungitox
Fungo
Fungo 50
Labilite
Methyl thiophamate
Methyl thiophanate
Methyl topsin
Methylthiofanate
Methylthiophanate
Mildothane
Neotopsin
NF 44
NF-44
Pelt 14
Sigma
Thiophanate methyl
Thiophanate-methyl dimethyl
Topsin WP methyl
Trevin

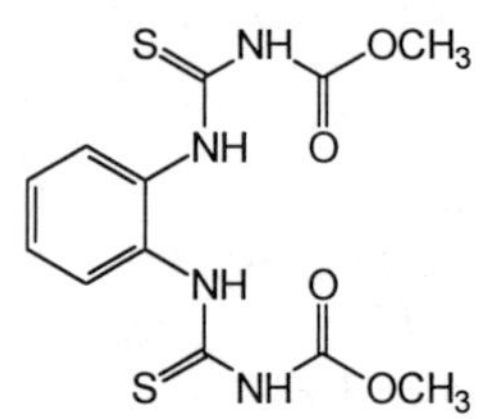

Merck Index No.
MP: 172° (dec)
BP:
Density:
Solubility: 26.6 mg/l water 20°, 58.1 g/kg acetone, 43 g/kg cyclohexanone, 29.2 g/kg methanol, 26.2 g/kg chloroform, 24.4 g/kg acetonitrile, 11.9 g/kg EtOAc, sl. sol. hexane
Octanol/water PC: 25
LD_{50}: rat orl 7500 mg/kg; rat orl 6640 mg/kg; rat ihl LC_{50} 1700 mg/m^3/4H; rat skn >10000 mg/kg; rat ipr 1140 mg/kg; mus orl 3510 mg/kg; mus orl 3510 mg/kg; mus skn >10000 mg/kg; mus ipr 790 mg/kg; dog orl LDLo 4000 mg/kg; rbt orl 2270 mg/kg; rbt skn >10000 mg/kg; gpg orl 3640 mg/kg; gpg skn >10000 mg/kg; qal skn >500 mg/kg.
40CFR180.371

CAS RN 137-26-8
$C_6H_{12}N_2S_4$

Thiram
Thioperoxydicarbonic diamide ($[(H_2N)C(S)]_2S_2$), tetramethyl- (9CI)
Disulfide, bis(dimethylthiocarbamoyl) (8CI)
α,α'-Dithiobis(dimethylthio)formamide
Aapirol
Aatiram
Accel TMT
Accelerator T
Accelerator Thiuram
Aceto TETD
Arasan
Arasan 42-s
Arasan 42-S
Arasan 42S
Arasan 70
Arasan 70-S Red
Arasan 75
Arasan-m
Arasan-sf-x
Arasan-M
Arasan-SF
Atiram
Bis(dimethyl-thiocarbamoyl)-disulfid (German)
Bis(dimethylthiocarbamoyl) disulfide
Bis(dimethylthiocarbamyl) disulfide
Bis(dimethylthiocarbamyl) sulfide
Bis[(dimethylamino)carbonothioyl] disulfide
Bis[dimethylthiocarbamoyl] disulfide
Cyuram DS
Delsan
Disolfuro di tetrametiltiourame (Italian)
Disulfure de tetramethylthiourame (French)
Ekagom tb
Ekagom TB
ENT-987
Falitiram
Fermide
Fernacol
Fernasan
Fernasan A
Fernide

$(CH_3)_2N-C(=S)-S-S-C(=S)-N(CH_3)_2$

Merck Index No. 9304
MP: 146° (technical), 155 - 156°
BP:
Density: 1.29^{20}
Solubility: 30 mg/l water 25°, <10 g/l ethanol, 80 g/l acetone, 230 g/l chloroform
Octanol/water PC:
LD_{50}: hmn ihl TCLo 0.03 mg/m^3/5Y-I; rat orl 560 mg/kg; rat orl 780-865 mg/kg; rat orl 640 mg/kg; rat ihl LC_{50} 500 mg/m^3/4H; rat skn >5000 mg/kg; rat ipr 138 mg/kg; rat scu 646 mg/kg; rat unr 740 mg/kg; mus orl 1350 mg/kg; mus orl 1500-2000 mg/kg; mus ipr 70 mg/kg; mus scu 1109 mg/kg; mus unr 1150 mg/kg; cat orl LDLo 230 mg/kg; rbt orl 210 mg/kg; rbt skn LDLo 1000 mg/kg; rbt unr 210 mg/kg; ckn unr 840 mg/kg; dom unr 225 mg/kg; mam unr 400 mg/kg; bwd orl 300 mg/kg.
40CFR180.132

CAS RN 88-82-4
$C_7H_3I_3O_2$

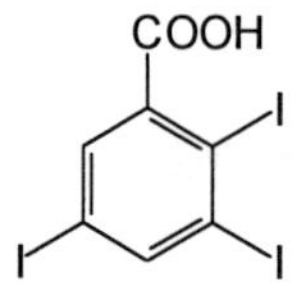

2,3,5-Triiodobenzoic acid (ACN)
Benzoic acid, 2,3,5-triiodo- (8CI9CI)
Floraltone
Johnkolor
Regim 8
Regin 8
Triiodobenzoic acid (VAN)
TIB
TIBA
2,3,5-TIBA

Merck Index No.
MP: 230.8 - 231.2°, 224 - 226°
BP:
Density:
Solubility: v. sol. alcohol, ether, sol. benzene, sp. sol. water
Octanol/water PC:
LD_{50}: rat orl 813 mg/kg; mus orl 700 mg/kg; mus ipr 562 mg/kg.
40CFR180.219

CAS RN 123-88-6
C_3H_7ClHgO

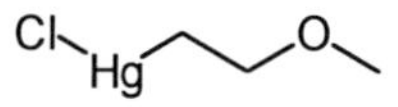

Triadimenol
Mercury, chloro(2-methoxyethyl)- (8CI9CI)
β-Methoxyethylmercury chloride
(β-Methoxyethyl)mercuric chloride
Agallol
Agallol '3'
Agallolat
Agalol
Aratan
Aretan 6
Atiran
Baytan
Celmer
Ceresan Universal Nazbeize
Ceresan Universal Wet
Ceresan-universal nassbeize
Cersan Universal Nazbeize
Chloro(2-methoxyethyl)mercury
Emisan 6
Falisan
Gramisan
Higosan
Merchlorate
Methoxyaethylquecksilbchlorid(German)
Methoxyaethylquecksilberchlorid (German)
Methoxyethyl mercuric chloride
Methoxyethylmercury chloride
MEMC
Sedresan
Tafasan
Tafasan 6W
Tayssato
2-Methoxyethylmercuric chloride
2-Methoxyethylmercury chloride

Merck Index No.
MP:
BP:
Density:
Solubility:
Octanol/water PC:
LD_{50}: rat orl 22 mg/kg; rat unr 50 mg/kg; mus orl 47 mg/kg; mus scu 88 mg/kg.
40CFR180.450

CAS RN 2303-17-5
$C_{10}H_{16}Cl_3NOS$

Triallate
Carbamothioic acid, bis(1-methylethyl)-, S-(2,3,3-trichloro-2-propenyl) ester (9CI)
Carbamic acid, diisopropylthio-, S-(2,3,3-trichloroallyl) ester (8CI)
Avadex BW
Carbamothioic acid, bis(1-methylethyl)-, S-(2,3,3-trichloro-2-propenyl) este
CP 23426
CP-23426
Dipthal
Far-go
N-Diisopropylthiocarbamic acid S-2,3,3-trichloro-2-propenyl ester
N,N-Diisopropyl-2,3,3-trichlorallyl-thiolcarbamat (German)
S-(1,2,3-Trichloroallyl) diisopropylthiocarbamate
S-(2,3,3-Trichloro-2-propenyl) bis(1-methylethyl) carbamo thioate
S-(2,3,3-Trichloroallyl) diisopropylthiocarbamate (ACN)
S-(2,3,3-Trichloroallyl) N,N-diisopropylthiolcarbamate
S-2,3,3-Trichloroallyl N,N-diisopropylthiocarbamate
Thiocarbamic acid, N-diisopropyl-, S-2,3,3-trichloroallyl ester

$(CH_3)_2CH$ $(CH_3)_2CH$ N S O Cl Cl Cl

Merck Index No. 9510
MP: 29 - 30°
BP: 117° $^{40\ mPa}$
Density: 1.273^{25}
Solubility: 4 mg/l water 25°, readily sol. in acetone, diethyl ether, ethyl acetate, ethanol, benzene, heptane
Octanol/water PC:
LD_{50}: rat orl 800 mg/kg; rat orl 1100 mg/kg; rat skn LDLo 3500 mg/kg; rat unr 1471 mg/kg; mus orl 930 mg/kg; mus unr 832 mg/kg; cat ihl LCLo 400 mg/m^3/4H; rbt skn 2225 mg/kg; qal orl 2251 mg/kg.
40CFR180.314

CAS RN 55335-06-3
$C_7H_4Cl_3NO_3$

Triclopyr (ACN)
Acetic acid, [(3,5,6-trichloro-2-pyridinyl)oxy]- (9CI)
(3,5,6-Trichloro-2-pyridinyl)oxy acetic acid (ACN)
[(3,5,6-Trichloro-2-pyridinyl)oxy]acetic acid
Acetic acid, [(3,5,6-trichloro-2-pyridyl)oxy]-
Dowco 233
Garlon
3,5,6-Trichloro-2-pyridyloxyacetic acid

Cl Cl Cl N O COOH

Merck Index No. 9572
MP: 148 - 150°
BP: dec 290°
Density:

Solubility: 440 mg/l water 25°, 989 g/kg acetone, 307 g/kg 1-octanol, 126 g/kg acetonitrile, 27.9 g/kg xylene, 27.3 g/kg benzene, 27.3 g/kg chloroform, 0.41 g/kg hexane
Octanol/water PC:
LD_{50}: rat orl 713 mg/kg; rat orl 630 mg/kg; rbt orl 550 mg/kg; rbt skn >2000 mg/kg; gpg orl 310 mg/kg; dck orl 1693mg/kg.
40CFR180.417

CAS RN 52-68-6
$C_4H_8Cl_3O_4P$

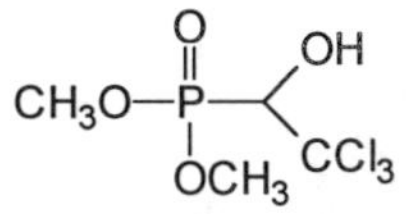

Trichlorfon
Phosphonic acid, (2,2,2-trichloro-1-hydroxyethyl)-, dimethyl ester (8CI9CI)
(1-Hydroxy-2,2,2-trichloroethyl)phosphonic acid, dimethyl ester
(2,2,2-Trichloro-1-hydroxyethyl)phosphonate, dimethyl ester
(2,2,2-Trichloro-1-hydroxyethyl)phosphonic acid dimethyl ester
(2,2,2-Trichloro-1-hydroxyethyl)phosphonic acid, dimethyl ester
[(2,2,2-Trichloro-1-hydroxyethyl) dimethylphosphonate]
Dimethyl (2,2,2-trichloro-1-hydroxyethyl)phosphonate (ACN)
Dimethyl 1-hydroxy-2,2,2-trichloroethylphosphonate
Agroforotox
Anthon
Bay 13/59
Bayer l13/59
Bayer L 13/59
Bayer L 1359
Bayer 2349
Bovinox
Briten
Briton
Britten
BAY 15922
Cekufon
Chlorofos
Chloroftalm
Chlorophthalm
Chloroxyphos
Chlorphos
Ciclosom
Clorofos (Russian)
Combat
Combot
Danex
Dipterax

Merck Index No. 9536
MP: 75 - 79°, 83 - 84°
BP: $100°^{0.1}$
Density: 1.73^{20}
Solubility: 120 g/l water 20°, 154 g/l water 25°, 152 g/kg benzene, 299 g/kg dichloromethane, 200 g/kg isopropanol, 30 g/kg toluene, 750 g/kg chloroform, 170 g/kg diethyl ether, 0.8 g/kg hexane
Octanol/water PC:
LD_{50}: hmn ihl TCLo 1.71 mg/kg/m^3/90D-I; rat orl 560 mg/kg; rat orl 630 mg/kg; rat ihl LC_{50} 1300 mg/m^3; rat skn 2000 mg/kg; rat ipr 160 mg/kg; rat scu 400 mg/kg; rat ims 395 mg/kg; rat unr 550 mg/kg; mus orl 300 mg/kg; mus ipr 196 mg/kg; mus scu 267 mg/kg; mus ivn 290 mg/kg; mus unr 445 mg/kg; dog orl 400 mg/kg; dog scu 269 mg/kg; dog ivn 150 mg/kg; cat orl 97 mg/kg; cat unr 98 mg/kg; rbt orl 160 mg/kg; rbt skn 1500 mg/kg; rbt ipr 60 mg/kg; rbt ims LDLo 500 mg/kg; pgp orl LDLo 420 mg/kg; gpg irp 300 mg/kg; ham irp 1000 mg/kg; ckn orl 75 mg/kg; ckn scu 125 mg/kg; qal orl 73.11 mg/kg; mam skn >2800 mg/kg; mam unr 295 mg/kg; bwd orl 37 mg/kg.
40CFR180.198

CAS RN 87-90-1
$C_3Cl_3N_3O_3$

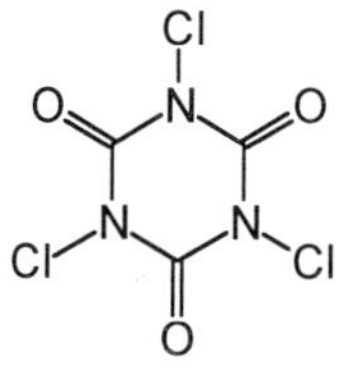

Symclosene (USAN)
Trichloroisocyanuric acid
1,3,5-Triazine-2,4,6(1H,3H,5H)-trione, 1,3,5-trichloro- (9CI)
s-Triazine-2,4,6(1H,3H,5H)-trione, 1,3,5-trichloro- (8CI)
s-Triazinetrione, trichloro-
ACL 85
Chloreal
Fi clor 91
Fi Clor 91
Fichlor 91
Isocyanuric chloride
Kyselina trichloisokyanurova (Czech)
N,N,N''-Trichloroisocyanuric acid
N,N',N -Trichloroisocyanuric acid
NSC-405124
Symclosen
Symelosene
Trichlorinated isocyanuric acid
Trichloro-*s*-triazine-2,4,6(1H,3H,5H)-trione
Trichloro-*s*-triazinetrione (ACN)
Trichloro-*s*-triazinetrione, dry (DOT)
Trichlorocyanuric acid
Trichloroisocyanic acid
Trichlorotriazinetrione
1,3,5-Triazine, 2,4,6(1H,3H,5H)-trione, 1,3,5-trichloro-
1,3,5-Trichloro-*s*-triazine-2,4,6(1H,3H,5H)-trione
1,3,5-Trichloro-*s*-triazine-2,4,6-trione
1,3,5-Trichloro-*s*-triazinetrione
1,3,5-Trichloro-1,3,5-triazine-2,4,6(1H,3H,5H)-trione
1,3,5-Trichloro-2,4,6-trioxohexahydro-*s*-triazine
1,3,5-Trichloroisocyanuric acid

Merck Index No. 8993
MP: 246 - 247° (dec)
BP:
Density:
Solubility: 0.2% water 25°, sol. chlorinated and highly polar solvents
Octanol/water PC:
LD_{50}: hmn orl LDLo 3570 mg/kg; rat orl 406 mg/kg.
40CFR180. pend 1978

CAS RN 41814-78-2
$C_9H_7N_3S$

Tricyclazole (ACN)
1,2,4-Triazolo[3,4-b]benzothiazole, 5-methyl- (9CI)
Beam
Bim
Blascide
EL 291
EL-291
S-Triazolo[3,4-b]benzothiazole, 5-methyl-
5-Methyl-1,2,4-triazole[3,4-b]benzothiazole
5-Methyl-1,2,4-triazolo[3,4-b]benzothiazole (ACN)

Merck Index No.
MP: 187 - 188°
BP:
Density:
Solubility: 1.6 g/l water 25°, 10.4 g/l acetone, 25 g/l methanol, 2.1 g/l xylene
Octanol/water PC: 25
LD_{50}: rat orl 250 mg/kg; rat orl 314 mg/kg; rat skn >5000 mg/kg; mus orl 245 mg/kg; dog orl >50 mg/kg; rbt orl LDLo 320 mg/kg; rbt skn >2000 mg/kg; qal orl >100 mg/kg; dck orl >100 mg/kg.
40CFR180. temp(1980)

CAS RN 24602-86-6
$C_{19}H_{39}NO$

Tridemorph
Morpholine, 2,6-dimethyl-4-tridecyl- (8CI9CI)
BAS 2203F
BAS 2205-F
BASF 220F
Calixin
Calixine
E-236
F 220 (fungicide)
Morpholine, 2,6-dimethyl-N-tridecyl-
N-Tridecyl-2,6-dimethylmorpholine
2,6-Dimethyl-4-tridecylmorpholine (ACN)
4-Tridecyl-2,6-dimethylmorpholine

Merck Index No. 9576
MP:
BP: $134^{\circ\ 0.5}$,$130 - 133^{\circ\ 0.7}$, $139 - 142^{\circ\ 1.3}$
Density: 0.86
Solubility: 11.7 mg/l water 20°, misc. ethanol, acetone, ethyl acetate, cyclohexane, diethyl ether, benzene, chloroform, olive oil
Octanol/water PC: 15800 pH7
LD_{50}: rat orl 480 mg/kg; rat orl 650 mg/kg; mus orl 1560 mg/kg; cat orl 540 mg/kg; rbt orl 750 mg/kg; gpg orl 1000 mg/kg.
40CFR180.372

CAS RN 1582-09-8
$C_{13}H_{16}F_3N_3O_4$

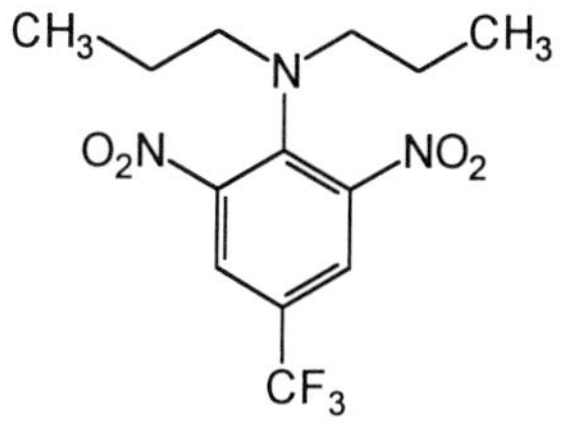

Trifluralin (ACN)
Benzenamine, 2,6-dinitro-N,N-dipropyl-4-(trifluoromethyl)- (9CI)
p-Toluidine, α,α,α-trifluoro-2,6-dinitro-N,N-dipropyl- (8CI)
α,α,α-Trifluoro-2,6-dinitro-N,N-dipropyl-*p*-toluidine
α,α,α-Trifluoro-2,6-dinitro-N,N-dipropyl-*p*-toluidine (ACN)
p-Toluidine, 2,6-dinitro-N,N-dipropyl-α,α,α-trifluoro-
Agreflan
Agriflan 24
Crisalin
Digermin
Elancolan
ENT 28203
L-36352
Lilly 36,352
N,N-Di-N-propyl-2,6-dinitro-4-trifluoromethylaniline
N,N-Dipropyl-4-trifluoromethyl-2,6-dinitroaniline
Nitran
Nitrofor
NCI-C00442
Olitref
Su seguro carpidor
Trefanocide
Treficon
Treflam
Treflan
Treflanocide
Triflan
Trifluoralin
Trifluraline
Trifurex
Trikepin
TRIM
2,6-Dinitro-N,N-di-n-propyl-α,α,α-trifluro-*p*-toluidine
2,6-Dinitro-N,N-dipropyl-4-(trifluoromethyl)aniline
2,6-Dinitro-N,N-dipropyl-4-(trifluoromethyl)benzenamine
2,6-Dinitro-4-trifluormethyl-N,N-dipropylanilin (German)
4-(Di-n-propylamino)-3,5-dinitro-1-(trifluoromethyl)benzene

CH_3 CH_3 N O_2N NO_2 CF_3

Merck Index No. 9598
MP: 48.5 - 49° (technical >95%), 42° (pure)
BP: 139 - 140° $^{4.2}$
Density:
Solubility: <1 mg/l water 27°, 400 g/l acetone, 580 g/l xylene
Octanol/water PC: 118000 (octanol/water pH 7, 25°)
LD_{50}: rat orl 1930 mg/kg; rat orl >10000 mg/kg; rat ihl LC_{50} 2800mg/m^3/1H; rat skn >5000 mg/kg; mus orl 3197 mg/kg; mus orl 500 mg/kg; mus ipr LDLo 1500 mg/kg; dog orl >10000mg/kg; rbt orl >2000 mg/kg; ckn orl >2000 mg/kg; mam orl 3700 mg/kg; mam unr LDLo 5000 mg/kg.
40CFR180.207

CAS RN 26644-46-2
$C_{10}H_{14}Cl_6N_4O_2$

Triforine (ACN)
Formamide, N,N'-[1,4-piperazinediylbis(2,2,2-trichloroethylidene)]bis- (8CI9CI)
Biformychlorazin
Biformylchlorazin
Compd. W
Compound W
CA 70203
CA 73021
CELA 50
CELA-W 524
CME 74770
CW 524
Funginex
N,N-[1,4-Piperazinediylbis(α,α,α-trichloroethylidene)]bis[formamide]
N,N'-[Piperazinediylbis(2,2,2-trichloroethylidene)]bis[formamide]
N,N'-[1,4-Piperazinediylbis(2,2,2-trichloroethylidene)]bis(formamide)
N,N'-[1,4-Piperazinediylbis(2,2,2-trichloroethylidene)]bisformamide
N,N'-Bis(1-formamido-2,2,2-trichloroethyl)piperazine
Piperazin-1,4-diyl-bis[1-(2,2,2-trichloroethyl)formamide]
Piperazine, 1,4-bis(1-formamido-2,2,2-trichloroethyl)-
Saprol
Triforin
W 524
1,4-Bis(1-formamido-2,2,2-trichloroethyl)piperazine
1,4-Bis(2,2,2-trichloro-1-formamidoethyl)piperazine

CCl3
NH CHO
N
N
OHC NH
CCl3

Merck Index No. 9602
MP: 155° (dec)
BP:
Density:
Solubility: 6 mg/l water 25°, 330 g/l dimethylformamide, 476 g/l dimethylsulfoxide, 476 g/l N-methylpyrrolidone, 11 g/l acetone, 1 g/l dichloromethane, 10 g/l methanol, 10 g/l dioxane, 10 g/l cyclohexanone, sol. tetrahydrofuran, insol. benzene, petroleum ether, cyclohexane
Octanol/water PC: 158
LD_{50}: rat orl 6000 mg/kg; rat orl >16000 mg/kg; rat ihl LC_{50} >4500 mg/m^3/1H; rat skn 10000 mg/kg; mus orl >6000 mg/kg; qal orl >5000 mg/kg.
40CFR180.382

CAS RN 13356-08-6
$C_{60}H_{78}OSn_2$

Fenbutatin oxide
Distannoxane, hexakis(2-methyl-2-phenylpropyl)- (9CI)
Distannoxane, hexakis(β,β-dimethylphenethyl)- (8CI)
Bendex
Bis(trineophyltin) oxide
Bis[tris(β,β-dimethylphenethyl)tin]oxide
Bis[tris(2-methyl-2-phenylpropyl)tin] oxide
Di[tri-(2,2-dimethyl-2-phenylethyl)tin]oxide
ENT 27738
Hexakis(β,β-dimethylphenethyl)distannoxane
Hexakis(2-methyl-2-phenylpropyl)distannoxane (ACN)
Neostanox
Stannane, oxybis[tris(β,β-dimethylphenethyl)]-
Torque
Torque (pesticide)
Trineophyltin oxide
Vendex

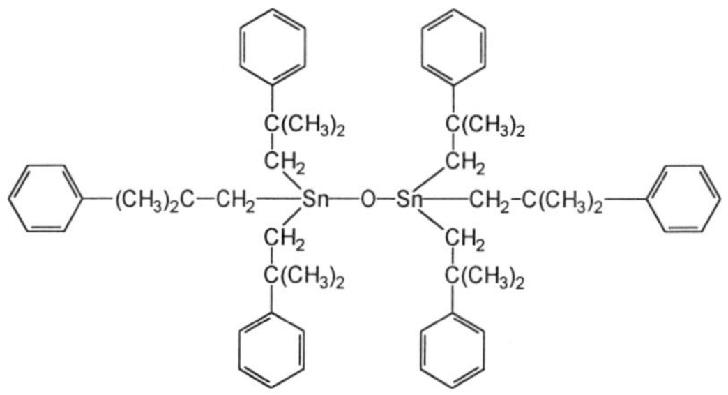

Merck Index No. 3907
MP: 138 - 139°
BP:
Density:
Solubility: 0.005 mg/l water 23°, 6 g/l acetone, 140 g/l benzene, 380 g/l dichloromethane, sl. sol. aliphatic hydrocarbons and mineral oils
Octanol/water PC:
LD_{50}: rat orl 2631 mg/kg; rat skn 1000 mg/kg; mus orl 1450 mg/kg; dog orl >1500 mg/kg; rbt skn >2000 mg/kg; qal orl 0.007 mg/kg.
40CFR180.362

CAS RN 2686-99-9 (2,3,5 isomer; 3971-89-9)
$C_{11}H_{15}NO_2$

Trimethacarb
3,4,5-trimethylphenyl methylcarbamate
Broot

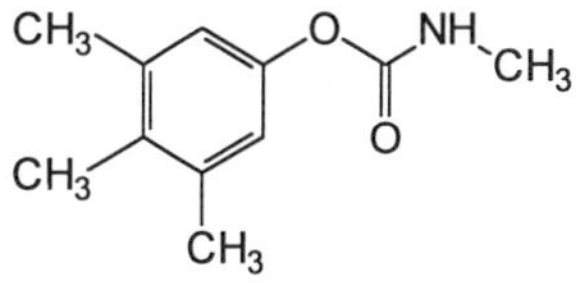

Merck Index No.
MP: 117 - 119°, 122 - 123°
BP:
Density:
Solubility:
Octanol/water PC:
LD_{50}: rat orl 178 mg/kg; rat skn >2000 mg/kg; rat ipr 94.4 mg/kg; rat ivn 32mg/kg; rat ims 283 mg/kg; mus orl 101 mg/kg; mus ipr 420 mg/kg; rbt skn >2500 mg/kg; rbt scu >2500 mg/kg; pgn orl 168 mg/kg; ckn orl 50.3 mg/kg; qal orl 71 mg/kg; dck orl 22 mg/kg; bwd orl 10 mg/kg.
40CFR180.305

CAS RN 1929-77-7
$C_{10}H_{21}NOS$

Vernolate 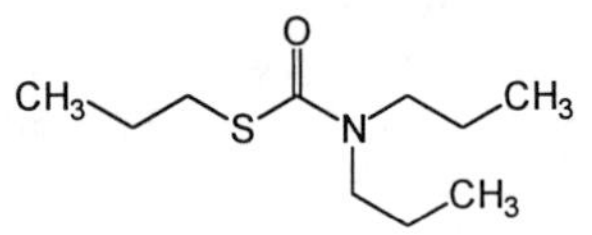
Carbamothioic acid, dipropyl-, S-propyl ester (9CI)
Carbamic acid, dipropylthio-, S-propyl ester (8CI)
n-Propyl-di-N-propyl thiolcarbamate
Dipropylthiocarbamic acid S-propyl ester
n-Propyl-di-N-propylthiolcarbamate
Propyl dipropylthiocarbamate
Propyl dipropylthiolcarbamate
Propyl N,N-dipropylthiolcarbamate
PPTC
R-1607
S-Propyl dipropylcarbamothioate
S-Propyl dipropylthiocarbamate (ACN)
S-Propyl N,N-dipropylthiocarbamate
Vanalate
Vernam
Vernnolaolate

Merck Index No. 9866
MP:
BP: $150°^{30}$
Density: 0.952^{20}, 0.9440^{30}
Solubility: 90 mg/l water 20°, 107 mg/l water 25°, misc. xylene, methyl isobutyl ketone, kerosene etc.
Octanol/water PC: 6918 (octanol/water 20)
LD_{50}: rat orl 1200 mg/kg; rat orl 1500 mg/kg; rat orl 1550 mg/kg; rat orl 1780 mg/kg; rbt skn >9000 mg/kg.
40CFR180.240

CAS RN 50471-44-8
$C_{12}H_9Cl_2NO_3$

Vinclozolin
2,4-Oxazolidinedione, 3-(3,5-dichlorophenyl)-5-ethenyl-5-methyl- (9CI)
Ronilan
3-(3,5-Dichlorophenyl)-5-ethenyl-5-methyl-2,4-oxazolidinedione
3-(3,5-Dichlorophenyl)-5-methyl-5-vinyl-2,4-oxazolidinedione

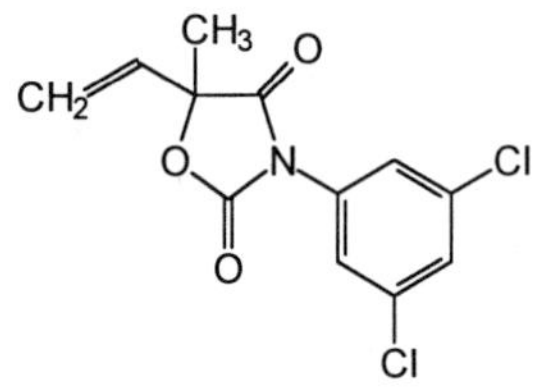

Merck Index No. 9890
MP: 108°
BP: $131°^{0.05}$
Density: 1.51
Solubility: 3.4 mg/l water 20°, 1 g/l water 20°, 14 g/kg ethanol, 435 g/kg acetone, 253 g/kg ethyl acetate, 9 g/kg cyclohexane, 63 g/kg diethyl ether, 146 g/kg benzene, 110 g/kg xylene, 540 g/kg cyclohexanone, 319 g/kg chloroform
Octanol/water PC: 1000
LD_{50}: rat orl >10000 mg/kg; rat ihl LC_{50} >29100 mg/m^3/4H; rat skn >2000 mg/kg; mus orl >10000 mg/kg; rbt ihl LC_{50} 1170 mg/m^3/4H; gpg orl 8000 mg/kg.
40CFR180.380

CAS RN 1314-84-7
P_2Zn_3

Zn=P—Zn—P=Zn

Zinc phosphide (ACN)(DOT)(VAN)
Zinc phosphide (Zn_3P_2) (8CI9CI)
Delusal
Kilrat
Mous-Con
Phosphure de zinc (French)
Phosvin
Rumetan
Trizinc diphosphide
Wuehlmaus-Koeder
Wuehlmaustod arvikol
Wuehlmaustod Arvikol
Zinc(phosphure de) (French)
Zinc-Tox
Zinco(fosfuro di) (Italian)
Zinkfosfide (Dutch)
Zinkphosphid (German)

Merck Index No. 10056
MP: >420°
BP: 1100°
Density: 4.55
Solubility: insol. water, alcohol, sol. benzene, CS2
Octanol/water PC:
LD_{50}: wmn orl LDLo 80 mg/kg; man unr LDLo 40 mg/kg; rat orl 12 mg/kg; rat orl 40.5-46.7 mg/kg; rat ipr 450 mg/kg; rat unr 45 mg/kg; mus orl 40 mg/kg; mus ipr 263 mg/kg; dog orl LDLo 40 mg/kg; cat orl LDLo 40 mg/kg; rbt orl LDLo 40 mg/kg; rbt skn 2000 mg/kg; qal orl 13.5 mg/kg; dck orl 37.5 mg/kg; dom orl 60 mg/kg; bwd orl 23.7 mg/kg.
40CFR180.284

CAS RN 7733-02-0
$H_2O_4S.Zn$

$^{\ominus}O-S(=O)(=O)-O^{\ominus}$ $Zn^{\oplus\oplus}$

Zinc sulfate, basic
Sulfuric acid, zinc salt (1:1) (8CI9CI)
component of Phenylzin
component of Zincfrin
Basic zinc sulfate
Bonazen
Bufopto zinc sulfate
Goslavite
Nuzinc
Op-Thal-Zin
Optraex
Sulfate de zinc (French)
Sulfuric acid, zinc salt (VAN)
Sulfuric acid, zinc salt(1:1)
Verazinc
White vitriol (VAN)
White Copperas
White Vitriol
Zinc sulfate (ACN)
Zinc vitriol (VAN)
Zincate
Zincomed
Zinkosite

Merck Index No. 10064
MP: loses water above 238°
BP:
Density: 1.97 (heptahydrate)
Solubility: sol. water, insol. alcohol, 1 g heptahydrate dissolves in 6 ml water
Octanol/water PC:
LD_{50}: hmn orl TDLo 45 mg/kg/7D-C; hmn orl TDLo 106 mg/kg; man orl TDLo 180 mg/kg/6W-I; wmn orl TDLo 3120 mg/kg/43W-I; rat orl 2949 mg/kg; rat ipr 258 mg/kg; rat scu LDLo 330 mg/kg; rat ivn LDLo 50 mg/kg; mus orl 57 mg/kg; mus ipr 71.75 mg/kg; mus scu LDLo 1.5 mg/kg; dog scu LDLo 78 mg/kg; dog ivn LDLo 66mg/kg; rbt orl LDLo 2000 mg/kg; rbt scu LDLo 300 mg/kg; rbt ivn LDLo 23 mg/kg.
40CFR180.244

CAS RN 12122-67-7
$C_4H_6N_2S_4Zn$

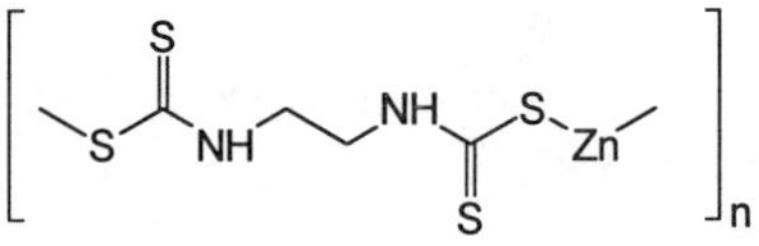

Zineb
Zinc, [[1,2-ethanediylbis[carbamodithioato]](2-)]- (9CI)
Zinc, [ethylenebis[dithiocarbamato]]- (VAN8CI)
[[1,2-Ethanediylbis(carbamodithioato)](2-)]zinc
[Ethylenebis(dithiocarbamate)]zinc
[Ethylenebis(dithiocarbamato)]zinc
Aaphytora
Aphytora
Aspor (VAN)
Aspor C
Asporum
Bercema
Blightox
Blitex
Blizene (VAN)
Carbadine (VAN)
Carbamodithioic acid, 1,2-ethanediylbis-, zinc salt
Chem zineb
Cineb
Crittox
Cyneb
Cynkotox
Daisen
Deikusol
Discon
Dithane z-78
Dithane Z
Dithane Z 78 (VAN)
Dithane Z-78
Dithane 65
Dithiane z-78
Ethyl zimate
Ethylenebis[dithiocarbamic acid], zinc salt
EBDC, zinc salt
ENT 14,874
Fungo-Pulvit
Funjeb

Merck Index No. 10071
MP: dec 157°
BP:
Density:
Solubility: 10 mg/l water 25°, sol. pyridine, carbon disulfide, insol. common organic solvents
Octanol/water PC: ,20 (octanol/water 20)
LD_{50}: hmn orl LDLo 5000 mg/kg; rat orl 1850 mg/kg; rat orl >5200 mg/kg; rat ihl LCLo >800 mg/m^3/4H; rat skn >10000 mg/kg; rat unr 1850 mg/kg; mus orl 7600 mg/kg; mus ipr 1940 mg/kg; rbt orl 4450 mg/kg; mam unr 1350 mg/kg.
40CFR180.115

CAS RN 137-30-4
$C_6H_{12}N_2S_4Zn$

Ziram
Zinc, bis(dimethylcarbamodithioato-S,S')-, (T-4)- (9CI)
Zinc, bis(dimethyldithiocarbamato)- (VAN8CI)
z-c Spray
Aaprotect
Aavolex
Aazira
Accelerator L
Aceto ZDED
Aceto ZDMD
Alcobam ZM
Bis(dimethylcarbamodithioato-S,S')zinc
Bis(dimethyldithiocarbamato)zinc
Bis(N,N-dimetil-ditiocarbammato)di zinco (Italian)
Bis-dimethyldithiocarbamate de zinc (French)
Carbamic acid, dimethyldithio-, zinc salt (2:1)
Carbamodithioic acid, dimethyl-, zinc salt
Carbazinc
Corona Corozate
Corozate
Cuman
Cuman L
Cymate
Dimethylcarbamodithioic acid, zinc complex
Dimethylcarbamodithioic acid, zinc salt
Dimethyldithiocarbamate zinc salt
Dimethyldithiocarbamic acid zinc salt & mercapto benzothiazole
Dimethyldithiocarbamic acid, zinc salt
Drupina 90
Eptac No. 1
Eptac 1
ENT 988
Fuclasin
Fuclasin ultra
Fuclasin-Ultra
Fuklasin
Fungostop
Hermat ZDM
Hexazir
Karbam White
Methasan

$[(CH_3)_2N\text{-}C(=S)\text{-}S^{\ominus}]$ $1/2Zn^{\oplus\oplus}$

Merck Index No. 10075
MP: 250°, 240°, 240 - 244° (technical)
BP:
Density: 1.66^{25}
Solubility: 65 mg/l water 25°, sol. chloroform, carbon disulfide, alkali, mod. sol acetone, insol. ethanol, diethyl ether
Octanol/water PC:
LD_{50}: rat orl 1400 mg/kg; rat orl 267 mg/kg; rat ihl LC_{50} 81 mg/m^3/4H; rat skn >6000 mg/kg; rat ipr 23 mg/kg; rat scu 1340 mg/kg; rat unr 1230 mg/kg; mus orl 480 mg/kg; mus ihl LC_{50} 1056 mg/m^3/2H; mus ipr 73mg/kg; mus scu 800 mg/kg; mus ivn 18 mg/kg; rbt orl 400 mg/kg; rbt skn >2000 mg/kg; rbt ipr LDLo 50 mg/kg; rbt scu 400 mg/kg; gpg orl 200 mg/kg; gpg ipr LDLo 30 mg/kg; mam unr 1400 mg/kg; bwd orl 100 mg/kg.
40CFR180.116

CAS RN 1646-88-4
$C_7H_{14}N_2O_4S$

Aldoxycarb (ACN)
Propanal, 2-methyl-2-(methylsulfonyl)-, O-[(methylamino)carbonyl]oxime (9CI)
Propionaldehyde, 2-methyl-2-(methylsulfonyl)-, O-(methylcarbamoyl)oxime (8CI)
Aldicarb sulfone
Aldoxycarb (2-methyl-2-(methylsulfonyl)propanal O-((methylamino)carbonyl)oxi
Ent ai3-29261
Ent 4.9
Standak
Sulfocarb
Temik sulfone
UC 21865
UC-21865
VC-21865
2-Methyl-2-(methylsulfonyl)propanal O-[(methylamino)carbonyl]oxime (ACN)
2-Methyl-2-(methylsulfonyl)propionaldehyde O-(methylcarbamoyl)oxime

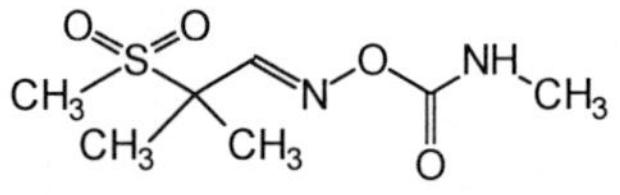

Merck Index No.
MP:
BP:
Density:
Solubility:
Octanol/water PC:
LD_{50}: rat orl 20 mg/kg; rat ihl LC_{50} 140 mg/m^3/4H; rat skn 1000 mg/kg; rat ipr 21 mg/kg; rat ivn 14.9 mg/kg; rbt skn 200 mg/kg; dck orl 33.5 mg/kg.
40CFR180. pending (1978)

CAS RN 7773-06-0
$H_3NO_3S.H_3N$

Ammonium sulfamate (ACN)
Sulfamic acid, monoammonium salt (8CI9CI)
Amcide
Amicide
Amidosulfate
Ammat
Ammate
Ammate X
Ammonium amidosulfate
Ammonium amidosulfonate
Ammonium amidotrioxosulfate
Ammoniumsalz der amidosulfonsaure (German)
AMS (VAN)
AMS (salt)
Fyran 206K
Ikurin
Monoammonium sulfamate
Sulfaminsaure (German)

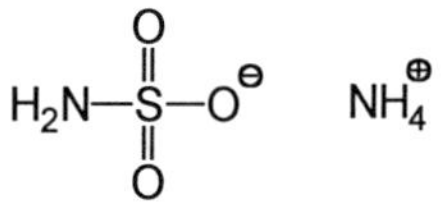

Merck Index No. 583
MP: 131° (dec 160°)
BP:
Density:
Solubility: sol. water, liq. ammonia, sl. sol. alcohol, mod. sol. glycerol, glycol, formamide
Octanol/water PC:
LD_{50}: rat orl 3000 mg/kg; rat orl 2000 mg/kg; rat ipr LDLo 800 mg/kg; mus orl 3100 mg/kg;
40CFR180.188

CAS RN 10311-84-9
$C_{14}H_{17}ClNO_4PS_2$

Dialifor (ACN)
Dialifos
Dialiphor
ENT 27320
Hercules 14503
Phosphorodithioic acid,
S-[2-chloro-l-(I,3-dihydro-1,3-dioxo-2H-isoindol-2-yl)ethyl]O,O-diethyl ester (9CI)
Phosphorodithioic acid, O,O-diethyl ester, S-ester with N-(2-chloro-l-mercaptoethyl)phthalimide (8CI)
O,O-Diethyl S-(2-chloro-l-phthalimidoethyl) phosphorodithioate (ACN)
Torak

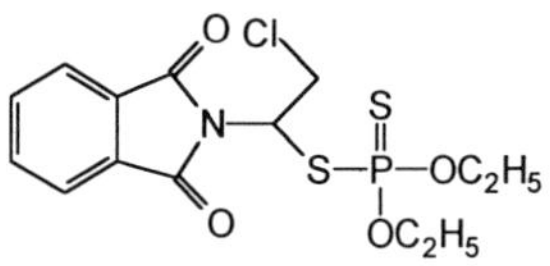

Merck Index No. 2949
MP: 67 - 69°
BP:
Density:
Solubility: Insoluble in water, soluble in aromatic hydrocarbons, ethers, esters, ketones, 760 g/kg acetone, 400 g/kg isophorone, 620 g/kg chloroform, 570 g/kg xylene, 500 g/kg diethyl ether, <10 g/l ethanol, <10 g/kg hexane

Octanol/water PC:
LD_{50}: rat orl 24 mg/kg; rat orl 5 mg/kg; rat orl 6 mg/kg; rat orl 43-53 mg/kg; rat orl 5 mg/kg; rat skn 28 mg/kg; mus orl 39 mg/kg; mus orl 65 mg/kg; dog orl 94 mg/kg; rbt orl 35 mg/kg; rbt skn 145 mg/kg; dck orl 940 mg/kg.
40CFRI80.326

CAS RN 67329-04-8
$C_9H_{17}ClN_3O_3PS$

Isazophos
Phosphorothioic acid, O-[2-chloro-l-(1-methylethyl)-IH-imidazol-4-yl] O,O-diethyl ester
Phosphorothioic acid, o-(2-chloro-l-isopropylimidazol-4-yl) O,O-diethyl ester
Phosphorothioic acid, O-(2-chloro-l-isopropylimidazol-4-yl) O,O-diethyl ester
CGA-12223
Miral

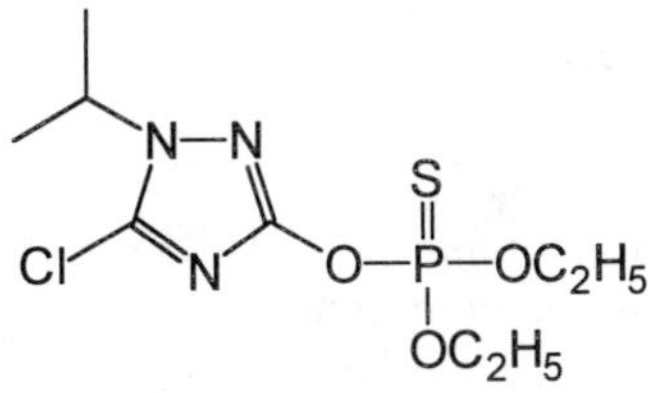

N—N
S
Cl
N
O—P—OC2H5
OC2H5

Merck Index No. 4988
MP:
BP: $100°^{0.001}$, $170°^{760}$
Density: 1.22^{20}
Solubility: 250 mg/l water 20°, misc. methanol, chloroform, benzene, hexane
Octanol/water PC:
LD_{50}: rat orl 60 mg/kg.
40CFRI80. temp (1989)

CAS RN 64902-72-3
$C_{12}H_{12}ClN_5O_4S$

Chlorsulfuron (ACN)
DPX 4189
I-[(*o*-Chlorophenyl)sulfonyl]-3-(4-methoxy-6-methyl-*s*-triazin-2-yl)urea
2-Chloro-N-[[(4-methoxy-6-methyl-1,3,5-triazin-2-yl)amino]carbonyl]be
nzenesulfonamide

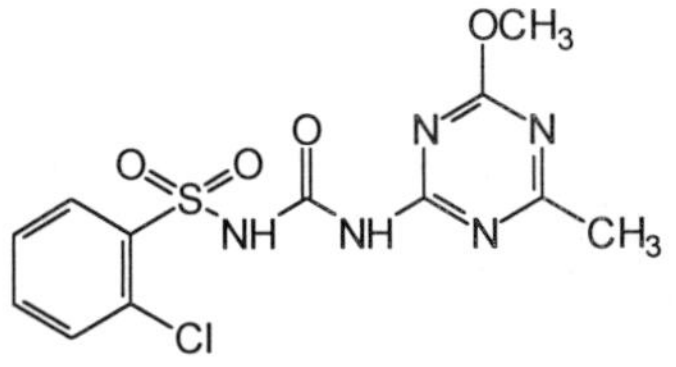

OCH3
O
N
N
O=S=O
S
NH
NH
N
CH3
Cl

Merck Index No. 2192
MP: 174 - 178°
BP:
Density:
Solubility: 0.3 g/l water pH 5 25°, 27.9 g/l water pH 7 25°, 102 g/l dichloromethane, 57 g/l acetone, 14 g/l methanol, 3 g/l toluene, <0.01 g/l hexane
Octanol/water PC:
LD_{50}: rat orl 5545 mg/kg; rat orl 6293 mg/kg; rat ihl LC_{50} >5900 mg/m^3; rbt skn 3400 mg/kg; dck unr 5000 mg/kg; brd unr 5000 mg/kg.
40CFRI 80.405

CAS RN 107534-96-3
$C_{16}H_{22}ClN_30$

tebuconazole
(RS)-l-*p*-chlorophenyl-4,4-dimethyl-3-(IH-1,2,4-triazol-1-ylmethyl)pentan-3-ol
(RS)-I-(4-chlorophenyl)-4,4-dimethyl-3-(IH-1,2,4-triazol-1-ylmethyl)pentan-3-ol
(±)α-[2-(4-chlorophenyl)ethyl]-α(I,I-dimethylethyl)-IH-1,2,4-triazole-l-ethanol

Merck Index No.
MP: 102.4°
BP:
Density:
Solubility: 32 mg/l water 20°, >200 g/l dichloromethane, <0.l g/l hexane, 50-100 g/l isopropanol, 50-100 g/l toluene
Octanol/water PC: 5000
LD_{50}: rat orl 4000 mg/kg; rat ihl LC_{50} >800 mg/m^3/4H; rat skn >5000 mg/kg; mus orl 2000 mg/kg; ckn orl 4488 mg/kg; brd orl >1000 mg/kg.
40CFRI80. pend. 1989

CAS RN 114369-43-6
$C_{20}H_{19}ClN_4$

fenbuconazole
4-(4-chlorophenyl)-2-phenyl-2-(l H-1,2,4-triazol-1-ylmethyl)butyronitrile
α[2-(4-chlorophenyl)ethyl]-α -phenyl- 1H- 1,2,4-triazole-1-propanenitrile
(R,S)-4-(4-chlorophenyl)-2-phenyl-2-(1H-1,2,4-triazol-1-ylmethyl)butyronitrile

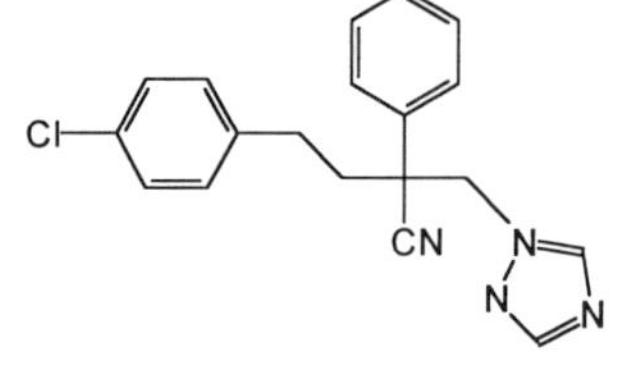

Merck Index No.
MP: 124 - 126°
BP:
Density:
Solubility: 0.2 mg/l water 25°, sol. ketones, esters, alcohols and aromatic hydrocarbons
Octanol/water PC: 1700
LD_{50}: rat orl >2000 mg/kg; rat ihl LC_{50} >2100 mg/m^3/4H; rat skn >5000 mg/kg.
40CFRI80. temp (1990)

CAS RN 81777-89-1
$C_{12}H_{14}ClNO_2$

clomazone
2-(2-chlorobenzyl)-4,4-dimethyl-1,2-oxazolidin-3-one
2-(2-chlorobenzyl)-4,4-dimethylisoxazolidin-3-one
2-[(2-chlorophenyl)methyl]-4,4-dimethyl-3-isoxazolidinone
dimethazone
FMC 57020
command

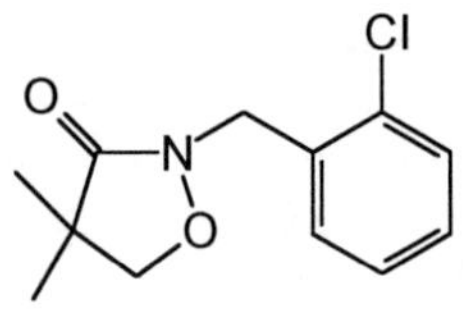

Merck Index No. 3203
MP:
BP:
Density: 1.192^{20}
Solubility: 1.1 g/l water, misc. acetone, chloroform, cyclohexanone, dichloromethane, methanol, toluene, heptane, dimethylformamide
Octanol/water PC: 350
LD_{50}: rat orl 2077 mg/kg; rat orl 1369 mg/kg; rat ihl LC_{50} 4800 mg/m^3/4H; rbt skn >2000 mg/kg; qal orl >2510 mg/kg; dck orl >2510 mg/kg.
40CFRI80.425

CAS RN 99129-21-2
$C_{17}H_{26}ClNO_3S$

clethodim
(±)-2-[(E)-1-[(E)-3-chloroallyloxyimino]propyl]-5-[2-(ethylthio)propyl]-3-hydroxy-cyclohex-2-enone
(E,E)-(±)-2-[1-[[(3-chloro-2-propenyl)oxy]imino]propyl]-5-[2-(ethylthio)propy11-3-hydroxy-2-cyclohex-1-one

Merck Index No.
MP:
BP:
Density: 1.14^{20}
Solubility: sol. in water, ph-dependent, sol. most. org. solvents
Octanol/water PC:
LD_{50}: rat orl 1630 mg/kg; rat orl 1360 mg/kg
40CFRI80.458

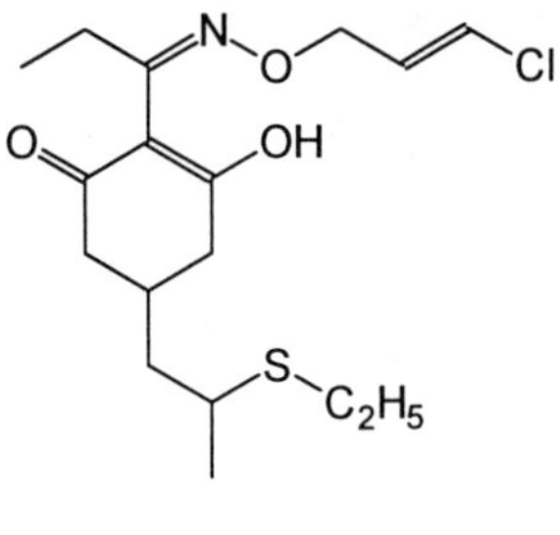

CAS RN 72178-02-0
$C_{15}H_{10}ClF_3N_2O_6S$

fomesafen
5-(2-chloro-ααα-trifluoro-*p*-tolyloxy)-N-methylsulfonyl-2-nitrobenzamide
5-(2-chloro-ααα-trifluoro-*p*-tolyloxy)-N-mesyl-2-nitrobenzamide
5-[2-chloro-4-(trifluoromethyl)phenoxy]-N-(methylsulfonyl)-2-nitrobenzamide

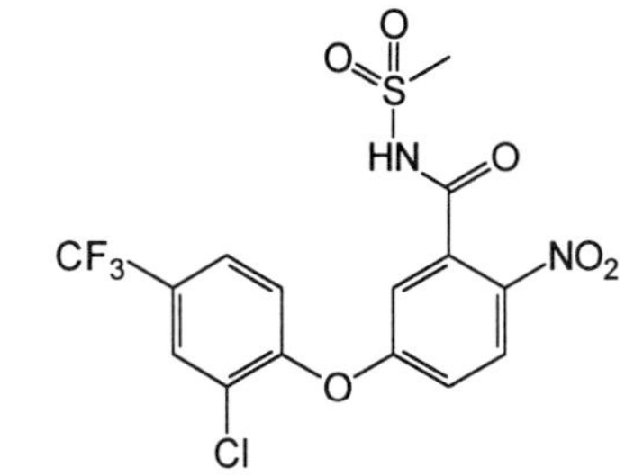

Merck Index No.
MP: 220 - 221°
BP:
Density: 1.28^{20}
Solubility: 50 mg/l water 20°, < 1 mg/l water 20° pH 1-2, 300 g/l acetone, 150 g/l cyclohexanone, 10 g/l dichloromethane, 1.9 g/l xylene, 0.5 g/l hexane
Octanol/water PC: 794 pH 1
LD_{50}: rat orl 1250-2000 mg/kg; rat orl 1600 mg/kg; rbt skn >1000 mg/kg; dck orl >5000 mg/kg.
40CFRI80.433

CAS RN 68694-11-1
$C_{15}H_{15}ClF_3N_3O$

triflumizole
(E)-4-chloro-α-trifluoro-N-(l-imidazol-1-yl-2-propoxyethylidene)-*o*-toluidine
(E)-1-[1-[[4-chloro-2-(trifluoromethyl)phenyl]imino]-2-propoxyethyl]-1H-imidazole

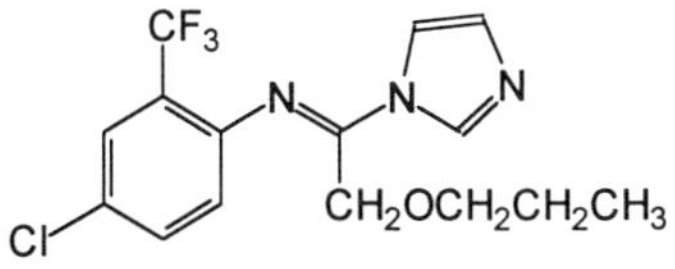

Merck Index No.
MP: 63.5°
BP:
Density:
Solubility: 12.5 g/l water 20°, 2220 g/l chloroform, 17.6 g/l hexane, 639 g/l xylene, 1440 g/l acetone, 496 g/lmethanol
Octanol/water PC: 25
LD_{50}: rat orl 715 mg/kg; rat orl 695 mg/kg; rat ihl LC_{50} >3200 mg/m^3/4H; rat skn >5000 mg/kg; rat ipr 710 mg/kg; rat scu 5000 mg/kg; mus orl 510 mg/kg; mus ihl LC_{50} >3200 mg/m^3; mus ipr 530 mg/kg; mus scu 5000 mg/kg; qal orl 2467 mg/kg; qal unr 2467 mg/kg.
40CFRI80. temp (1988)

CAS RN 74115-24-5
$C_{14}H_8Cl_2N_4$

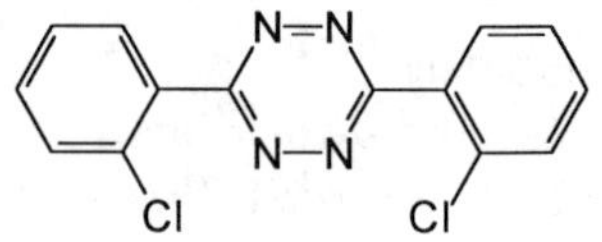

clofentezine
3,6-bis(2-chlorophenyl)-1,2,4,5-tetrazine
bisclofentezin
acaristop
apollo
bisclofentezine
clofentexine
panatac

Merck Index No. 2373
MP: 182.3°, 179 - 182°
BP:
Density:
Solubility: <1 mg/l water 20°, 50 g/l chloroform, 37 g/l dichloromethane, 9.3 g/l acetone, 2.5 g/l benzene, 0.5 g/l ethanol, 1 g/l hexane
Octanol/water PC: 12600
LD_{50}: rat orl >5200 mg/kg; rat orl >3200 mg/kg; rat ihl LC_{50} 9000 mg/m^3/4H; rat skn >2400 mg/kg; mus orl >3200 mg/kg; qal orl >750 mg/kg; dck orl >3000 mg/kg; dck unr >3000 mg/kg.
40CFRl 80.446

CAS RN 1702-17-6
$C_6H_3Cl_2NO_2$

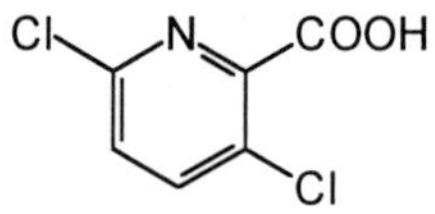

clopyralid
3,6-dichloropyridine-2-carboxylic acid
3,6-dichloro-2-pyridinecarboxylic acid
cirtoxin
cyronal
Dowco 290
lontrel
matrigon
reclaim
shield stinger

Merck Index No. 2398
MP: 151 - 152°
BP:
Density:
Solubility: 9 g/l water 25°, 153 g/kg acetone, 387 g/l cyclohexanone, 6.5 g/l xylene
Octanol/water PC:
LD_{50}: rat orl >4300 mg/kg; rat ipr 900 mg/kg; mus orl >5000 mg/kg; rbt skn >2000 mg/kg; dck orl 1465 mg/kg.
40CFRI80.431

CAS RN 68359-37-5
$C_{22}H_{18}Cl_2FNO_3$

cyfluthrin
(RS)-α-cyano-4-fluoro-3-phenoxybenzyl (IRS)-*cis-trans*-3-(2,2-dichlorovinyl)-2,2-dimethylcyclopropanecarboxylate
(RS)-α-cyano-4-fluoro-3-phenoxybenzyl (1RS, 3RS:1RS,3SR)-3-(2,2-dichlorovinyl)-2,2-dimethylcyclopropanecarboxylate
cyano(4-fluoro-3-phenoxyphenyl)methyl
3-(2,2-dichloroethenyl)-2,2-dimethylcyclopropanecarboxylate
cyfluthrine
cyfoxylate

Merck Index No. 2764
MP: ca. 60°
BP:
Density:
Solubility: 0.002 mg/l water 20°, >1000 g/l dichloromethane,
Octanol/water PC:
LD_{50}: rat orl 900 mg/kg; rat orl 400 mg/kg (in dimethylsulfoxide); rat ihl LC_{50} 469 mg/m^3/4H; rat skn >5000 mg/kg; mus orl 300 mg/kg; dog orl 500 mg/kg; ckn orl 5000 mg/kg; qal orl >5000 mg/kg; brd orl 250 mg/kg.
40CFRI80.436

CAS RN 64257-84-7
$C_{22}H_{23}NO_3$

fenpropathrin
(RS)-α-cyano-3-phenoxybenzyl 2,2,3,3-tetramethylcyclopropanecarboxylate
cyano(3-phenoxyphenyl)methyl 2,2,3,3-tetramethylcyclopropanecarboxylate
danitol
herald
kilumal
meothrin
rody
tame

Merck Index No.
MP: 45 - 50°
BP:
Density: 1.15^{25}
Solubility: 0.33 mg/l water 25°, 1000 g/kg xylene, 1000 g/kg cyclohexanone, 337 g/kg methanol
Octanol/water PC: 1000000
LD_{50}: rat orl 70.6 mg/kg; rat orl 66.7 mg/kg; rat ivn 2.586 mg/kg.
40CFRI80. temp (1986)

CAS RN 102851-06-9
$C_{26}H_{22}ClF_3N_2O_3$

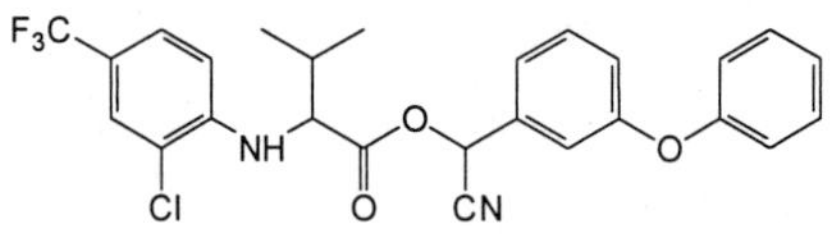

Fluvalinate
Mavrik
N-(2-Chloro-α,α,α-trifluoro-*p*-tolyl)-DL-valine (±)-α-cyano-*m*-phenoxybenzyl ester
N-[2-Chloro-4-(trifluoromethyl)phenyl]-DL-valine (±)-cyano(3-phenoxyphenyl) methyl ester (ACN)
ZR-3210
2-(2-Chloro-α,α,α-trifluoro-*p*-toluidino)-3-methylbutyric acid, (±)-α-cyano-*m*-phenoxybenzyl ester
2-[[2-Chloro-4-(trifluoromethyl)phenyl]amino]-3-methylbutanoic acid, (±)-cyano(3-phenoxyphenyl)methyl ester

Merck Index No.
MP:
BP: >450°
Density: 1.29^{25}
Solubility: 0.002 mg/l water, sol. aromatic hydrocarbons, alcohols, diethyl ether, dichloromethane
Octanol/water PC: 7000
LD_{50}: rat orl >3000 mg/kg.
CFR: 40CFR180.427

CAS RN 91465-08-6
$C_{23}H_{19}ClF_3NO_3$

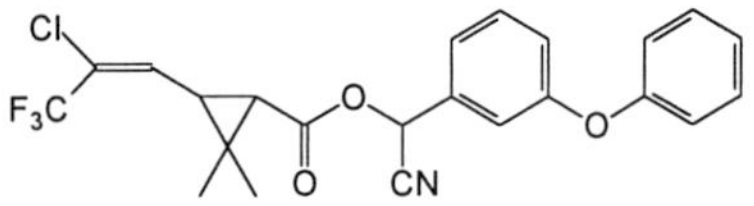

Cyhalothrin
(RS)-α-Cyano-3-phenoxybenzyl (Z)-(1RS,3RS)-3-(2-chloro-3,3,3-trifluoropropenyl)-2,2-dimethylcyclopropanecarboxylate
PP 563

Merck Index No.
MP: 49.2°
BP:
Density:
Solubility: 0.005 mg/l water pH 6.5 20°, >500 g/l acetone, methanol, toluene, hexane, ethyl acetate
Octanol/water PC: 10000000
LD_{50}: rat orl 79 mg/kg; rat orl 56 mg/kg; rat skn 632 mg/kg; rat ivn 1.951 mg/kg; dck orl >3950 mg/kg; mam ihl LC_{50} 60 mg/m^3/4H.
CFR: 40CFR180. (pend. 1986)

CAS RN 52918-63-5
$C_{22}H_{19}Br_2NO_3$

Deltamethrin
Cyclopropanecarboxylic acid, 3-(2,2-dibromoethenyl)-2,2-dimethyl -, cyano(3-phenoxyphenyl) methyl ester, [1R-[1α(S*),3α]]- (9CI)
α-Cyano-3-phenoxybenzyl *cis*-3-(2,2-dibromovinyl)-2,2-dimethylcyclopropanecarboxylate
α-Cyano-3-phenoxybenzyl-3-(2,2-dibromovinyl)-2,2-dimethylcyclopropanecarboxylate
Cyano(3-phenoxyphenyl)methyl-3-(2,2-dibromoethenyl)-2,2-dimethylcyclopropanecarboxylate
Cyclopropanecarboxylic acid, 3-(2,2-dibromoethenyl)-2,2-dimethyl -,cyano(3-phenoxyphenyl) methyl ester, [1R-[1α(S*),3α]]-
Cyclopropanecarboxylic acid, 3-(2,2-dibromoethenyl)-2,2-dimethyl-, cyano(3-phenoxyphenyl) methyl ester, 1R-*trans*-
Decamethrin
Decamethrine
Decis
Decis EC-25
FMC 45498
Nrdc 158
Nrdc 161
NRDC 161
RU 22974
RU 2974

Br
Br
O
O
NC
O

Merck Index No. 2869
MP: 98 - 101°
BP:
Density:
Solubility: <0.002 mg/l water 20°, 900 g/l dioxane, 750 g/l cyclohexanone, 700 g/l dichloromethane, 500 g/l acetone, 450 g/l benzene, 450 g/l dimethylsulfoxide, 250 g/l xylene, 15 g/l ethanol, 6 g/l isopropanol, sol. ethanol, acetone, dioxane, insol. water
Octanol/water PC: 270000
LD_{50}: rat orl 128 mg/kg; rat orl 139 mg/kg; rat orl 31 mg/kg; rat orl 9.36 mg/kg; rat ihl LC_{50} 785 mg/m^3/2H; rat skn >2000 mg/kg; rat ivn 2.526 mg/kg; rat ivr 4 mg/kg; mus orl 20 mg/kg; mus ice 26.1 mg/kg; dog orl >300 mg/kg; dog ivn 3.44 mg/kg; rbt skn 2000 mg/kg; dck orl >4640 mg/kg.
CFR: 40CFR180.435

CAS RN 60-51-5
$C_5H_{12}NO_3PS_2$

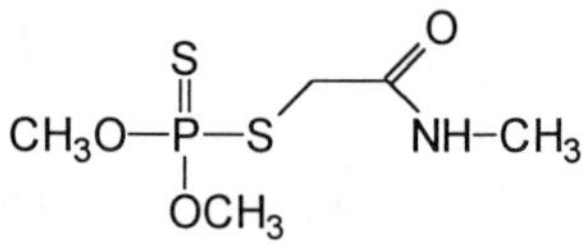

Cygon
Phosphorodithioic acid, O,O-dimethyl S-[2-(methylamino)-2-oxoethyl] ester (9CI)
Phosphorodithioic acid, O,O-dimethyl ester, S-ester with 2-mercapto-N-methyl acetamide (O,O-Dimethyl-S-(N-methyl-carbamoyl-methyl)-dithiophosphat) (German)
Acetic acid, O,O-dimethyldithiophosphoryl-, N-monomethylamide salt
American Cyanamid 12,880
AC-18682
BI 58
BI 58 EC
Cygon Insecticide
Cygon 4E
CL 12880
Dithiophosphate de O,o-dimethyle et de s(-N-methylcarbamoyl-methyle) (French)
Dithiophosphate de O,O-dimethyle et de s(-N-methylcarbamoyl-methyle) (French)
Experimental insecticide 12,880
Experimental Insecticide 12,880
Daphene
De-fend
Demos-L40
Dimetate
Dimethioate
Dimethoaat (Dutch)
Dimethoat (German)
Dimethoat technisch (German)
Dimethoate (ACN)
Dimethoate-267
Dimethogen
Dimeton
Dimevur
EI-12880
ENT 24,650
ENT-24650
Ferkethion
Fortion NM
Fosfamid
Fosfatox R
Fosfotox
Fosfotox R
Fosfotox R 35
Fostion MM

Merck Index No. 3209
MP: 49°, 52 - 52.5°
BP: $117^{\circ\ 0.1}$
Density: 1.277^{65}
Solubility: 25 g/l water 21°, >300 g/kg alcohols, ketones, benzene, toluene, dichloromethane, chloroform, sl. sol. xylene, carbon tetrachloride, aliphatic hydrocarbons, freely sol. org. solvents except hydrocarbons
Octanol/water PC: 5
LD_{50}: man orl TDLo 286 mg/kg; hmn orl 30 mg/kg; man orl TDLo 300 mg/kg; rat orl 60 mg/kg; rat orl 250 mg/kg; rat orl 290-325 mg/kg; rat skn 353 mg/kg; rat ipr 100 mg/kg; rat scu 350 mg/kg; rat ivn 450 mg/kg; rat unr 215 mg/kg; mus orl 60 mg/kg; mus orl 160 mg/kg; mus skn 1000 mg/kg; mus ipr 45 mg/kg; mus scu 60 mg/kg; dog orl 400 mg/kg; cat orl 100 mg/kg; rbt orl 300 mg/kg; rbt skn 1000 mg/kg; gpg orl 350 mg/kg; gpg skn 965 mg/kg; gpg ipr 390 mg/kg; ham orl 200 mg/kg; ham ipr 160 mg/kg; ham scu 60 mg/kg; ckn orl 25 mg/kg; qal orl 25 mg/kg; dck orl 42 mg/kg; bwd orl 6.6 mg/kg.
CFR: 40CFR180.204

CAS RN 1861-32-1
$C_{10}H_6Cl_4O_4$

Dimethyl tetrachloroterephthalate (ACN)
1,4-Benzenedicarboxylic acid, 2,3,5,6-tetrachloro-,dimethyl ester (9CI)
Terephthalic acid, tetrachloro-, dimethyl ester (8CI)
Chlorothal
Chlorthal-dimethyl
Chlorthal-methyl
Chlorthal-Dimethyl
Dacthal
Dacthalor
Dimethyl ester of tetrachloroterephthalic acid
Dimethyl
2,3,5,6-tetrachloro-1,4-benzenedicarboxylate
Dimethyl 2,3,5,6-tetrachloroterephthalate
DAC 4
DAC-893
DCP
DCPA (VAN)
Fatal
Terephthalic acid, 2,3,5,6-tetrachloro-, dimethyl ester
Tetrachloroterephthalic acid dimethyl ester
Tetrachloroterephthalic acid, dimethyl ester
2,3,5,6-Tetrachloroterephthalic acid, dimethyl ester

$COOCH_3$ Cl Cl Cl Cl $COOCH_3$

Merck Index No. 2830
MP: 155 - 156°
BP:
Density:
Solubility: 0.5 mg/l water 25°, 250 g/kg benzene, 170 g/kg toluene, 140 g/kg xylene, 120 g/kg dioxane, 100 g/kg acetone, 70 g/kg carbon tetrachloride, <5% water, >5% acetone, cyclohexanone, xylene
Octanol/water PC:
LD_{50}: rat orl >3000 mg/kg; rat ihl LC_{50} >5700 mg/m^3/4H; mus ivn 320 mg/kg; mus unr 3500 mg/kg; rbt skn 10000 mg/kg.
CFR: 40CFR180.185

CAS RN 75-99-0
$C_3H_4Cl_2O_2$

Dalapon (ACN)
Propanoic acid, 2,2-dichloro- (9CI)
Propionic acid, 2,2-dichloro- (8CI)
α-Dichloropropionic acid
α,α-Dichloropropionic acid
Alatex
Basinex P
Bh dalapon
Crisapon
D-Granulat
Dalapon 85
Dawpon-Rae
Ded-Weed
Dowpon M
DPA (VAN)
Gramevin
Kenapon
Liropon
Proprop
Radapon
Sys-Omnidel
Unipon
2,2-DPA

Cl CH_3 COOH Cl

Merck Index No. 2806
MP: 166.5° (dec)
BP: 98 - 99° [20] (sodium salt)
Density: 1.4014^{20}
Solubility:
Octanol/water PC:
LD_{50}: rat orl 7126 mg/kg; rat orl 6936 mg/kg; rat skn >5000 mg/kg.
CFR: 40CFR180.150

CAS RN 127-20-8
$C_3H_4Cl_2O_2.Na$

Dalapon, sodium salt (ACN)
Propanoic acid, 2,2-dichloro-, sodium salt (9CI)
Propionic acid, 2,2-dichloro-, sodium salt (8CI)
α,α-Dichloropropionic acid, sodium salt
Antigramigna
Basfapon
Dalacide
Dalaphon
Dalapon
Dalapon sodium
Dalapon sodium salt
Delapon
Dikopan
Doowpon
Dowpon
Gramevin
Natriumsalz der 2,2-dichlorpropionsaeure (German)
Omnidel Spezial
Propinate
Radapon
Sodium α,α-dichloropropionate
Sodium dichloropropionate
Sodium Dalapon
Sodium 2,2-dichloropropionate
Sodium 2,2-dichloropropionic acid
SYS 67 Omnidel
Tafapon
Unipon
2,2-Dichloropropionic acid sodium salt
2,2-Dichloropropionic acid, sodium salt (ACN)

Cl
CH_3 O
$O^{\ominus}$ $Na^{\oplus}$
Cl

Merck Index No. 2806
MP: 174 - 176°
BP:
Density:
Solubility: 45 g/100 ml water 25°, 900 g/kg water 25°, 185 g/kg ethanol, 179 g/kg methanol, 0.14 g/kg acetone, 0.02 g/kg benzene0.16 g/kg diethyl ether
Octanol/water PC:
LD_{50}: rat orl 9330 mg/kg; rat orl 7570 mg/kg; rat orl 3860 mg/kg; rat par >6000 mg/kg; mus orl >4600 mg/kg; mus unr 3650 mg/kg; rbt orl 3400 mg/kg; rbt skn >2000 mg/kg; gpg orl 3400 mg/kg; ckn orl 5600 mg/kg; mam orl 4000 mg/kg; mam ihl LC_{50} >20000 mg/m^3/8H; mam unr 6600 mg/kg.
CFR: 40CFR180.150

CAS RN 50-29-3
$C_{14}H_9Cl_5$

DDT
Dichlorodiphenyltrichloroethane (ACN)
Benzene, 1,1'-(2,2,2-trichloroethylidene)bis[4-chloro- (9CI)
α,α-Bis(*p*-chlorophenyl)-β,β,β-trichlorethane
α,α-Bis(*p*-chlorophenyl)-β,β,β-trichloroethane
p,p'-Dichlorodiphenyltrichloroethane
p,p'-DDT
Aavero-extra
Agritan
Anofex
Arkotine
Azotox
Azotox M-33
Bosan supra
Bovidermol
Cesarex
Chlorophenothan
Chlorophenothane
Chlorophenothanum
Chlorophenothanum Technicum
Chlorophenotoxum
Chlorphenothan
Chlorphenothanum
Chlorphenotoxum
Citox
Clofenotan
Clofenotane
Clofenotane Technique
Clofenotano
Clofenotanum
D.D.T. Technique
Dedelo
Deoval
Detox
Detoxan
Dibovan
Dibovin
Dicophane
Didigam
Didimac

Cl, Cl, CCl_3

Merck Index No. 2832
MP: 108.5 - 109°
BP: 185 - 187° $^{0.05}$
Density:
Solubility: Insol. water, 580 g/l acetone, 780 g/l benzene, 420 g/l benzyl benzoate, 450 g/l carbon tetrachloride, 740 g/l chlorobenzene, 1160 g/l cyclohexanone, 20 g/l 95% alcohol, 280 g/l diethyl ether, 100 g/l gasoline, 30 g/l isopropanol, 80-100 g/l kerosene, 750 g/l morpholine, 610 g/l tetralin, 110 g/l peanut oil, 100-160 g/l pine oil, 500 g/l tributyl phosphate, freely sol. pyridine, dioxane
Octanol/water PC:
LD_{50}: inf orl LDLo 150 mg/kg; man orl TDLo 6 mg/kg; hmn orl TDLo 16 mg/kg; hmn orl LDLo 500 mg/kg; hmn orl TDLo 5 mg/kg; man unr LDLo 221 mg/kg; rat orl 113 mg/kg; rat orl 118 mg/kg; rat orl 87 mg/kg; rat skn 1931 mg/kg; rat ipr 9.1 mg/kg; rat scu 1500 mg/kg; rat ivn 68 mg/kg; rat unr 300 mg/kg; mus orl 135 mg/kg; mus ipr 32 mg/kg; mus ivn 68.5 mg/kg; dog orl 150 mg/kg; dog ivn LDLo 75 mg/kg; mky orl 200 mg/kg; mky ivn LDLo 50 mg/kg; cat orl LDLo 250 mg/kg; cat ivn LDLo 40 mg/kg; rbt orl 250 mg/kg; rbt skn 300 mg/kg; rbt scu 250 mg/kg; rbt ivn LDLo 50 mg/kg; gpg orl 150 mg/kg; gpg skn 1000 mg/kg; gpg scu 900 mg/kg; ham orl >5000 mg/kg; ckn orl LDLo 300 mg/kg; frg orl 7.6 mg/kg; frg par 24.1 mg/kg; dom orl LDLo 300 mg/kg; mam unr 200 mg/kg.
CFR: 40CFR180.147 (removed 12/24/86)

CAS RN 919-86-8
$C_6H_{15}O_3PS_2$

Demeton S methyl
Phosphorothioic acid, S-[2-(ethylthio)ethyl] O,O-dimethyl ester
Bayer 25/154
BAY 18436
Demeton-S-methyl
Demeton-S-methyl sulfide
Demeton-S-metile (Italian)
Dimethyl S-(2-eththioethyl) thiophosphate
Dimethyl S-(2-eththioethyl)thiophosphate
Ethanethiol, 2-(ethylthio)-, S-ester with O,O-dimethyl phosphorothioate
Isometasystox
Isomethylsystox
Meta-isosystox
Metaisoseptox
Metaisosystox
Metasystox
Metasystox (i)
Metasystox forte
Metasystox I
Metasystox J
Metasystox 55
Methyl demeton thioester
Methyl isosystox
Methyl-mercaptofos teolovy
Methylthionodemeton
O,o-Dimethyl-S-(2-aethylthio-aethyl)-monothiophosphat (German)
O,o-Dimethyl-S-(2-ethylthio-ethyl)-monothiofosfaat (Dutch)
O,o-Dimetil-S-(2-etiltio-etil)-monotiofosfato (Italian)
O,O-Dimethyl S-(ethylmercapto)ethyl thiophosphate
O,O-Dimethyl S-(2-aethylthio-aethyl)-monothiophosphat (German)
O,O-Dimethyl S-(2-ethethioethyl)phosphorothioate
O,O-Dimethyl S-[2-ethylthio)ethyl]-monothiofosfaat (Dutch)
O,O-Dimethyl S-2-(ethylthio)ethyl phosphorothioate
O,O-Dimethyl 2-(ethylmercapto)ethyl thiophosphate, thiolo isomer

O,O-Dimethyl-S (3-thia-pentyl)-monothiophosphat (German)
O,O-Dimetil-S (2-etilitio-etil)-monotiofosfato (Italian)
Phosphorothioic acid, O,O-dimethyl S-[2-(ethylthio)ethyl] ester

$CH_3O-P(=O)(OCH_3)-S-CH_2CH_2-S-CH_2CH_3$

Merck Index No.
MP:
BP: $89^{\circ\,0.15}$, $118^{\circ\,1}$
Density: 1.207^{20}
Solubility: 3.3 g/l water 20°, readily sol. alcohols, ketones, chlorinated hydrocarbons
Octanol/water PC:
LD_{50}: rat orl 40-106 mg/kg; rat orl 30 mg/kg; rat ihl LC_{50} 500 mg/m^3/4H; rat skn 85 mg/kg; rat ipr 7.5 mg/kg; rat ivn 65 mg/kg; gpg orl 110 mg/kg; gpg ipr 12.5 mg/kg; mam unr 40 mg/kg.
CFR: 40CFR180.105

CAS RN 13684-56-5
$C_{16}H_{16}N_2O_4$

Desmedipham (ACN)
Carbamic acid, [3-[[(phenylamino)carbonyl]oxy]phenyl]-, ethyl ester (9CI)
Carbanilic acid, *m*-hydroxy-, ethyl ester, carbanilate (ester) (8CI)
Bentanex
Betanal AM
Betanal-475 (contested)
Betanex
Carbamic acid, [[(phenylcarbamoyl)oxy]phenyl]-, ethyl ester
Carbanilic acid, *m*-carbaniloyloxy-, ethyl ester
Ethyl *m*-hydroxycarbanilate carbanilate (ACN)
Ethyl *m*-hydroxycarbanilate carbanilate (ester)
Ethyl [[(phenylcarbamoyl)oxy]phenyl]carbamate
Ethyl [3-[[(phenylamino)carbonyl]oxy]phenyl]carbamate
Ethyl N-[3-(N-phenylcarbamoyloxy)phenyl]carbamate
Ethyl N-[3-(N'-phenylcarbamoyloxy)phenyl]carbamate
Ethyl 3-phenylcarbamoyloxyphenylcarbamate
EP 475
EP-475
Schering 38107
Sn-038107
Sn-475
SN 38107
3-(Aethoxycarbonylaminophenyl)-N-phenyl-carbamat (German)
3-[(Ethoxycarbonyl)amino]phenyl phenylcarbamate
3-[[(Ethoxycarbonyl)amino]phenyl]-N-phenylcarbamate
3-Ethoxycarbnylaminophenyl N-phenylcarbamate
3-Ethoxycarbonylaminophenyl N-phenylcarbamate
3-Ethoxycarbonylaminophenylphenylcarbamate

NH O NH OC_2H_5 O O

Merck Index No.
MP: 120°
BP:
Density:
Solubility: 7 mg/l water 20°, 400 g/l acetone, 180 g/l methanol, 400 g/l isophorone, 149 g/l ethyl acetate, 80 g/l chloroform, 0.5 g/l hexane, 1.2 g/l toluene, 1.6 g/l benzene
Octanol/water PC: 2455 (octanol/water pH 3.9)
LD_{50}: rat orl >10,250 mg/kg; rat orl 9600 mg/kg; mus orl >500 mg/kg; rbt skn >2000 mg/kg; qal orl 2480 mg/kg.
CFR: 40CFR180.353

CAS RN 520-45-6
$C_8H_8O_4$

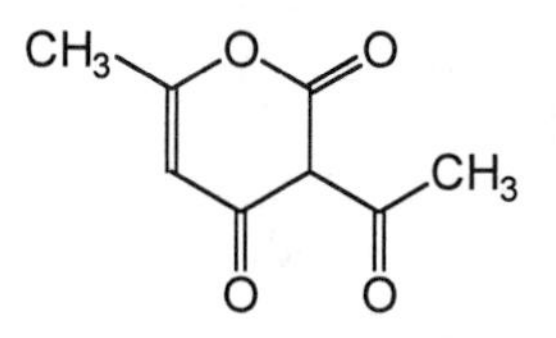

Dehydroacetic acid (ACN)(VAN)
DHA (VAN)
2H-Pyran-2,4(3H)-dione, 3-acetyl-6-methyl- (8CI9CI)
Acetic acid, dehydro-
Dehydracetic acid (VAN)
DHS
Methylacetopyranone
Methylacetopyronone
2-Acetyl-5-hydroxy-3-oxo-4-hexenoic acid, δ-lactone
2H-Pyran-2-one, 3-acetyl-4-hydroxy-6-methyl-
2H-Pyran-2-one,3-acetyl-4-hydroxy-6-methyl-
3-Acetyl-4-hydroxy-6-methyl-2H-pyran-2-one
3-Acetyl-6-methyl-2,4-pyrandione
3-Acetyl-6-methyl-2H-pyran-2,4(3H)-dione
3-Acetyl-6-methyl-2H-pyran-2,4(3H)-dione. enol form
3-Acetyl-6-methyldihydropyrandione-2,4
3-Acetyl-6-methylpyrandione-2,4
4-Hexenoic acid, 2-acetyl-5-hydroxy-3-oxo-, δ-lactone

Merck Index No. 2855
MP: 109 - 111°
BP: 269.9°
Density:
Solubility: <0.1% water 25°, 22% acetone, 18% benzene, 5% methanol, 3% carbon tetrachloride, 3% ethanol, <0.1% glycerol, 0.7% heptane, 1.6% olive oil, 1.7% propylene glycol
Octanol/water PC:
LD_{50}: rat orl 1000 mg/kg; rat orl 500 mg/kg; mus orl 1330 mg/kg; mus ipr 922 mg/kg; rbt skn LDLo 5000 mg/kg.
CFR: 40CFR180.159

CAS RN 2303-16-4
$C_{10}H_{17}Cl_2NOS$

Diallate
Carbamothioic acid, bis(1-methylethyl)-, S-(2,3-dichloro-2-propenyl) ester (9CI)
Carbamic acid, diisopropylthio-, S-(2,3-dichloroallyl) ester (8CI)
Avadex
Bis(1-methylethyl) carbamothioic acid, S-(2,3-dichloro-2-propenyl) ester
Bis(1-methylethyl)carbamathioic acid S-(2,3-dichloro-2-propenyl) ester
Bis(1-methylethyl)carbamothioic acid, S-(2,3-dichloro-2-propenyl) ester
Carbamothioic acid, bis(1-methylethyl)-, S-2,3-dichloro-2-propenyl ester
CP 15,336
CP 15336
Di-allate
Di-isopropylthiolocarbamate de S-(2,3-dichloro allyle) (French)
Diallaat (Dutch)
Diallat (German)
Dichloroallyl diisopropylthiocarbamate
Diisopropylthiocarbamic acid, S-(2,3-dichloroallyl) ester
Diisopropylthiocarbamic acid, S-2,3-dichloroallyl ester
DATC
S-(2,3-Dichlor-allyl)-N,N-diisopropyl-monothiocarbamaat (Dutch)
S-(2,3-Dichloro-allil)-N,N-diisopropil-monotiocarbammato (Italian)
S-(2,3-Dichloroallyl) diisopropylthiocarbamate (ACN)
S-2,3-Dichloroallyl diisopropylthiocarbamate
2-Propene-1-thiol, 2,3-dichloro-, diisopropylcarbamate
2,3-Dichloro-2-propene-1-thiol, diisopropylcarbamate
2,3-Dichloroallyl diisopropylthiolcarbamate
2,3-Dichloroallyl N,N-diisopropylthiolcarbamate
2,3-Dichloroallyl-N,N-diisopropylthiocarbamate
2,3-DCDT

Cl O CH(CH3)2 N Cl S CH(CH3)2

Merck Index No. 2950
MP:
BP: 150°[9]
Density:
Solubility: 40 ppm water 25°, sol. acetone, benzene, chloroform, diethyl ether, heptane
Octanol/water PC:
LD_{50}: rat orl 395 mg/kg; rat skn 2124 mg/kg; rat unr 393 mg/kg; dog orl 510 mg/kg; rbt skn 2000 mg/kg; gpg orl 420 mg/kg; mam unr 395 mg/kg.
CFR: 40CFR180.277

CAS RN 117-80-6
$C_{10}H_4Cl_2O_2$

Dichlone (ACN)
1,4-Naphthalenedione, 2,3-dichloro- (9CI)
1,4-Naphthoquinone, 2,3-dichloro- (8CI)
Algistat
Compound 604
Dichloronaphthoquinone
Dichloronapthoquinone
Diclone
ENT 3,776
ENT-3776
Phygon
Phygon Seed Protectant
Phygon XL
Quintar
Sanquinon
U.S. Rubber 604
USR 604
2,3-Dichlor-1,4-naphthochinon (German)
2,3-Dichloro-α-naphthoquinone
2,3-Dichloro-1,4-naphthalenedione
2,3-Dichloro-1,4-naphthoquinone (ACN)
2,3-Dichloronaphthoquinone
2,3-Dichloronaphthoquinone-1,4
2,3-Dichloronapthoquinone

Merck Index No. 3032
MP: 193°
BP:
Density:
Solubility: 0.1 ppm water 25°, 4% xylene, o-dichlorobenzene, mod. sol. acetone, diethyl ether, benzene, dioxane
Octanol/water PC:
LD_{50}: rat orl 1300 mg/kg; rat orl 160 mg/kg; mus orl 440 mg/kg; mus ipr 30 mg/kg; rbt skn 5000 mg/kg.
CFR: 40CFR180.118

CAS RN 542-75-6
$C_3H_4Cl_2$

1,3-Dichloropropene (ACN)
1-Propene, 1,3-dichloro- (9CI)
Propene, 1,3-dichloro- (8CI)
α-Chloroallyl chloride
α,γ-Dichloropropylene
γ-Chloroallyl chloride
D-d Mixture
Dichloropropene
Nemex
NCI-C03985
Propylene, 1,3-dichloro-
Telone (VAN)
Telone C
Telone II
Vidden D
1,3-Dichloro-1-propene
1,3-Dichloro-2-propene
1,3-Dichloropropene-1
1,3-Dichloropropene, E,Z
1,3-Dichloropropylene
3-Chloroallyl chloride
3-Chloropropenyl chloride

Merck Index No. 3064
MP:
BP: 108°, 104.3° (*cis*), 112.0° (*trans*)
Density: 1.220^{25}, 1.224^{20} (*cis*), 1.217^{20} (*trans*)
Solubility:
Octanol/water PC:
LD_{50}: rat orl 713 mg/kg; rat orl 470 mg/kg; rat ihl LC_{50} 500 ppm; rat skn 775 mg/kg; rat ipr 175 mg/kg; mus orl 640 mg/kg; mus ihl LC_{50} 4650 mg/m^3/2H; rbt skn 504 mg/kg; gpg ihl LCLo 400 ppm/7H.
40CFR180. (temp. 1980)

CAS RN 333-41-5
$C_{12}H_{21}N_2O_3PS$

Diazinon
Phosphorothioic acid, O,O-diethyl O-[6-methyl-2-(1-methylethyl)-4-pyrimidinyl] ester (9CI)
Phosphorothioic acid, O,O-diethyl O-(2-isopropyl-6-methyl-4-pyrimidinyl) ester (8CI)
o,o-Dietil-o-(2-isopropil-4-metil-pirimidin-il)-monotiofosato (Italian)
Alfa-Tox
Antigal (VAN)
AG-500
Basudin
Basudin spectracide
Basudin 10 G
Bazuden
Bazudin
Ciazinon
Dacutox
Dassitox
Dazzel
Dianon
Diater
Diazajet
Diazide
Diazinon AG 500
Diazinone
Diazitol
Diazol
Dicid
Diethyl 2-isopropyl-4-methyl-6-pyrimidinyl phosphorothionate
Diethyl 2-isopropyl-4-methyl-6-pyrimidyl thionophosphate
Diethyl 4-(2-isopropyl-6-methylpyrimidinyl) phosphorothionate
Dipofene
Dizinon
Dyzol
Exodin (VAN)
ENT 19,507
Flytrol
G 301

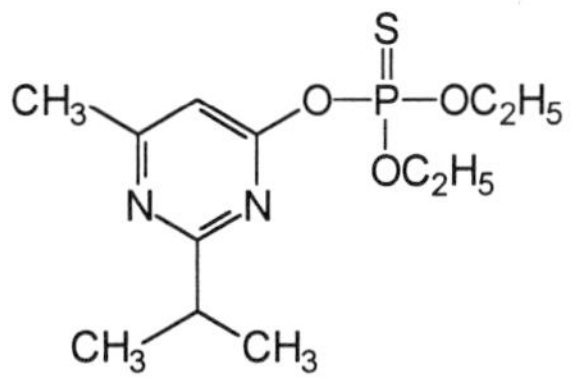

Merck Index No. 2978
MP:
BP: $83\text{-}84^{\circ\ 0.0002}$, $125^{\circ\ 1}$
Density: $1.116\text{-}1.118^{20}$
Solubility: 40 mg/l water 20°, completely misc. common org. solvents such as ethers, alcohols, benzene, toluene, hexane, cyclohexane, dichloromethane, acetone, petroleum oils
Octanol/water PC:
LD_{50}: hmn orl TDLo 214 mg/kg; rat orl 66 mg/kg; rat orl 300-400 mg/kg; rat ihl LC_{50} 3500 mg/m^3/4H; rat skn 180 mg/kg; rat ipr 65 mg/kg; mus orl 17 mg/kg; mus orl 80-135 mg/kg; mus ihl LC_{50} 1600 mg/m^3/4H; mus skn 2750 mg/kg; mus ipr 33 mg/kg; mus scu 58 mg/kg; mus ivn 180 mg/kg; rbt orl 143 mg/kg; pig orl 320 mg/kg; pig skn 633 mg/kg; gpg orl 250 mg/kg; gpg ihl LC_{50} 5500 mg/m^3/4H; pgn orl 3.16 mg/kg; ckn orl 8.4 mg/kg; qal orl 4.21 mg/kg; dck orl 3.50 mg/kg; mam unr 76 mg/kg; bwd orl 2 mg/kg.
40CFR180.153

CAS RN 43222-48-6
$C_{17}H_{17}N_2.CH_3O_4S$

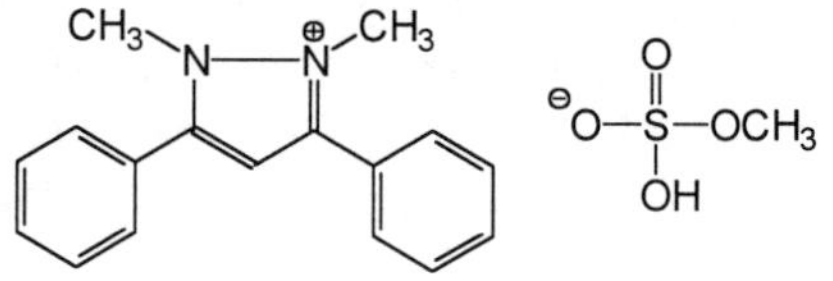

Difenzoquat methyl sulfate (ACN)
1H-Pyrazolium, 1,2-dimethyl-3,5-diphenyl-, methyl sulfate (9CI)
1,2-Dimethyl-3,5-diphenyl-1-(H)-pyrazolium methyl sulfate
1,2-Dimethyl-3,5-diphenyl-1H-pyrazolinium methyl sulfate (ACN)
1,2-Dimethyl-3,5-diphenyl-1H-pyrazolium methyl sulfate
1,2-Dimethyl-3,5-diphenylpyrazolium methyl sulfate
Avenge
AC 84777
AC 84777 Finaven
Difenzoquat
Finaven
Mataven

Merck Index No.
MP: 150 - 160°
BP:
Density:
Solubility: 765 g/l water 25°, 360 g/l dichloromethane, 500 g/l chloroform, 588 g/l methanol, 9.8 g/l acetone, <0.01 g/l xylene, <0.01 g/l heptane, sl. sol. petroleum ether, benzene, dioxane
Octanol/water PC: 0.24 pH 7.0
LD_{50}: rat orl 270 mg/kg; rat orl 470 mg/kg; mus orl 31 mg/kg; mus orl 44 mg/kg; rbt skn 470 mg/kg.
40CFR180.369

CAS RN 55290-64-7
$C_6H_{10}O_4S_2$

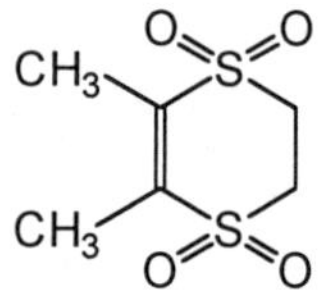

dimethipin
2,3-dihydro-5,6-dimethyl-1,4-dithi-ine, 1,1,4,4-tetraoxide
2,3-dihydro-5,6-dimethyl-1,4-dithiin, 1,1,4,4-tetraoxide
oxydimethin
harvade
N252

Merck Index No.
MP: 162 - 167°
BP:
Density:
Solubility: 3 g/l water 25°, 180 g/kg acetone, 10 g/kg xylene
Octanol/water PC: 0.66
LD_{50}: rat orl 1180 mg/kg; rat orl 1150 mg/kg; rat ihl LC_{50} >20000 mg/m^3/1H; mus orl 440 mg/kg; mus orl 600 mg/kg; rbt skn 8000 mg/kg.
CFR: 40CFR180.406

CAS RN 81334-34-1
$C_{13}H_{15}N_3O_3$

imazapyr
2-(4-isopropyl-4-methyl-5-oxo-2-imidazolin-2-yl)nicotinic acid
2-[4,5-dihydro-4-methyl-4-(1-methylethyl)-5-oxo-1H-imidazol-2-yl]-3-pyridinecarboxylic acid
AC 252,925
arsenal
assault
chopper
CL 252-925
contain
pivot

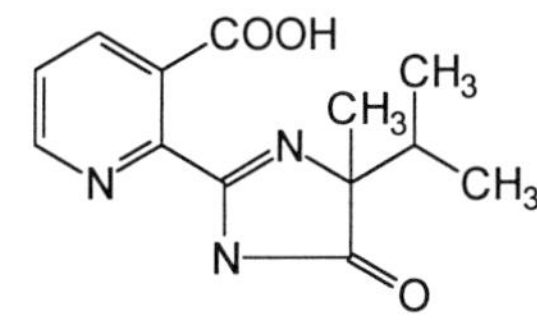

Merck Index No.
MP: 169 - 173°
BP:
Density:
Solubility: 10-15 g/l water 25°
Octanol/water PC: 1.3
LD_{50}: rat orl >5000 mg/kg; mus orl >2000 mg/kg; rbt skn >2000 mg/kg; qal orl >2150 mg/kg; dck orl >2150 mg/kg.
CFR: 40CFR185.....

CAS RN 32809-16-8
$C_{13}H_{11}Cl_2NO_2$

procymidone
N-(3,5-dichlorophenyl)-1,2-dimethylcyclopropane-1,2-dicarboximide
3-(3,5-dichlorophenyl)-1,5-dimethyl-3-azabicyclo[3.1.0]hexane-2,4-dione
S-7131
Sialex
Sumiboto
Sumilex
Sumisclex

Merck Index No.
MP: 166 - 166.5°
BP:
Density: 1.452^{25}
Solubility: 4.5 mg/l water 25°, sl. sol. alcohols, 180 g/l acetone, 180 g/l xylene, 210 g/l chloroform, 230 g/l dimethylformamide
Octanol/water PC: 1380
LD_{50}: rat orl 6800 mg/kg; rat orl 7700 mg/kg; rat orl 7000 mg/kg; rat skn >2500 mg/kg; rat ipr 730 mg/kg; rat scu >10000 mg/kg; mus per 7800 mg/kg; mus orl 9100 mg/kg; mus skn >2500 mg/kg; mus ipr 1650 mg/kg; mus scu >10000 mg/kg; rbt orl 10000 mg/kg; qal unr LDLo >6637 mg/kg; dck unr >4000 mg/kg; mam orl >10000 mg/kg.
CFR: 40CFR180.455

CAS RN 83657-24-3 (76714-88-0, 83657-18-5, 83657-19-6)
$C_{15}H_{17}Cl_2N_3O$

diniconazole
(E)-(RS)-1-(2,4-dichlorophenyl)-4,4-dimethyl-2-(1H-1,2,4-triazol-1-yl)pent-1-en-3-ol
(E)-(±)-β-[(2,4-dichlorophenyl)methylene]-α-(1,1-dimethylethyl)-1H-1,2,4-triazole-1-ethanol
Mixor
Ortho Spotless
Spotless
Sumi-8
Sumi-eight
XE-779

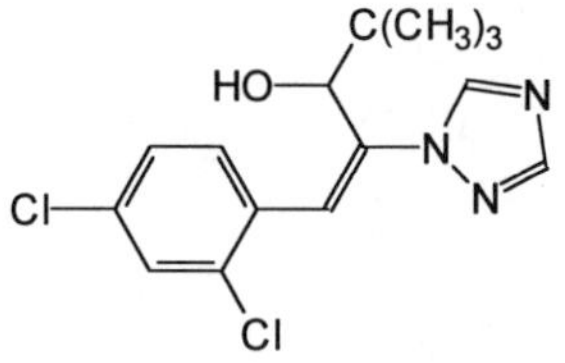

Merck Index No.
MP: 134 - 156°
BP:
Density:
Solubility: 4 mg/l water 25°, 95 g/kg acetone, 95 g/kg methanol, 14 g/kg xylene, 0.7 g/kg hexane
Octanol/water PC: 20000
LD_{50}: rat orl 639 mg/kg; rat orl 474 mg/kg; rat skn >5000 mg/kg; qal unr 1490 mg/kg.
CFR: 40CFR180. (pend. 1992)

CAS RN 60207-90-1
$C_{15}H_{17}Cl_2N_3O_2$

propiconazole
(±)-1-[2-(2,4-dichlorophenyl)-4-propyl-1,3-dioxalan-2-ylmethyl]-1H-1,2,4-triazole
1-[[2-(2,4-dichlorophenyl)-4-propyl-1,3-dioxolan-2-yl]methyl]-1H-1,2,4-triazole
Banner
CGA 64250
Desmel
Orbit
Radar
Spire
Tilt
Alamo
Practis

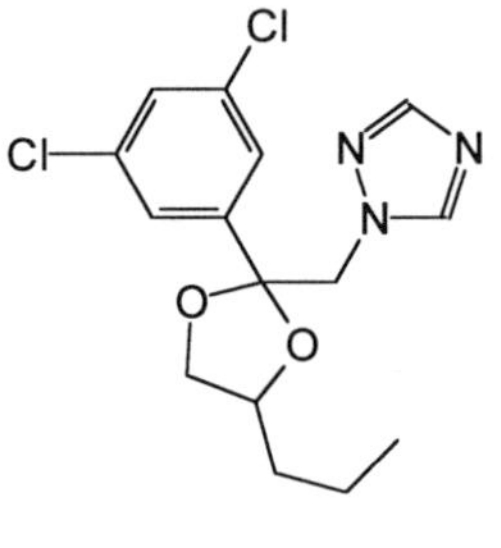

Merck Index No.
MP:
BP: $180^{\circ\ 0.1}$
Density: 1.27^{20}
Solubility: 110 mg/l water 20°, 60 g/kg hexane, completely misc. with acetone, methanol, isopropanol
Octanol/water PC:
LD_{50}: rat orl 1517 mg/kg; rat skn >4000 mg/kg.
CFR: 40CFR180.434

CAS RN 58138-08-2
$C_{10}H_7Cl_5O$

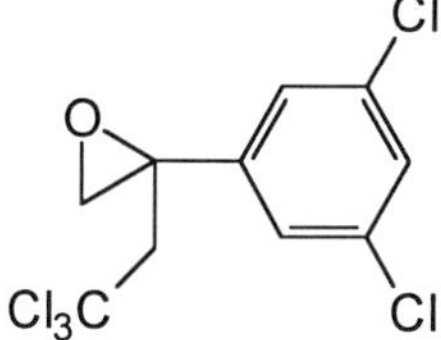

tridiphane
(RS)-2-(3,5-dichlorophenyl)-2-(2,2,2-trichloroethyl)oxirane
(±)-2-(3,5-dichlorophenyl)-2-(2,2,2-trichloroethyl)oxirane
Dowco 356
Nelpon
Tandem

Merck Index No.
MP: 42.8°
BP:
Density:
Solubility: 1.8 mg/l water 25°, 0.98 kg/kg methanol, 9.1 kg/kg acetone, 4.6 kg/kg xylene, 5.6 kg/kg C6H5Cl, 7.1 kg/kg dichloromethane
Octanol/water PC:
LD_{50}: rat orl 1743-1918 mg/kg; rat orl 1500 mg/kg; mus orl 740 mg/kg; rbt skn 3536 mg/kg; dck orl >2510 mg/kg.
CFR: 40CFR180.424

CAS RN 84087-01-4
$C_{10}H_5Cl_2NO_2$

quinclorac
3,7-dichloroquinoline-8-carboxylic acid
3,7-dichloro-8-quinolinecarboxylic acid
BAS 514 H
Facet

Merck Index No.
MP: 274°
BP:
Density: 1.75
Solubility: 0.065 mg/kg water 20°, 2 g/kg ethanol, 2 g/kg acetone, insol. other solvents
Octanol/water PC: 0.07 pH 7
LD_{50}: rat orl >2610 mg/kg; rat orl 2190 mg/kg; rat ihl LC_{50} >5170 mg/m^3; rat skn >2000 mg/kg; mus orl 5000 mg/kg; qal unr >2000 mg/kg.
CFR: 40CFR180. (temp. 1991)

CAS RN 119446-68-3
$C_{19}H_{17}Cl_2N_3O_3$

difenoconazole
cis,trans-3-chloro-4-[4-methyl-2-(1H-1,2,4-triazol-1-ylmethyl)-1,3-dioxolan-2-yl]phenyl 4-chlorophenyl ether
1-[2-[4-(4-chlorophenoxy)-2-chlorophenyl]-4-methyl-1,3-dioxolan-2-ylmethyl]-1H-1,2,4-triazole
CGA 169374
Geyser
Score

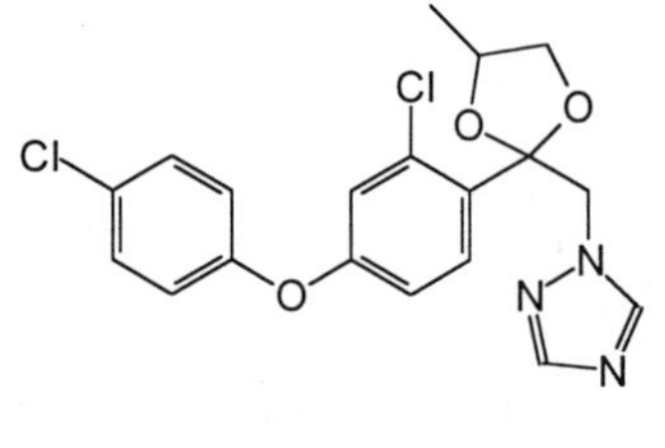

Merck Index No.
MP: 76°
BP: 220° $^{0.03}$
Density:
Solubility: 3.3 mg/l water 20°, v. sol. most org. solvents
Octanol/water PC: 20000
LD_{50}: rat orl 1453 mg/kg; rat ihl LC_{50} > 45 mg/m^3/4H; rbt skn 2010 mg/kg; dck orl >2150 mg/kg;
CFR: 40CFR180. (pend. 1992)

CAS RN 97886-45-8
$C_{15}H_{16}F_5NO_2S_2$

dithiopyr
S,S'-dimethyl-2-difluoromethyl-4-isobutyl-6-trifluoromethylpyridine-3,5-dicarbothioate
S,S'-dimethyl-2-difluoromethyl-4-(2-methylpropyl)-6-(trifluoromethyl)-3,5-pyridine-dicarbothioate
Dimension
MON 15100
MON 7200
Stakeout

Merck Index No.
MP: 65°
BP:
Density:
Solubility: 1.4 mg/l water 20°
Octanol/water PC:
LD_{50}: rat orl >5000 mg/kg; rat ihl LC_{50} >6000 mg/m^3/4H; rat skn >5000 mg/kg; mus orl >5000 mg/kg; qal orl >2250 mg/kg.
40CFR180. (pend. 1991)

CAS RN 111991-09-4
$C_{15}H_{18}N_6O_6S$

nicosulfuron
2-(4,6-dimethoxypyrimidin-2-ylcarbamoylsulfamoyl)-N,N-dimethylnicotinamide
1-(4,6-dimethoxypyrimidin-2-yl)-3-(3-dimethylcarboxamoyl-2-pyridylsulfonyl)urea
2-[[[[(4,6-dimethoxy-2-pyrimidinyl)amino]carbonyl]amino]sulfonyl]-N,N-dimethyl-3-pyridine-carboxamide
Accent
DPX-V9360
MU-495
SL-950

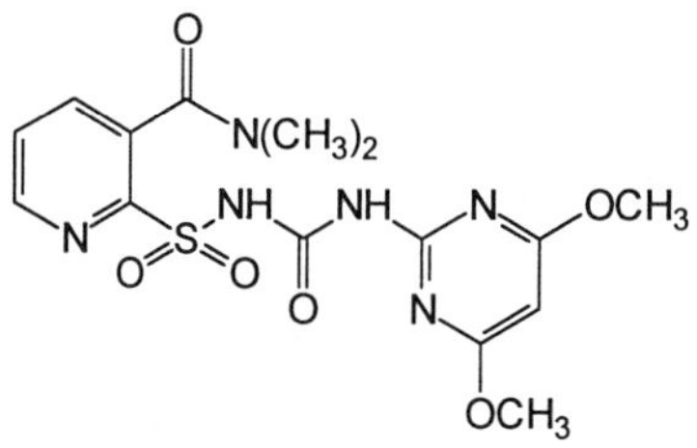

Merck Index No.
MP: 172 - 173°
BP:
Density:
Solubility: 22 g/l water pH 7 25°, 44 mg/l water pH 3.5 25°, 18 g/kg acetone, 4.5 g/kg ethanol, 64 g/kg chloroform, 64 g/kg dimethylformamide, 370 g/kg toluene, 160 g/kg dichloromethane
Octanol/water PC: 0.44 (octanol/water pH 5)
LD_{50}: rat orl >5000 mg/kg; mus orl >5000 mg/kg; rbt skn >2000 mg/kg; qal orl >2250 mg/kg.
40CFR180.454

CAS RN 71626-11-4
$C_{20}H_{23}NO_3$

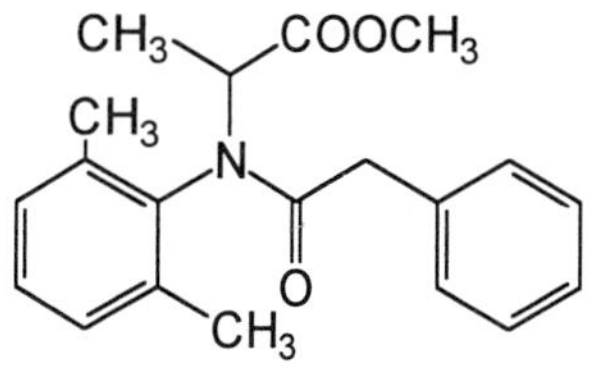

benalaxyl
methyl N-phenylacetyl-N-2,6-xylyl-DL-alaninate
methyl N-(2,6-dimethylphenyl)-N-(phenylacetyl)-DL-alaninate
Galben
M 9834
Tairel

Merck Index No.
MP: 78 - 80°
BP:
Density: 1.27^{25}
Solubility: 37 mg/l water 25°, >500 g/kg acetone, chloroform, dichloromethane, dimethylformamide, >400 g/kg cyclohexanone, >300 g/kg hexane
Octanol/water PC: 2500
LD_{50}: rat orl 4200 mg/kg; rat skn >5000 mg/kg; rat ipr 1100 mg/kg; mus orl 680 mg/kg; rbt skn >5000 mg/kg; ckn orl 4600 mg/kg; qal orl 3700 mg/kg
40CFR......

CAS RN 87237-48-7
$C_{17}H_{15}ClF_3NO_4$

haloxyfopethoxyethyl
(RS)-2-[4-(3-chloro-5-trifluoromethyl-2-pyridyloxy)phenoxy]propionic acid, ethyl ester
(±)-2-[4-[[3-chloro-5-(trifluoromethyl)-2-pyridinyl]oxy]phenoxy]propanoic acid, ethylester
Dowco 453
Gallant

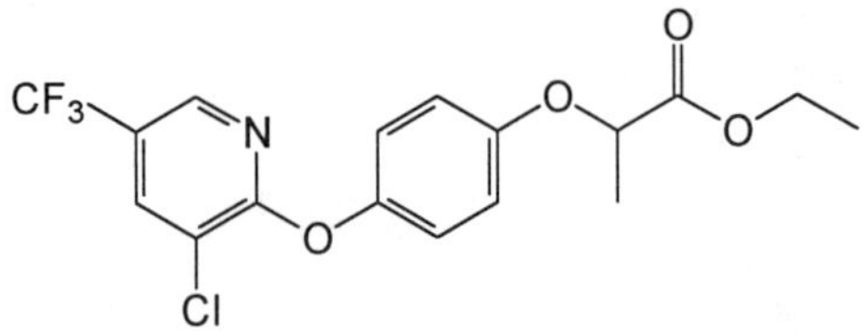

Merck Index No.
MP: 58 - 61°
BP:
Density: 1.34
Solubility: (ethoxyethyl ester) 2.7 mg/l water 20°, 2760 g/l dichloromethane, 1250 g/l xylene, >1000 g/l acetone, >1000 g/l EtOAc, >1000 g/l toluene, 233 g/l methanol, 52 g/l isopropanol, 44 g/l hexane
Octanol/water PC: 29500
LD_{50}: rat orl 531 mg/kg (ethoxyethyl ester); rat orl 518 mg/kg (ethoxyethyl ester); rat skn >2000 mg/kg; rbt skn >5000 mg/kg; dcl orl >2150 mg/kg.
CFR: 40CFR180. (pend. 1985)

CAS RN 77-06-5
$C_{19}H_{22}O_6$

Gibberellic acid
gibberellin A3
GA3

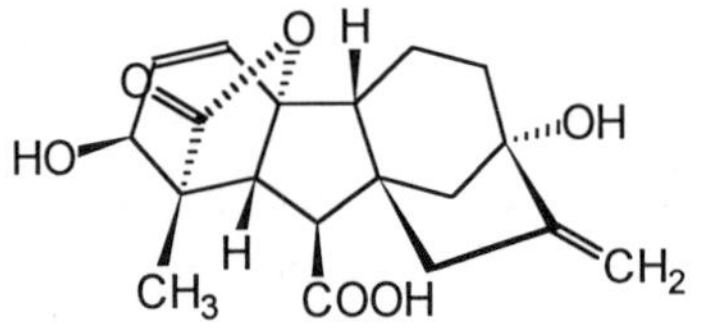

Merck Index No. 4313
MP: 223 - 225°
BP:
Density:
Solubility: 5 g/l water 25°, sol. methanol, ethanol, acetone, sl. sol. diethyl ether, ethyl acetate, insol. chloroform
Octanol/water PC:
LD_{50}: rat orl >15000 mg/kg; rat orl 6300 mg/kg; mus orl >15000 mg/kg; mus orl 8500 mg/kg; rbt skn >2000 mg/kg.
CFR: 40CFR180.224

CAS RN 39148-24-8
$C_6H_{18}AlO_9P_3$

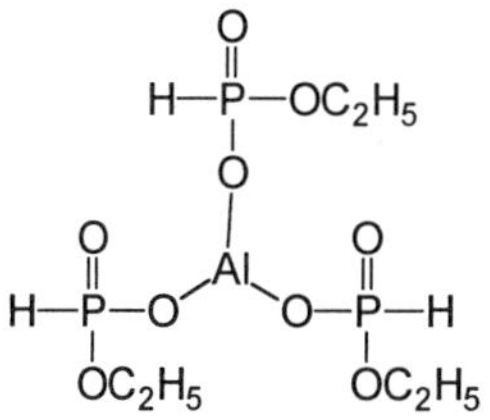

fosetyl-Al
ethyl hydrogen phosphonate, aluminum salt
Aliette
LS 74783

Merck Index No. 4167
MP: >300°
BP:
Density:
Solubility: 120 g/l water 20°, 920 mg/l methanol, 13 mg/l acetone, 80 mg/l propylene glycol, 5 mg/l ethyl acetate, 5 mg/l hexane
Octanol/water PC: 0.002 (octanol/water pH 4)
LD_{50}: rat orl 5800 mg/kg; rat orl 5000 mg/kg; rat skn >3200 mg/kg; rat ipr 550 mg/kg; mus orl 3700 mg/kg; mus skn LD >3200 mg/kg; dog orl LDLo >2140 mg/kg; rbt orl 2680 mg/kg; rbt skn >2000 mg/kg; gpg orl 2780 mg/kg; qal orl 4997 mg/kg.
CFR: 40CFR180.415

CAS RN 70124-77-5
$C_{26}H_{23}F_2NO_4$

flucythrinate
(RS)-α-cyano-3-phenoxybenzyl (S)--2-(4-difluoromethoxyphenyl)-3-methylbutyrate
cyano(3-phenoxyphenyl)methyl 4-(difluoromethoxy)-α-(1-methylethyl)benzeneacetate
AC 222705
CL 222705
Cybolt
Cythrin
Pay-Off

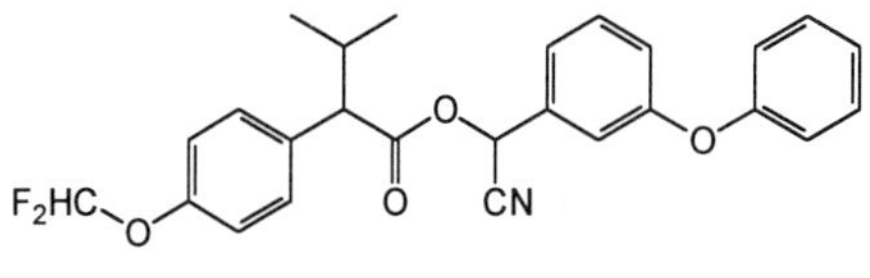

Merck Index No. 4055
MP:
BP: 108° $^{0.35}$
Density: 1.189^{22}
Solubility: 0.5 mg/l water 21°, >820 g/l acetone, 1810 g/l xylene, >780 g/l propanol, >560 g/l corn oil, >300 g/l cottonseed oil, >300 g/l soybean oil, 90 g/l hexane
Octanol/water PC: 120
LD_{50}: rat orl 81 mg/kg; rat orl 67 mg/kg; mus orl 76 mg/kg; rbt skn >1000 mg/kg; qal orl 2708 mg/kg; dck orl >2510 mg/kg.
CFR: 40CFR180.400

CAS RN 69806-50-4
$C_{19}H_{20}F_3NO_4$

fluazipop-butyl
(RS)-2-[4-(5-trifluoromethyl-2-pyridyloxy)phenoxy]propionic acid, butyl ester
(±)-2-[4-[[5-(trifluoromethyl)-2-pyridinyl]oxy]phenoxy]propanoic acid, butyl ester
Fusilade
Hache Uno Super
ICIA0009
IH 773B
Onecide
Ornamec
PP009
TF1169

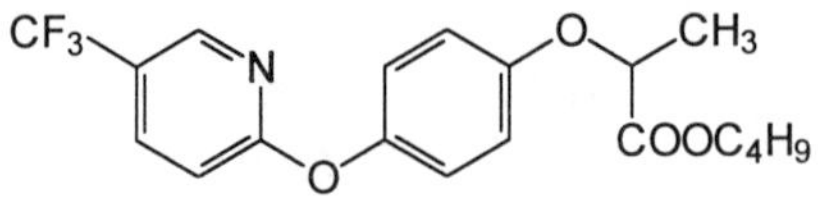

Merck Index No. 4049
MP: 13°
BP: 167° $^{0.05}$, 170° $^{0.5}$
Density: 1.21^{20}
Solubility: 2 mg/l water 20°, misc. acetone, cyclohexanone, hexane, methanol, dichloromethane, xylene, 24 g/l propylene glycol
Octanol/water PC: 31620
LD_{50}: rat orl 3328 mg/kg; rat orl 2910 mg/kg; rat ihl LC_{50} >5240 mg/m^3; rat skn >6050 mg/kg; rat ipr 1620 mg/kg; rat scu >5000 mg/kg; mus orl 1490 mg/kg; mus orl 1770 mg/kg; mus ipr 1240 mg/kg; mus scu >2000 mg/kg; rbt orl 621 mg/kg; rbt skn >2420 mg/kg; gpg orl 2659 mg/kg; dck orl 17000 mg/kg.
CFR: 40CFR180.411

CAS RN 79241-46-6
$C_{19}H_{20}F_3NO4$

fluazipop-P-butyl
(R)-2-[4-(5-trifluoromethyl-2-pyridyloxy)phenoxy]propionic acid, butyl ester
(R)-2-[4-[[5-(trifluoromethyl)-2-pyridinyl]oxy]phenoxy]propanoic acid, butyl ester
Fusilade 2000, Fusilade 5
Fusilade Super
ICI0005
PP005

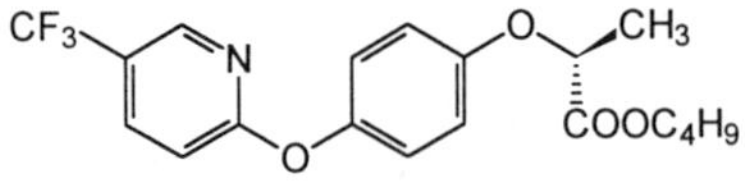

Merck Index No.
MP: 5°
BP:
Density:
Solubility: 1 mg/l water, misc. acetone, hexane, methanol, dichloromethane, EtOAc, toluene, xylene
Octanol/water PC: 31620
LD_{50}: rat orl 3680 mg/kg; rat orl 2451 mg/kg; rat orl 2712 mg/kg; rbt skn >2000 mg/kg.
CFR: 40CFR180.411

CAS RN 79127-80-3
$C_{17}H_{19}NO_4$

fenoxycarb
ethyl 2-(4-phenoxyphenoxy)ethylcarbamate
ethyl [2-(4-phenoxyphenoxy)ethyl]carbamate
ACR-2807 B
ACR-2913 A
Insegar
Logic
Pictyl
Ro 13-5223/000
Torus
Varikill

Merck Index No.
MP: 53 - 54°
BP:
Density:
Solubility: 6 mg/l water 20°, >250 g/kg in most org. solvents, 5 g/kg hexane
Octanol/water PC: 20000
LD_{50}: rat orl >10000 mg/kg.
CFR: 40CFR180. (temp. 1985)

CAS RN 98886-44-3
$C_9H_{18}NO_3PS_2$

fosthiazate
(RS)-S-*sec*-butyl O-ethyl 2-oxo-1,3-thiazolidin-3-ylphosphonothioate
(RS)-S-*sec*-butyl O-ethyl-2- oxothiazolidin-3-ylphosphonothioate
(RS)-3-[*sec*-butyl(ethoxy)phosphinoyl]-1,3-thiazolidin-2-one
(RS)-3-[*sec*-butylthio(ethoxy)phosphinoyl]thiazolidin-2-one
O-ethyl S-(1-methylpropyl) (2-oxo-3-thiazolidinyl)phosponothioate
IKI-1145

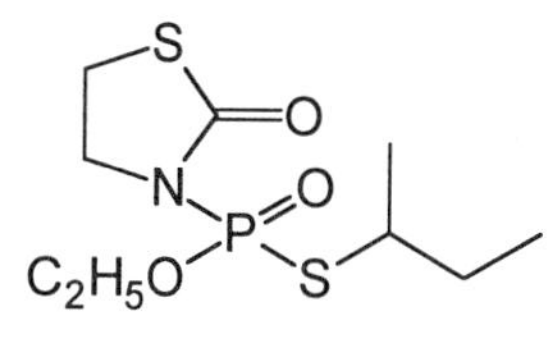

Merck Index No.
MP:
BP:
Density:
Solubility:
Octanol/water PC:
LD_{50}:
CFR: 40CFR180. (temp. 1991)

CAS RN 90982-32-4
$C_{15}H_{15}ClN_4O_6S$

chlorimuron
2-(4-chloro-6-methoxypyrimidin-2-ylcarbamoylsulfamoyl) benzoic acid, ethyl ester
2-[[[[(4-chloro-6-methoxy-2-pyrimidinyl)aimno]carbonyl]amino]sulfonyl]
benzoic acid, ethyl ester
Classic
DPX-F6025

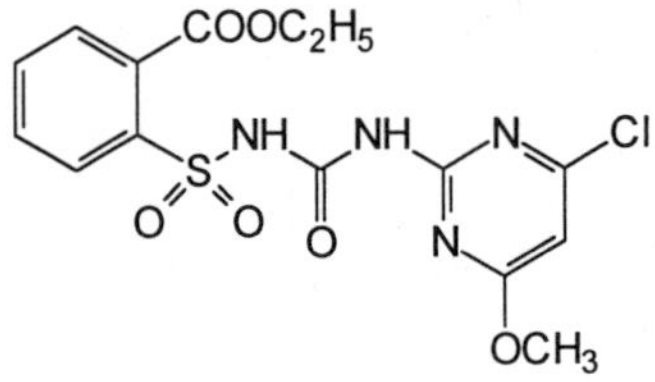

Merck Index No. 2092
MP: 185 - 187°
BP:
Density:
Solubility: 11 mg/l water pH 5, 25°, 1200 mg/l pH 7, low. sol. org. solvents
Octanol/water PC: 2.3 pH 7
LD_{50}: rat orl 4102 mg/kg; rat ihl LC_{50} >5000 mg/m^3/4H; rbt skn >2000 mg/kg; dck unr >2510 mg/kg.
CFR: 40CFR180.429

CAS RN 74051-80-2
$C_{17}H_{29}NO_3S$

sethoxydim
(±)-(EZ)-2-(1-ethoxyiminobutyl)-5-[2-(ethylthio)propyl]-3-hydroxycyclohex-2-enone
(±)-2-[1-(ethoxyimino)butyl]-5-[2-(ethylthio)propyl]-3-hydroxy-2-cyclohexen-1-one
sethoxydime
BAS 90520H
Checkmate
Expand
Fervinal
Grasidim
Nabu
NP-55
Poast
SN 81742

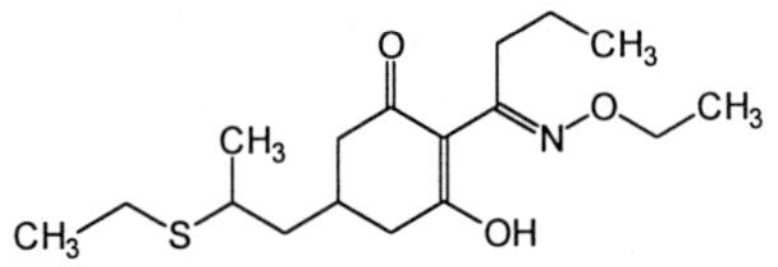

Merck Index No. 8424
MP:
BP: >90° $^{0.00003}$
Density: 1.043^{25}
Solubility: 25 mg/l water 20° pH 4, 4700 mg/l pH 7, >1 kg/kg in acetone, benzene, ethyl acetate, hexane, methanol
Octanol/water PC: 32360 pH 5, 44.7 pH 7
LD_{50}: rat orl 3200 mg/kg; rat orl 2676 mg/kg; rat ihl LC_{50} >6280 mg/m^3/4H; rat skn 5000 mg/kg; rat ipr 1493 mg/kg; rat ivn 505 mg/kg; mus orl 5600 mg/kg; mus orl 6300 mg/kg; mus skn 5000 mg/kg; mus ivn 485 mg/kg; rbt orl 4600 mg/kg; qal unr >5000 mg/kg.
CFR: 40CFR180.412

CAS RN 81335-37-7
$C_{17}H_{17}N_3O_3$

imazaquin
(RS)-2-(4-isopropyl-4-methyl-5-oxo-2-imidazolin-2-yl)quinoline-3-carboxylic acid
(±)-2-[4,5-dihydro-4-methyl-4-(1-methylethyl)-5-oxo-1H-imidazol-2-yl]-3-quinolinecarboxylic acid
imazaquine
AC 252,214
CL 252,214
Image
Scepter

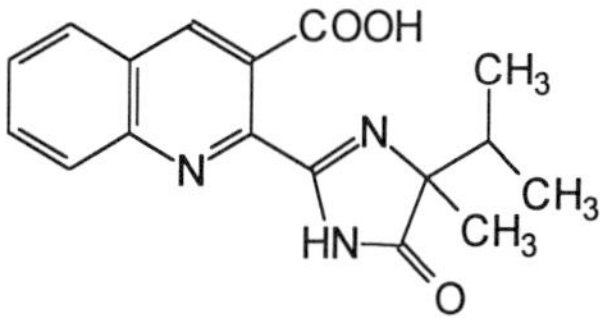

Merck Index No. 4826
MP: 219 - 224°
BP:
Density:
Solubility: 60-120 mg/l water 25°, 0.4 g/l toluene, 68 g/l dimethylformamide, 159 g/l dimethylsulfoxide, 14 g/l dichloromethane
Octanol/water PC: 2.2
LD_{50}: rat orl 5000 mg/kg; mus orl 2353 mg/kg; rbt skn 2000 mg/kg; qal orl >2150 mg/kg; dck orl >2150 mg/kg.
CFR: 40CFR180.426

CAS RN 81335-77-5
$C_{15}H_{19}N_3O_3$

imazethapyr
(RS)-5-ethyl-2-(4-isopropyl-4-methyl-5-oxo-2-imidazolin-2-yl)nicotinic acid
(±)-2-[4,5-dihydro-4-methyl-4-(1-emthylethyl)-5-oxo-1H-imidazol-2-yl]-5-ethyl-3-pyridinecarboxylic acid
AC 263,499
CL 263,499
Event
Pivot
Pursuit
Overtop

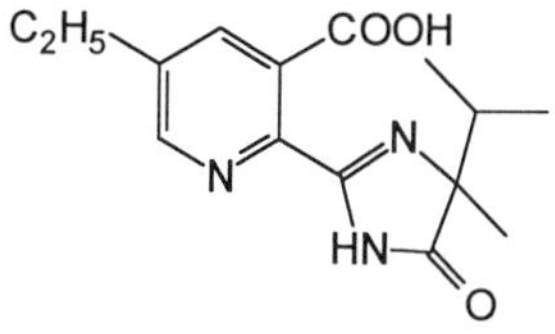

Merck Index No.
MP: 172 - 175°
BP:
Density:
Solubility: 1.4 g/l water 25°, 48.2 g/l acetone, 105 g/l methanol, 5 g/l toluene, 185 g/l dichloromethane, 422 g/l dimethylsulfoxide, 17 g/l isopropanol
Octanol/water PC: 11 (octanol/water pH 5)
LD_{50}: rat orl >5000 mg/kg; rat ihl LC_{50} 3270 mg/m^3; mus orl >5000 mg/kg; rbt skn >2000 mg/kg; qal orl >2150 mg/kg; dck orl >2150 mg/kg.
CFR: 40CFR180.447

CAS RN 77501-63-4
$C_{19}H_{15}ClF_3NO_7$

lactofen
ethyl O-[5-(2-chloro-α,α,α-trifluoro-*p*-tolyloxy)-2-nitrobenzoyl]-DL-lactate
(±)-2-ethoxy-1-ethyl-2-oxoethyl 5-[2-chloro-4-(trifluoromethyl)phenoxy]-2-nitrobenzoate
Cobra
PPG-844

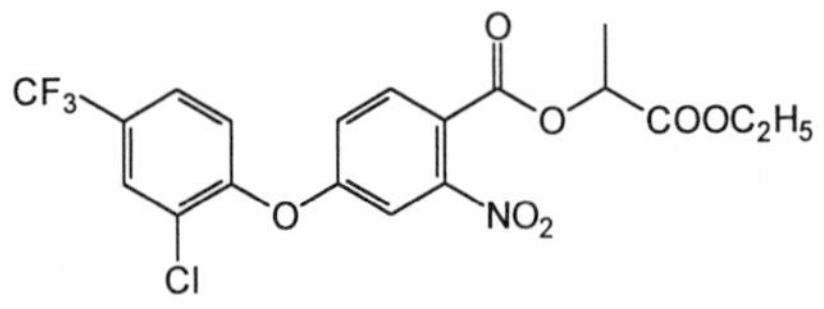

Merck Index No.
MP:
BP:
Density:
Solubility: <1 mg/l water 20°
Octanol/water PC:
LD_{50}: rat orl >5000 mg/kg; rat skn 2000 mg/kg.
CFR: 40CFR180.432

CAS RN 82097-50-5
$C_{14}H_{16}ClN_5O_5S$

triasulfuron
1-[2-(2-chloroethoxy)phenylsulfonyl]-3-(4-methoxy-6-methyl-1,3,5-triazin-2-yl)urea
2-(2-chloroethoxy)-N-[[(4-methoxy-6-methyl-1,3,5-triazin-2-yl)amino]carbonyl]benzene sulfonamide
Amber
CGA 131036
Logran

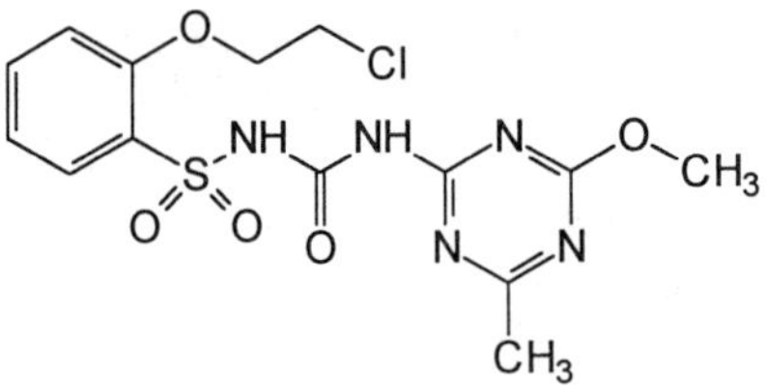

Merck Index No.
MP: 186°
BP:
Density:
Solubility: 1.5 g/l water 20° pH 7, 3.4 g/l methanol, 16 g/l acetone,
17 g/l cyclohexanone, 15 g/l dichloromethane, 180 g/l n-octanol, 166 g/l xylene, 0.2 g/l hexane
Octanol/water PC: 0.11 (octanol/water pH 7)
LD_{50}: rat orl >5000 mg/kg; rat orl >5050 mg/kg; rat ihl LC_{50} >2320 mg/m^3/4H; rat skn 2000 mg/kg; rbt skn 2000 mg/kg; qal orl >2150 mg/kg.
CFR: 40CFR180.459

CAS RN 77732-09-3
$C_{14}H_{18}N_2O_4$

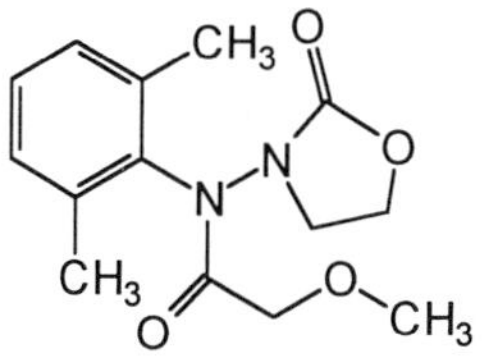

oxadixyl
2-methoxy-N-(2-oxo-1,3-oxazolidin-3-yl)acet-2',6'-xylidide
N-(2,6-dimethylphenyl)-2-methoxy-N-(2-oxo-3-oxazolidinyl)acetamide
Anchor
SAN 371F
Sandofan

Merck Index No.
MP: 104 - 105°
BP:
Density:
Solubility: 3.4 g/kg water 25°, 344 g/kg acetone, 390 g/kg dimethylsulfoxide, 50 g/kg ethanol, 17 g/kg xylene, 6 g/kg diethyl ether
Octanol/water PC: 4.5 - 6.3
LD_{50}: rat orl 3480 mg/kg; rat orl 1860 mg/kg; rat skn >2000 mg/kg; rat ipr 278 mg/kg; rat scu 611 mg/kg; mus orl 693 mg/kg; mus orl 1860 mg/kg; mus orl 2150 mg/kg; mus ipr 184 mg/kg; mus scu 560 mg/kg; rbt ihl LC_{50} >6000 mg/m^3; dck orl >2510 mg/kg.
CFR: 40CFR180.456

CAS RN 82657-04-3
$C_{23}H_{22}ClF_3O_2$

bifenthrin
2-methylbiphenyl-3-ylmethyl (Z)-(1RS,3RS)-3-(2-chloro-3,3,3-trifluoroprop-1-enyl)- 2,2-dimethylcyclopropanecarboxylate
[1α,3α(Z)]-(±)-(2-methyl[1,1'-biphenyl]-3-yl)methyl 3-(2-chloro-3,3,3-trifluoro-1-propenyl)-2,2-dimethylcyclopropanecarboxylate
bifenthrine
Brigade
Capture
FMC 54800
Talstar

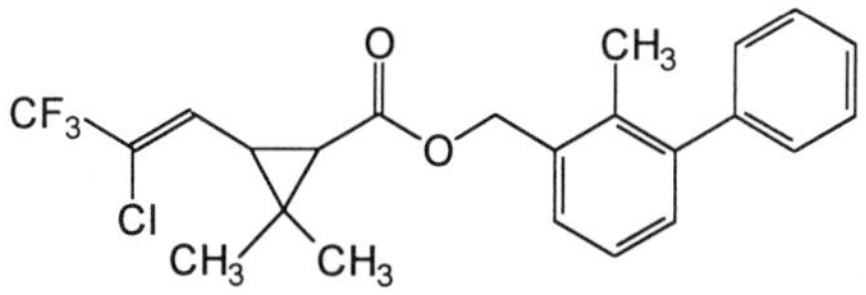

Merck Index No. 1229
MP: 68 - 70.6°
BP:
Density: 1.2^{125}
Solubility: 0.1 mg/l water 25°, sol. acetone, chloroform, dichloromethane, diethyl ether, toluene, sl. sol. heptane, methanol
Octanol/water PC: 1000000
LD_{50}: rat orl 54.5 mg/kg; rbt skn >2000 mg/kg; qal orl 1800 mg/kg; dck orl >4450 mg/kg.
CFR: 40CFR180.442

CAS RN 83055-99-6
$C_{16}H_{18}N_4O_7S$

bensulfuron-methyl
α-(4,6-dimethoxypyrimidin-2-ylcarbamoylsulfamoyl)-*o*-toluic acid, methyl ester
2-[[[[[(4,6-dimethoxypyrimidin-2-yl)amino]carbonyl]amino]sulfonyl]methyl]benzoic acid, methyl ester
DPX-F 5384
Londax
Mariner

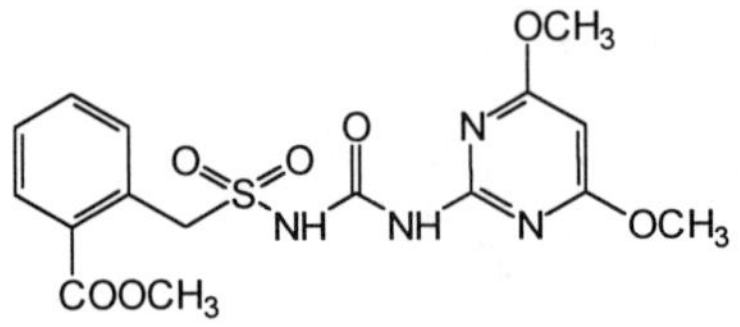

Merck Index No.
MP: 185 - 188°
BP:
Density:
Solubility: 2.9 mg/l water pH 5, 12 mg/l pH 6, 120 mg/l pH 7, 1200 mg/l pH 8, sol. dichloro-methane, rel. sol. acetonitrile, acetone, methanol, ethyl acetate
Octanol/water PC: 4.1 pH 7
LD_{50}: rat orl >5000 mg/kg; rat ihl LC_{50} >5000 mg/m^3; rat skn 2000 mg/kg; mus orl >10985 mg/kg; rbt skn >2000 mg/kg; dcl orl >2510 mg/kg.
40CFR180.445

CAS RN 55702-49-3
$C_{10}H_{13}N_3O_4$

N-*sec*-butyl-2,6-dinitroaniline
benzeneamine, N-(1-methylpropyl)-2,6-dinitro

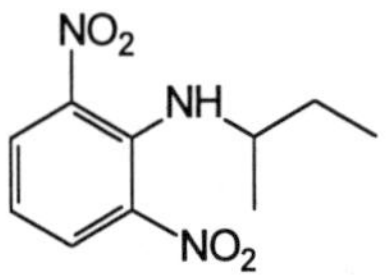

Merck Index No.
MP:
BP:
Density:
Solubility:
Octanol/water PC:
LD_{50}:
CFR: 40CFR180. temp (1973)

CAS RN 6086-22-2
$C_6H_{11}N_3$

1H-1,2,4-triazole, 1-butyl
1-*n*-butyl-1,2,4-triazole
1-butyl-*s*-triazole
1-butyl-1,2,4-triazole

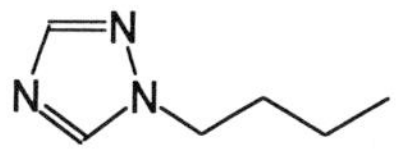

Merck Index No.
MP:
BP:
Density:
Solubility:
Octanol/water PC:
LD_{50}:
CFR: 40CFR180. pend. (1977)

CAS RN 10103-61-4
$AsH_3O_4.xCu$

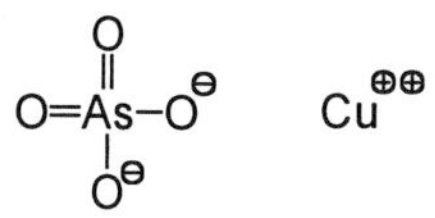

copper arsenate (ACN)
Arsenic acid (H_3AsO_4), copper salt
basic copper arsenate

Merck Index No.
MP:
BP:
Density:
Solubility:
Octanol/water PC:
LD_{50}:
CFR: 40CFR180. revoked 5-4-88

CAS RN 78-48-8
$C_{12}H_{27}OPS_3$

S,S,S,-tributyl phosphorotrithioate (ACN)
phosphorotrithioic acid, S,S,S-tributyl ester
B-1,776
Butifos
Butiphos
butyl phosphorotrithioate
Chemagro B-1776
Chemagro 1,1776
De-Green
DEF
DEF defoliant
E-Z-Off D
Fos-Fall A
Fosfall
orthophosphate defoliant
S,S,S-tributyl trithiophosphate
TBTP

SC_4H_9
$C_4H_9S—P=O$
SC_4H_9

Merck Index No.
MP: < -25°
BP: 150° $^{0.3}$
Density: 1.057^{20}
Solubility: 2.3 mg/l water 20°, sol. aliphatic, aromatic and chlorinated hydrocarbons
Octanol/water PC: 1700
LD_{50}: rat orl 233 mg/kg; rat orl 200 mg/kg; rat orl 150 mg/kg; rat skn 168 mg/kg; rat ipr 210 mg/kg; mus orl 77 mg/kg; mus ihl LCLo 3804 mg/m^3/1H; mus ipr 285 mg/kg; rbt orl 242 mg/kg; rbt skn 97 mg/kg; gpg orl 140 mg/kg; gpg ipr 150 mg/kg; ckn skn LDLo 1000 mg/kg; mam unr 170 mg/kg.
CFR: 40CFR180.272

CAS RN 80-13-7
$C_7H_5Cl_2NO_4S$

***p*-(N,N-dichlorosulfamoyl)benzoic acid** (ACN)
benzoic acid, 4-[(dichloroamino)sulfonyl]-
benzoic acid, *p*-(dichlorosulfamoyl)-
p-(dichlorosulfamoyl)benzoic acid
p-(N,N-dichlorosulfamyl)benzoic acid
p-carboxybenzenesulfondichloramide
p-dichlorosulfamoylbenzoic acid
p-sulfondichloroamidobenzoic acid
benzoic acid, *p*-(N,N-dichlorosulfonamide)
Halazone
Pantocid
Pantocide
Pantosid
Pantoside
Pantotsid
Pentocid
Zeptabs
4-(N,N-dichlorosulfamoyl)benzoic acid

O O
S
N Cl
Cl
HOOC

Merck Index No. 4503
MP: 195° (dec)
BP:
Density:
Solubility: sl. sol. water, chloroform
Octanol/water PC:
LD_{50}: rat orl LDLo 3500 mg/kg; rat ivn 300 mg/kg.
CFR: 40CFR180. pend.

CAS RN 136-45-8
$C_{13}H_{17}NO_4$

Dipropyl isocinchomeronate (ACN)
2,5-pyridinedicarboxyl acid, dipropyl ester
di-*n*-propyl isocinchomeronate
di-*n*-propyl 2,5-pyridinedicarboxylate
di-propylisocinchomeronate
Di-*N*-propyl-isocinchomeronate (German)
dipropyl 2,5-pyridinedicarboxylate
ENT 17591
isocinchomeronic acid, dipropyl ester
Mgk r-326
Mgk Repellent 326
pyridini-2,5-dicarbonsaeure-di-N-propylester (German)

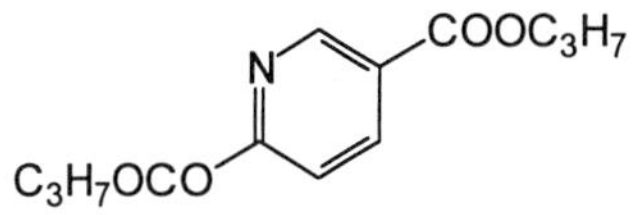

Merck Index No.
MP:
BP:
Density:
Solubility:
Octanol/water PC:
LD_{50}: rat orl 5230 mg/kg; rat skn 9400 mg/kg; mus orl 1600 mg/kg; mus ipr 330 mg/kg; rbt skn 9500 mg/kg.
CFR: 40CFR180.143

CAS RN 8018-01-7
$C_4H_6MnN_2S_4.C_4H_6N_2S_4Zn$

mancozeb
manganese, [[1,2-ethanediylbis[carbamodithioato]](2-)]-, mixt. with [[1,2-ethanediylbis [carbamodithioato]](2-)]zinc
Carbamic acid, ethylenebis[dithio, manganese zince complex
carbamic acid, ethylenebis(dithio-, manganese zinc complex
dithane ultra
Zinc ion coordinated manganese ethylenebis[dithiocarbamate]
zinc manganese ethylenebis[dithiocarbamate]
Dithane M45
Dithane M 45
Dithane SPC
F 2966
Fore
Greem diasen M
Green-daisen M
Manoseb
Manzate 200
Manzeb
Marzin
Vondozeb
Zimanat

Merck Index No. 5598
MP: 192 - 194° (dec)
BP:
Density:
Solubility: insol. water, most org. solvents
Octanol/water PC:
LD50: rat orl 5000 mg/kg; rat skn >15000 mg/kg; rbt skn >5000 mg/kg.
CFR: 40CFR180.176

INDEXES

CAS REGISTRY NUMBER INDEX

MOLECULAR FORMULA INDEX

COMPOUND NAME INDEX

A

B

D

E

F

G

H

I

J

K

L

M

N

O

P

Q

R

S

T

U

V

W

X

Y

Z